From Overtourism to Sustainability Governance

Insightful and international in scope, this significant volume explores the transition from overtourism to sustainability governance and elaborates perspectives for developing resilient destinations.

The book is split into three parts and comprises interdisciplinary contributions from renowned authors and scholars in the field, with each part including case studies to illustrate real-world applications of the topics and issues discussed. Part I provides an overview of current academic discussion on overtourism, presents theoretical concepts and illustrates the impact of the COVID-19 pandemic, as well as the development from overtourism to no-tourism. Part II discusses approaches towards a new sustainability through the lens of current research and future trends, including a possible new understanding of tourism in a post-pandemic world. Part III presents strategies to deal with overtourism, including management strategies and governance theories. Equipped with a wide range of examples and insights from across the globe, the book is intended to facilitate the ongoing journey towards a more sustainable tourism industry that is increasingly resilient and less vulnerable to crises.

This will be of pivotal interest to academics, researchers and practitioners in the fields of tourism, over- and mass-tourism, as well as sustainability governance.

Harald Pechlaner is Head of the Center for Advanced Studies at Eurac Research, Chair of Tourism and founding Dean of the School of Transformation and Sustainability at the Catholic University of Eichstaett-Ingolstadt. His research area covers sustainable destination development and selected questions about global governance combined with economics and politics. Since 2014 he has been Adjunct Research Professor at Curtin University (Perth, Australia) and President of AIEST (Association Internationale d'Experts Scientifiques du Tourisme), the oldest tourism experts' association, based in the St. Gallen University.

Elisa Innerhofer was a Senior Researcher at the Center for Advanced Studies at Eurac Research Bozen-Bolzano, Italy. She studied international economics, business science, and political science at the University of Innsbruck (Austria) and the Marquette University in Milwaukee (WI, USA) and holds a PhD in economics from the Catholic University of Eichstätt-Ingolstadt in Germany.

Julian Philipp is a Consultant at PKF Munich with a focus on public transport, and a former Consultant for Regional Development in South-East Germany. He is a PhD student at the Chair of Tourism / Center for Entrepreneurship, Catholic University of Eichstaett-Ingolstadt, Germany. His dissertation focuses on the role of local and regional identity in the governance of spatial and regional development. He completed his BSc at University of Wuppertal, Germany, and his MSc at Oxford Brookes University, United Kingdom. His main fields of interest encompass regional development, regional ecosystems, sustainable destination development, overtourism and entrepreneurship.

From Overtourism to Sustainability Governance

A New Tourism Era

Edited by
Harald Pechlaner, Elisa Innerhofer and Julian Philipp

Routledge
Taylor & Francis Group
LONDON AND NEW YORK

First published 2024
by Routledge
4 Park Square, Milton Park, Abingdon, Oxon OX14 4RN

and by Routledge
605 Third Avenue, New York, NY 10158

Routledge is an imprint of the Taylor & Francis Group, an informa business

This funding for this Open Access book has been provided by Eurac Research.

British Library Cataloguing-in-Publication Data
A catalogue record for this book is available from the British Library

Library of Congress Cataloging-in-Publication Data
Names: Pechlaner, Harald, editor. | Innerhofer, Elisa, editor. | Philipp, Julian, editor.
Title: From overtourism to sustainability governance : a new tourism era / edited by Harald Pechlaner, Elisa Innerhofer, and Julian Philipp.
Description: New York : Routledge, 2024. | Includes bibliographical references and index.
Identifiers: LCCN 2024005019 (print) | LCCN 2024005020 (ebook) | ISBN 9781032431369 (hbk) | ISBN 9781032431376 (pbk) | ISBN 9781003365815 (ebk)
Subjects: LCSH: Overtourism. | Sustainable tourism—Management.
Classification: LCC G156.5.O94 F76 2024 (print) | LCC G156.5.O94 (ebook) | DDC 910.68—dc23/eng/20240507
LC record available at https://lccn.loc.gov/2024005019
LC ebook record available at https://lccn.loc.gov/2024005020

ISBN: 978-1-032-43136-9 (hbk)
ISBN: 978-1-032-43137-6 (pbk)
ISBN: 978-1-003-36581-5 (ebk)

DOI: 10.4324/9781003365815

Contents

Figures and Tables

Figures

Tables

Contributors

Pietro Beritelli studied Business Administration at the University St. Gallen (HSG), Switzerland, with an emphasis on travel and transport. Since 2003 he has been an Associate Professor at the University of St. Gallen and Vice-Director of the Institute for Systemic Management and Public Governance (IMP-HSG), Research Centre for Tourism and Transport. Since 2004 he has also held the role of Director of the Master Program in Marketing Management at the University of St. Gallen. He advises tourist enterprises and public institutions in questions regarding destination management and marketing, tourism policy, and strategic management and is actively involved in the industry through mandates as a board member of several tourism organisations.

Elena Borin, PhD in Economics (The University of Ferrara, Italy) and Doctor Europaeus, is currently an Associate Professor in Business Administration for Link Campus University in Rome (Italy). She is member of the Board of ENCATC – the European Network on Cultural Management and Policy (Brussels, Belgium) – and Interim Editor in Chief of the *European Journal of Cultural Management and Policy*. Her research interests include governance of cultural and creative industries, PPPs, business models and entrepreneurial ecosystems, and sustainability accounting and reporting in the cultural and creative sector. Her scientific contributions include books and around 50 papers and book chapters published in international journals and book series.

Rosa Codina is a Senior Lecturer in Tourism and Events Management at Oxford Brookes University, United Kingdom, where she also obtained a PhD in Tourism Management. Her doctoral research examined the role of tourism in local power relations through an ethnographic study of the town of Pisac, Peru. She actively conducts research on the socio-cultural and political impacts of tourism and events in developing nations, with a focus on tourism host communities and traditionally marginalised groups. She is particularly interested in the challenges and opportunities tourism presents for women, as well as the role of informal actors in tourism development in the Global South.

Antonia Correia is Professor of Tourist Behaviour and Tourism Economics, University of Algarve, and the President of the first and unique collaborative laboratory in tourism, KIPT Colab, both in Portugal. Research areas include consumer behaviour, tourism economics and modelling. She has published more than 100 papers in tourism, leisure and economics journals. She is a member of the editorial boards of several leading journals including *Journal of Travel Research, Current Issues in Tourism* and *Anatolia*, amongst others.

Greta Erschbamer is a Researcher at the Center for Advanced Studies at Eurac Research in Italy and a doctoral student at the Chair of Tourism and the Center for Entrepreneurship of the Catholic University of Eichstaett-Ingolstadt in Germany. She holds a master's degree in Strategic Tourism Management from SKEMA Business School in France and a Master of Science in Tourism and Regional Planning from the Catholic University Eichstaett-Ingolstadt. Her current research focuses on tourism governance, participation and design studies. As a project manager, she was responsible for several projects in the field of tourism and regional development, with a particular focus on the development of mountain areas.

Martin Fontanari studied Business Administration and Political Science at the Universities of Innsbruck and Trier. After completing his doctorate with summa cum laude, he held top-management positions in tourism consultancy businesses (KPMG, Schitag Ernst & Young, European Tourism Institute) and a global NPO. In 2008 he returned to the academic sphere and research. His area of academic interest lies in the fields of tourism management, sustainable development, resilience in tourism, crisis management, cooperation and partnership.

Dirk Glaesser is the Director of the Sustainable Development of Tourism Department at the United Nations World Tourism Organization (UNWTO). Under his supervision, the programme addresses the challenges and opportunities of sustainable tourism development, including climate change, sustainable consumption and production, health emergencies, biodiversity and travel facilitation. He is a banker by profession and a Colonel of the Reserve of the German Armed Forces. He obtained his PhD from the University of Lüneburg (Germany) and was awarded the ITB scientific award for his work on Crisis Management. He has authored several publications, which have been widely translated.

Mirjam Gruber is a political scientist at Eurac Research (Italy) and a PhD candidate at the University of Leipzig (Germany). She holds a master's degree in Political Science (minor: Sustainable Development) from the University of Bern (Switzerland) and a bachelor's degree in International Development and Cooperation from the University of Bologna (Italy).

Stefan Hartman is the Head of Department at the European Tourism Futures Institute (ETFI), NHL Stenden University, Leeuwarden, The Netherlands. His research focuses on tourism futures, strategic foresight, transition management, resilience, (smart) destination governance and adaptive capacity building. He uses his knowledge in applied science projects to help actors in the leisure and tourism industry to develop strategies and actions that allow them to manage continually changing business environments.

Raúl Hernández-Martín is Senior Lecturer of Tourism Economics at the Department of Applied Economics and Quantitative Methods, University of La Laguna, Spain. He obtained his PhD in Economics after a research stay at the University of Social Science of Toulouse, France. He currently holds the Chair in Tourism at the University of La Laguna. His research focuses on tourism economic impacts, tourism satellite accounts, measuring the sustainability of tourism, and island tourism development, and has been published in *Tourism Management*, *Tourism Economics*, *Current Issues in Tourism* and the *Journal of Sustainable Tourism*.

Marcus Herntrei is Professor for International Tourism Management at the Deggendorf Institute of Technology, Germany. His research interests include participatory destination development, quality of life and resilience, social impacts of tourism and tourism acceptance.

Jasper Heslinga is a Senior Researcher at the European Tourism Futures Institute (ETFI) at NHL Stenden University, Leeuwarden, The Netherlands, and a Programme Manager for CELTH (Centre of Expertise Leisure, Tourism and Hospitality). He has a background in human geography and spatial planning at the University of Groningen and successfully completed his PhD research there on synergies between tourism and nature. His research focuses on how destinations can become more sustainable, inclusive and smart.

Donagh Horgan is an academic and practitioner in the area of regenerative placemaking and socio-spatial transformation. He is presently a Lab Lead at the Urban Leisure and Tourism Lab in Rotterdam, Netherlands. Trained as an architect and service designer, he is an expert on social innovation and resilience in the built environment. Outside of academia he consults on working on urban transformation with cities in the European neighbourhood and around the world for UNDP and BCPI. He is based at the Inholland University of Applied Sciences and Erasmus University Rotterdam, where he is helping to build a robust ecosystem looking at sustainable urban transitions.

Elisa Innerhofer was a Senior Researcher at the Center for Advanced Studies at Eurac Research Bozen-Bolzano, Italy. She studied international economics, business science and political science at the University of Innsbruck (Austria) and the Marquette University in Milwaukee (WI, USA) and holds a PhD in economics from the Catholic University of Eichstätt-Ingolstadt in Germany.

Veronika Jánová is a Research Associate at the Deggendorf Institute of Technology, European Campus Rottal-Inn, Germany. She studied Tourism Management at the Munich University of Applied Sciences and International Tourism Development at the Deggendorf Institute of Technology. Her research interests include (urban) tourism planning and development, tourism acceptance, and overtourism.

Szymon Kielar is a master's student at the Munich University of Applied Sciences, Germany, where he previously completed a bachelor's degree in Tourism

Management. His field of study is Strategy and Innovation in Tourism and his research interests include sustainability, mobility, destination management and development, and communication issues.

Ko Koens has been the Professor of the New Urban Tourism Research Group at Inholland University of Applied Sciences, Netherlands, since February 2020. Considering the increasing pressure on cities from urbanisation and fast-growing urban tourism, his research focuses on how we can find a new balance between visitors and residents in such a way that tourism and the social transformation of city neighbourhoods can go hand in hand. He aims to find and analyse good examples of projects that have been carried out in other countries, and to test new local initiatives in Inholland's living labs. With the Expertise Network Sustainable Urban Tourism, he plans to work at various levels of scale to provide solutions in the field of sustainable urban tourism.

Karin Malacarne is a PhD student at Auckland University of Technology (New Zealand). She received her master's degree in Management of Tourism and Sustainability and her bachelor's degree in Management and Economics at the University of Trento (Italy). She has worked for the New Zealand Tourism Research Institute and Cook Islands Tourism Corporation. Her research interests include rural tourism, rural entrepreneurship and the value of networks.

Paulo Martins is a psychologist and a Researcher at KIPT COLAB, the first and only collaborative laboratory in tourism in Portugal. His research areas include human resources, social sciences, cognition and human behaviour.

Hans Müller's passion for world travel was sparked during his studies in Political Science and Sociology in Trier, Germany. Over the past 35 years, he has organised land services in popular destinations like the Balearic and Canary Islands, Thailand, Romania, Greece, Tunisia, London, Malta and Turkey. With over two decades of experience in hotel contracting, Hans has successfully coordinated hotel procurement for renowned international tour operators such as the Thomas Cook Group and DER Touristik, covering countries like Spain, Portugal, the Caribbean and Eastern Europe. As the Chief Commercial Officer (CCO) of the Loro Parque Group, he oversees sales and marketing for their parks, hotels and restaurants.

Natalie Olbrich is an Executive Assistant at the School of Transformation and Sustainability at the Catholic University of Eichstaett-Ingolstadt, Germany, and a PhD student at the Chair of Tourism / Center for Entrepreneurship of the Catholic University of Eichstaett-Ingolstadt. Previously, she was a Research Assistant at the Chair of Tourism / Center for Entrepreneurship. She completed her bachelor's degree in Tourism Management at the Technical University of Deggendorf, Germany, and holds a master's degree (MSc) in Tourism and Regional Development from the Ernst-Moritz-Arndt-University Greifswald, Germany. Before coming to the Catholic University of Eichstaett-Ingolstadt, she worked as an Account Support Manager at HRS – HOTEL RESERVATION

SERVICE. Her main fields of interest encompass regional development, culture, experiences in tourism and gastronomy.

Hugo Padrón-Ávila is a Lecturer of Economics and Business Management at the Faculty of Social Sciences of Universidad Europea de Canarias, Spain. He obtained his PhD at the University of La Laguna after spending part of his research formation at the National Chung Hsing University (Taiwan), Università degli Studi di Bari Aldo Moro (Italy) and the University of Central Florida (USA). His research focuses on analysing the impacts of tourism, tourists' behaviour and tourism sustainable development. Some of his studies have been published in prestigious journals such as *International Journal of Tourism Research* and *Tourism Review*.

Harald Pechlaner is Head of the Center for Advanced Studies at Eurac Research, Chair of Tourism and founding Dean of the School of Transformation and Sustainability at the Catholic University of Eichstaett-Ingolstadt. His research area covers sustainable destination development and selected questions about global governance combined with economics and politics. Since 2014 he has been Adjunct Research Professor at Curtin University (Perth, Australia) and President of AIEST (Association Internationale d'Experts Scientifiques du Tourisme), the oldest tourism experts' association, based in the St. Gallen University.

Mike Peters is Professor at the Department of Strategic Management, Marketing and Tourism at the Faculty of Business and Management, University of Innsbruck, Austria. His research interests include small business development, family businesses in tourism, destination management, entrepreneurship and innovation.

Christof Pforr is Discipline Leader (Tourism, Hospitality & Events) with the School of Management & Marketing, Faculty of Business & Law, Curtin University (Western Australia). His past and current research is inter- and multidisciplinary. In essence, his activities have concentrated on four interconnected research areas, sustainability, tourism public policy, destination governance and special interest tourism – all fields he has frequently published in. Pforr has contributed to more than 200 publications (including ten books) and numerous national and international research projects.

Julian Philipp is a Consultant at PKF Munich with a focus on public transport, and former Consultant for Regional Development in South-East Germany. He is a PhD student at the Chair of Tourism / Center for Entrepreneurship, Catholic University of Eichstaett-Ingolstadt, Germany. His dissertation focuses on the role of local and regional identity in the governance of spatial and regional development. He completed his BSc at University of Wuppertal, Germany, and his MSc at Oxford Brookes University, United Kingdom. His main fields of interest encompass regional development, regional ecosystems, sustainable destination development, overtourism and entrepreneurship.

Albert Postma is Professor of Strategic Foresight and Scenario Planning at the European Tourism Futures Institute (ETFI), NHL Stenden University in The

Netherlands. He holds an MSc and PhD in spatial planning. Postma's current research focuses on strategic foresight, scenario planning and tourism community relations (overtourism). Postma is a respected speaker at business conferences, has authored dozens of technical reports and articles, and is co-editor of the *Journal of Tourism Futures* and the books 'The Future of European Tourism' (2013) and 'Scenario Planning and Tourism Futures' (2024).

Yurena Rodríguez-Rodríguez is a Lecturer at the Department of Business Administration and Economic History, University of La Laguna, Spain. She obtained her PhD from the University of La Laguna. Her research focuses on measuring the economic impacts of tourism, the delineation of tourism destinations, tourism statistics and sustainability, and has been published in journals such as *Current Issues in Tourism*, *European Journal of Tourism Research* and *Journal of Place Management and Development*.

Julia Schiemann studied Tourism Management (BSc) and Management and Planning of Tourism (MSc) at the University of La Laguna on Tenerife, Spain, with distinction. She is the Head of the Sustainability Department of the City of Gersthofen, Germany, and former Research Assistant at the Catholic University of Eichstaett-Ingolstadt and the University of La Laguna, Spain. Her research focuses on sustainable transformation, measuring sustainability on a local scale, and on female entrepreneurship enhancing competitiveness of a destination.

Sarah Schönherr is a Post-Doc University Assistant at the Department of Strategic Management, Marketing and Tourism at the University of Innsbruck, Austria. Her research interests include residents' attitudes towards tourism, quality of life in tourism, as well as responsible and sustainable tourism.

Anna Scuttari is Professor of Empirical Research in Tourism at the Munich University of Applied Sciences (Germany) and Senior Researcher at Eurac Research (Italy). She is the coordinator of the UNWTO-INSTO sustainable tourism observatory in South Tyrol (Italy). Her research is focused on sustainable tourism, destination management, mobilities and affective science in tourism.

Elina Störmann is a Research Associate and PhD student at the Catholic University of Eichstaett-Ingolstadt (Germany), Chair of Tourism / Center for Entrepreneurship. She completed her master's degree in Tourism and Regional Planning at the Catholic University of Eichstaett-Ingolstadt. Her main fields of research are destination management, regional and spatial development, resilience and circular economy.

Hannes Thees is Senior Analyst at the National Competence Center for Tourism in Germany. He obtained a doctoral degree at the Catholic University of Eichstaett-Ingolstadt, where he was a Researcher at the Chair of Tourism / Center for Entrepreneurship. He has working experience across consultancy, tour operation and destination management. His main research fields are sustainable development in tourism and logistics, destination governance, technological implementation and international regional cooperation through multi-level governance.

Anastasia Traskevich studied International Tourism and Economics and worked as a Project Manager, a Management Consultant and a Lecturer in the fields of tourism, hospitality and event management, as well as in the spa and wellness industry. In 2016, she completed her doctorate degree and is researching and publishing in the fields of tourism management, resilience in tourism, the spa industry and health & wellness tourism.

Jan van der Borg teaches Tourism Economics and Management at KU Leuven (Belgium) and Ca' Foscari University, Venice, Italy, where he coordinates the master's degree courses in tourism. He obtained a PhD in Economics from the Erasmus University Rotterdam, Netherlands, in 1991, on a dissertation discussing overtourism in Venice. Building further on this dissertation, the conditions to achieve sustainable urban tourism have been a recurring theme in most of his publications and teaching.

Maximilian Walder is a sociologist at Eurac Research (Italy) with a focus on media, discourses, sport and culture. He studied sociology at the Leopold-Franzens-University in Innsbruck (Austria) and holds a master's degree in Sociology of Media, Culture and the Arts from the Erasmus University Rotterdam, Netherlands.

Felix Windegger is a Researcher at the Center for Advanced Studies of Eurac Research (Italy). He studied Philosophy (BA), Economics (BSc) and Socio-Ecological Economics and Policy (MSc) in Vienna (Austria) and Edinburgh (United Kingdom). His research interests lie in ecological conflicts and social inequalities as well as social-ecological transformation more broadly.

Daniel Zacher is a Research Associate at the Catholic University of Eichstaett-Ingolstadt, Germany. He is currently working on a project establishing transfer structures between science and regional practice. Prior to this, he obtained his doctoral degree at the Chair of Tourism / Center for Entrepreneurship at the Catholic University of Eichstaett-Ingolstadt and conducted numerous destination and regional development studies at the municipal and state levels. His main research areas include community resilience, regional tourism, and destination development, as well as sustainability and transformation research in spatial systems.

Peter Zellmann has been the Director of the Vienna Institute for Leisure and Tourism Research (IFT) since 1987. Within the framework of this activity, he devotes himself primarily to social research and future research, including the areas of lifestyles, work and leisure. He also works as an economic and political consultant. In the context of his research work, he has given numerous lectures and has published a considerable number of papers and books. He is bearer of the Golden Decoration of Merit of the Republic of Austria.

Foreword

Dirk Glaesser

In today's travel and tourism world, addressing the negative impacts of tourism and identifying sustainable solutions have become more critical than ever. The dimensions of contemporary tourism are vast, both in terms of arrival numbers and economic significance, and this trend is set to continue in the coming decades. We are witnessing a transformation in the landscape of tourism destinations, including a shift toward a more responsible, inclusive and conscious travel ethos.

Numerous reasons exist for today's discussions on overtourism. Overtourism may relate to excessive visitor numbers, potentially exacerbated by seasonality or pertaining to excessive adverse visitor impacts. These impacts include issues such as noise disturbance, rowdiness or other disruptions attributed to visitors, and may cause a too strong physical impact on the visitor economy, such as the over-proliferation of hotels, facilities and retail geared to visitors rather than the local population.

Effectively addressing overtourism involves challenging and correcting prevailing myths. It is vital to convey that tourism congestion is not solely a product of visitor volume but is also intimately connected to the capacity to manage it efficiently.

What must concern us most in this context is the lack of proficient management, evidence-based policy approaches and monitored development. While many textbooks highlight these concepts as essential for tourism development, there remains a significant implementation gap that we urgently need to address.

Achieving sustainability in tourism can only occur if its development and management are executed with a well-informed, holistic focus on all three dimensions of sustainability (social, economic and environmental) and if both visitors and the local population are considered equally in the development of attractive destinations. This commitment involves initiatives like engaging local communities, effectively managing congestion, minimising seasonality, adhering to meticulous planning concerning capacity limitations and destination uniqueness, and diversifying products.

Sustainability is not only an aspiration but also a pivotal axis upon which the future of tourism must revolve. As the COVID-19 pandemic has also demonstrated, sustainability is a crucial investment in the resilience of our sector and our societies.

Learning our lessons from these discussions about overtourism will assist us in paving the way for a tourism sector that is economically vibrant, sustainable and resilient, thoughtfully attuned to both visitors and local communities for many generations ahead.

Part I

An Introduction to Overtourism

1 Introduction

Harald Pechlaner, Elisa Innerhofer and Julian Philipp

Overtourism is a phenomenon in many popular tourist destinations around the world. In the affected destinations, the excessive number of tourists is causing infrastructure overload, becoming a major burden on the environment, and impacting the quality of life of local people. Against this background, Harald Pechlaner and Elisa Innerhofer, together with Greta Erschbamer, published an edited volume in 2020 that shows a series of possible solutions and management strategies for dealing with overtourism and the various issues overtourism can impose. The success of the book and the developments of recent years have prompted the publisher and the editors of this volume to publish this new edition on the subject.

Recent developments such as the COVID-19 pandemic, the climate crisis, geopolitical conflicts and inflation have impacts on travel behaviour and tourism. Of these, one of the most significant developments in recent years has been the COVID-19 pandemic, which had a profound impact on the tourism sector worldwide and hit the industry hard. The considerable growth rates of the years prior to the pandemic ended abruptly. The pandemic-related measures led to an unprecedented decline in international travel. The world came to a standstill, and overtourism was no longer an issue. The impact was devastating as tourism, a major source of income, disappeared in many countries. The pandemic forced the travel industry to adapt and develop new strategies in the short term. Virtual tourism and digital platforms have seen increased use to offer travellers an away-from-home experience. In addition, strict hygiene measures and safety protocols were introduced to regain travellers' confidence.

But this unexpected standstill also allowed destinations to think about and reflect on their future development, their sustainability, their dependence on mass tourism and even the kind of target groups they want to attract.

Sustainability has become a central challenge in today's fast-living world. Discussions on the topic encompass the unprecedented rate of acceleration of the climate crisis and environmental change, pollution and the consequences of the COVID-19 pandemic. In this context, tourism as a globally impactful industry has been at the forefront of discussions. In particular, topics such as overtourism, overcrowding and mass tourism developments have received increasing attention recently, amplified by the COVID-19 crisis. What is needed are holistic approaches

DOI: 10.4324/9781003365815-2

and modern, innovative strategies for establishing sustainable and resilient tourism futures. Despite its momentous impacts on the global tourism industry, COVID-19 as a global health crisis represents an opportunity to rethink tourism completely.

The first impression usually is that overtourism is a negative extremity of increasing masses of tourists at particular places. This makes sense, as overcrowded places, overloaded infrastructures or exceeded capacities are not among the implications or effects that destination managers and tourism development strategies aim for. Overtourism certainly needs to be tackled, but the solutions and strategies to deal with overtourism are varied, and there is no "one size fits all" approach or blueprint as every region, destination or attraction has different characteristics, surroundings and stakeholders. Therefore, we argue that overtourism can be a starting point and an engine for transformation processes. In a certain way, all those destinations which are more or less affected by overcrowding and overtourism tendencies should interpret this as a chance to change the circumstances of their tourism model. Integration, collaboration and consensus among stakeholders and decision-makers are vital components of any discussion on more sustainable practices and strategies in the tourism industry. To find a balance between growth and sustainability and to focus on responsible, accountable, reliable and ethical practices, sustainability governance is needed.

About the Book

The volume includes original contributions from renowned authors and scholars in the field. The volume is interdisciplinary in coverage and international in scope. It includes three sections that describe the transition from overtourism to sustainability governance and elaborate perspectives for developing resilient destinations. Each section (Part I, Part II, Part III) will include Excursus (case studies) to show applications of the topics and issues discussed.

The chapters in Part I provide an overview of the current academic discussion on overtourism, outline the developments related to overtourism, present theoretical concepts such as carrying capacities, and illustrate the impact of the COVID-19 pandemic on tourism as well as the development from overtourism to no-tourism. The section excursus presents two case studies that strengthen the understanding of overtourism.

Part II discusses approaches towards a new sustainability, including a possible new understanding of tourism in a post COVID-19 world. Sustainability is discussed in the context of various aspects, covering current research as well as future trends. The section excursus presents two case studies that describe destinations and their measures and efforts on the pathway to sustainable tourism.

The book continues with a section on strategies to deal with overtourism (Part III). This section goes more deeply into management strategies and governance theories. A wide range of examples taken from various countries explore the interface between tourism, overtourism and sustainability. The section excursus presents a case study focusing on the concept of destination resilience.

The book targets academics and researchers in the fields of tourism, over- and mass-tourism, as well as sustainability governance. It will also be of interest for practitioners. The aim is to give practitioners insights into how tourism flows and how perceived overtourism may be managed to design sustainable and resilient destinations and to avoid negatively impacting the attitudes of residents towards tourists. The book is intended to help make the entire industry more sustainable and resilient and less vulnerable to crises.

2 From Carrying Capacity to Overtourism

The Changing Perspective in the Course of Time

Marcus Herntrei and Veronika Jánová

1 Introduction

Traditionally, the success of the tourism industry has been measured in terms of visitor numbers, following the growth-focused mindset of tourism policymakers and tourism service providers (Dodds & Butler, 2019). The focus in most destinations has been predominantly on creating unique visitor experiences and increasing visitor satisfaction (German Tourism Association [DTV] & German Institute for Tourism Research [DIFT], 2022). Accordingly, the perception of what a tourism destination is has been strongly shaped by the customer- and market-oriented perspective (Pechlaner, 2003) and the tourism destination has been regarded as a competitive unit (Herntrei, 2014) in which the well-being of *local communities* represents merely an efficiency disadvantage (Cracolici, Nijkamp & Rietveld, 2006). In recent years, a new direction can be observed in tourism – alongside tourism providers on the one side and visitors on the other side, local communities as a new stakeholder group are increasingly coming to the attention of destination planners and managers (Becken & Simmons, 2019).

The shift of attention in the destination work outlined above was sparked by a surprisingly high number of public anti-tourism protests globally in 2017/2018, indicating sharply decreasing *tourism acceptance* (Herntrei, 2019). The term *overtourism* was subsequently coined and has received considerable attention in the global media (Gössling, McCabe & Chen, 2020). Since then, overtourism also gradually gained increased attention from the scientific community and became the focus of numerous studies (Eckert, Zacher, Pechlaner, Namberger & Schmude, 2019; Goodwin, 2019, 2021; Koens, Postma & Papp, 2018; Pechlaner, Innerhofer & Erschbamer, 2019; Peeters et al., 2018; World Tourism Organisation [UNWTO], 2018).

Despite the topicality of the term overtourism, the phenomenon itself is not new (Dredge, 2017). Its underlying issues have been the subject of academic consideration since the 1970s (Gössling, McCabe & Chen, 2020). Even though the phenomenon as such is well known, capturing its essence is challenging given its existing diversity of meanings. The proximity and partially synonymous use of terms such as *overcrowding* underline the resulting ambiguity and overtourism remains open to multiple interpretations (Bauer, Gardini & Skock, 2020). Moreover, in social

DOI: 10.4324/9781003365815-3

science tourism research, the opinion is increasingly appearing that the complexity of the phenomenon cannot be adequately expressed solely in numbers (e.g. through the calculation of *carrying capacities*) and that it is defined by destination-specific acceptance of tourism by local communities (Carvalho, Guerreiro & Matos, 2020; Herntrei, Pillmayer, Scherle & Nikitsin, 2022). However, the concept of carrying capacity is considered the precursor to the current concerns with overtourism (Wall, 2020) and continues to be used as a framework to interpret the phenomenon (Benner, 2020; Peeters et al., 2018; Postma, Koens & Papp, 2020).

In pursuit of contributing to the contemporary understanding of overtourism, this article explores its origins and development, reviews and evaluates the (in) appropriateness of using the concept of carrying capacity in relation to overtourism, and defines overtourism while addressing its distinction from overcrowding. Possible implications for tourism practice and research are consequently discussed.

2 The Origin and Development of Overtourism

The deteriorating quality of life of local inhabitants is central to the protests against tourism in recent years (Milano, Novelli & Cheer, 2019). The ongoing debates on overtourism thus draw on the negative social/socio-cultural impacts of tourism on host communities – one of the traditional areas of tourism research since the second half of the twentieth century.

2.1 The 1960s: Tourism as a Driver of Economic Growth. Relatively Positive and Uncritical Approaches to Tourism

Early approaches to tourism in the 1960s were relatively uncritical. Recognising that international tourism trade generates foreign exchange earnings and stimulates employment, the importance of tourism for economic development was widely accepted (Davis, 1967). In addition to its economic value, tourism was seen as providing an important contribution to cultural understanding (McIntosh, 1964) and local and national tourism authorities encouraged its development and promotion. Jafari (2005) speaks about *the advocacy platform* during which tourism studies emphasised the positive economic impacts of tourism to justify the industry's focus on growth. Although the first voices critical of the negative tourism-induced socio-cultural impacts on local communities were raised in the 1960s (Forster, 1964), tourism-related criticism was initially rather sporadic, only appearing regularly in later decades.

2.2 The 1970s: Limits to Growth Recognised. Tourism Is Increasingly Coming Under Critical Scrutiny

In the 1970s, the negative consequences of tourism, such as economic dependence on a single industry, employment fluctuation, loss of cultural identity and environmental degradation, were increasingly recognised (Pizam, 1978). It was acknowledged that growth has its limits (Meadows, Meadows, Randers & Behrens, 1972) and tourism development should not focus solely on economic progress and prosperity while neglecting the social and environmental costs. Tourism studies shifted from the positive economic impacts of tourism to its negative socio-cultural and environmental impacts, and the 1970s were subsequently termed *the cautionary platform* (Jafari, 2005).

Young's "Tourism: Blessing or Blight?" (1973), Krippendorf's "Die Landschaftsfresser" (The Landscape Eaters) (1975), and Turner's and Ash's "The Golden Hordes" (1975) rank among the many influential studies from the 1970s with distinctly articulated warnings about the ambivalent implications of constantly growing tourism. Further notable scientific considerations from this period pointing to the concerns that excessive tourism harms the environment, leads to negative attitudes of local communities towards tourism, and causes social tensions, include those of Butler (1974), Doxey (1975) and Pizam (1978). In the late 1970s, Rosenow and Pulsipher (1979) coined the term "visitor overkill" and recognised seasonal visitor pressure, inappropriate visitor behaviour and adverse environmental impacts as the major negative impacts of tourism. Boissevain (1979) analysed the impacts of tourism on the Mediterranean island of Gozo, observing the increasing resentment of local inhabitants towards Maltese tourists – (also) due to pollution and the growing dependence of the Gozitans on tourism from Malta. Without specifically referring to the term overtourism, the aforementioned works already approached the contemporary understanding of the phenomenon (see section 4).

2.3 The 1980s: Alternative Tourism Concepts, Tourism Acceptance, Participatory Planning Approaches and Carrying Capacity at the Forefront

In the 1980s, tourism continued to grow strongly due to ongoing prosperity in the Western world. In response to the two previously mentioned platforms for and against tourism, the 1980s saw the emergence of *the adaptancy platform* (Jafari, 2005) in tourism studies, exploring alternative and sustainable forms of tourism development. The concept of soft tourism (Jungk, 1984) falls into this period, followed by ecotourism, which was further examined in the 1990s (Wood, 1991).

Furthermore, in the 1980s (and 1990s), examining the local communities' perceptions of the impacts of tourism formed a traditional area of tourism research and tourism acceptance became a central focus of numerous studies (Ap, 1992; Brougham & Butler, 1981; Dogan, 1989; Getz, 1994; King, Pizam & Milman, 1993; Murphy, 1981; Perdue, Long & Allen, 1990). Moreover, the adaptancy platform approach acknowledged the need for all stakeholders to benefit from tourism (Postma, Koens & Papp, 2020). Local communities were recognised as key actors in the tourism industry (Ap, 1992) and participatory tourism planning approaches were increasingly encouraged (Haywood, 1988; Jamal & Getz, 1995; Keogh, 1990; Murphy, 1983; Simmons, 1994; Timothy, 1999; Tosun, 1999).

Additionally, in the 1980s, the attention of many studies was centred on the carrying capacity of a destination – be it in the form of tourism area life cycle (TALC) (Butler, 1980), the tourism saturation point (UNWTO, 1983) or the interrelated dimensions of tourism carrying capacity (O'Reilly, 1986). A more detailed overview of the concept of carrying capacity is provided in section 3.

2.4 The 1990s: Contradictions Within Science: A Holistic View of Tourism Complexity or Destinations as Mere Competitive Units?

Along with ongoing tourism impact studies and academic calls for participatory approaches in tourism (see section 2.3), the 1990s were marked by the formation of *the knowledge platform* (Jafari, 2005). Tourism was acknowledged as a global industry with both desirable and undesirable impacts. Focusing on the specific impacts of tourism was found to be insufficient to provide a holistic view of its complexity (Postma, Koens & Papp, 2020). Rather than simply seeking a maximum number of visitors to set limits on tourism growth (see section 3), the emphasis shifted to recognising the importance of local conditions and management objectives as a prerequisite for addressing the appropriate levels of use. Using theories from disciplines such as ecology, economics and system dynamics, comprehensive alternative planning and management frameworks were established. The Limits of Acceptable Change (LAC) (McCool, 1994) exemplify this development.

However, in parallel with the knowledge platform, the discussion about the competitiveness of tourism destinations inspired by Porter (1979, 1980) entered the field of tourism science in the 1990s (Herntrei, 2014). With the view that destinations should be managed as competitive units, the needs of markets (profit and efficiency) and the needs of visitors (satisfaction and experience) shaped the prevailing paradigm. Consequently, little emphasis was placed on the needs of local communities, whose well-being was seen as a mere constraint to efficiency (Cracolici, Nijkamp & Rietveld, 2006). Following the liberal and globalised economic order, the government played an indirect role in guiding the economy and creating the preconditions for sustainable growth (Boughton, 2022). Since the 1990s, greater responsibility was given to tourism stakeholders who often lack the necessary skills and knowledge to act sustainably (Koens, Postma & Papp, 2018). As such, the 1990s saw numerous public protests as a response to constant tourism growth – e.g. in Spain, Italy, France and Malta (Boissevain, 1996). "If in the 1960s tourists were welcomed with pride and native hospitality, by the beginning of the 1990s the welcome seems less enthusiastic" (Boissevain, 1996, p. 7).

2.5 The 21st Century: Social (Urban) Unrest and the Advent of Overtourism

In the twenty-first century, the growth of tourism has been driven by increasing global welfare, decreasing travel costs (low-cost carriers), the spread of the sharing economy, and the power of social media (Dodds & Butler, 2019; Goodwin, 2019). Concurrently, economic and political crises ranging from the Arab Spring to terrorist attacks in Islamic countries to Turkey's coup attempt have led to a spatial shift in international tourism flows to European destinations that had traditionally been perceived as safe (Herntrei, 2019). Particularly in the second half of the 2010s, many European destinations such as Spain, Italy and Germany experienced record visitor numbers (UNWTO, 2019). However, unlike the tourism industry, which celebrated its continuous growth as a sign of success (McKinsey & Company & World Travel & Tourism Council [WTTC], 2017), local inhabitants have been expressing less enthusiasm. As the negative socio-cultural and socio-economic impacts of tourism have become a more prominent concern in a wider variety of destinations, public resentment has been increasingly manifested in various forms – from public initiatives, social movements, and activist groups to radical forms of protests such as vandalism (Herntrei, 2019). In 2017/2018, such developments were reported from Barcelona, Palma de Mallorca, Venice, Rome, Dubrovnik, Lisbon, Amsterdam, Berlin and Prague (Milano, Novelli & Cheer, 2019).

While public protests against tourism are not new (see section 2.4), protests leading up to the COVID-19 pandemic were more "organised, vocal, and politically active" (Gössling, McCabe & Chen, 2020, p. 1). Particularly in Europe, several urban social movements have arisen, including the Assembly of Neighbourhoods for Sustainable Tourism in Barcelona, the No Big Ships Committee in Venice, the Naplesland – Rights in the Age of Tourism in Naples, and the community group People Live Here in Lisbon. Out of these (and other) movements, the Network of Southern European Cities against Touristification was formed in 2018, through which 16 cities have joined their forces in engaging in protest actions against the growth-centred model of urban tourism development (Milano, Cheer & Novelli, 2019). The increasing anti-tourism sentiments have been demonstrated by the simultaneous appearance of signs, involving graffiti with slogans such as "No more rolling suitcases" and "Tourists f*** off" in Berlin (Novy & Colomb, 2017), "Tourists go home" and "Tourist: your luxury trip, my daily misery" in Barcelona (Burgen, 2018), or "Tourism kills the city" in Palma de Mallorca (Zeit Online, 2017). Consequently, public discontent with tourism development has been associated with terms such as tourismphobia, anti-tourism movements, overcrowding (McKinsey & Company & WTTC, 2017; Peeters et al., 2018) and – since 2017 – particularly overtourism (Kagermeier & Erdmenger, 2019a).

The term overtourism has proven to be very marketable and was trademarked by the market research company Skift in 2018 (Koens, Postma & Papp, 2018). Having become particularly popular among the global mass media (Novy & Colomb, 2019; UNWTO, 2018), overtourism has developed into an overused and under-conceptualised buzzword. Nevertheless, overtourism has received considerable parallel attention from the scientific community. Several edited books (e.g. Dodds & Butler, 2019; Milano, Cheer & Novelli, 2019; Pechlaner, Innerhofer & Erschbamer, 2019) and numerous journal articles (e.g. Arlt, 2018; Bauer, Gardini & Skock, 2020; Eckert, Zacher, Pechlaner, Namberger & Schmude, 2019; Goodwin, 2019, 2021; Gössling, McCabe & Chen, 2020; Kagermeier & Erdmenger, 2019a, 2019b; Koens, Postma & Papp, 2018; Mihalic, 2020; Milano, Novelli & Cheer, 2019; Wall, 2020) underline the academic interest in the phenomenon. Furthermore, a notable amount of literature has been published by institutions such as the European Parliament (Peeters et al., 2018), the UNWTO (2018) and the McKinsey & Company and WTTC (2017).

As stated previously, overtourism is not a new phenomenon. Its potentially disruptive occurrences and risks associated with the excessive development of tourism to host communities have been increasingly recognised by the scientific community since the 1970s (see section 2.2). Dredge (2017) argues that focusing on overtourism as a *new* concern is thus merely "resetting the clock on well-established debates" (para. 4). Nevertheless, shaped by contemporary developments and reflecting wider societal, technological and mobility advancements, overtourism is a constantly evolving phenomenon that remains subject to multiple interpretations (Bauer, Gardini & Skock, 2020).

The following two sections focus on providing a more detailed insight into the topics of carrying capacity and overtourism – among others, to evaluate the frequent use of the first concept to interpret the latter.

3 Carrying Capacity

Carrying capacity has a long research tradition, reaching back to the 1930s (Dhondt, 1988). The concept was initially applied in relation to domesticated and wild herbivores (Caughley, 1976 in Dhondt, 1988) based on the notion that the availability of suitable conditions for living determines the number of organisms that can exist in a given environment (Carey, 1993). Concerns about large increases in visitation to national parks and protected areas in the US during the 1950s led to calls for finding a rationale to limit use to protect quality recreation (McCool & Lime, 2001; Wagar, 1974). The intuitively appealing concept of numerical carrying capacity has therefore shifted from wildlife management to recreation management, where it has been widely applied since the 1960s (Frissell & Stankey, 1972; Wall, 2020) and, subsequently, to tourism management, where it has been widely applied since the 1980s (Butler, 1980; O'Reilly, 1986; UNWTO, 1983). However, attempts to transfer carrying capacity to socio-economic sectors, comprising recreation and tourism management, have not been successful (Saarinen, 2006; Seidl & Tisdell, 1998). Some of the arguments against carrying capacity are listed in the following sections.

3.1 Carrying Capacity in Recreation Management: From the Focus on Numbers to the Focus on Objectives

Early definitions of recreational carrying capacity addressed the impacts of visitation on an area from environmental (biophysical) and experiential (social) points of view. Carrying capacity was defined as the maximum level of use that will not adversely affect the quality of the environment or the recreational experience. The underlying oversimplified assumption, thus, was that with growing numbers of users, environmental damage increases and the quality of experience decreases (Frissell & Stankey, 1972; McCool & Lime, 2001; Wall, 2020). However, already in the 1960s, research started to reveal that high levels of use do not necessarily lead to decreased recreational experience and carrying capacity depends on human needs and value judgments (Wagar, 1964). In the 1970s, it was recognised that appropriate levels of use vary with the objectives that are established for an area (Frissell & Stankey, 1972). Correspondingly, each area has multiple carrying capacities. The same area might be designated as a nature reserve, golf course or theme park, thus having different implications for appropriate levels of use (McCool & Lime, 2001;

Wall, 2020). Arising from this recognition, carrying capacity was approached as a relative condition rather than a specific number (Washburne, 1982). In addition, it was acknowledged that the relationship between the level of use and amount of impact is intervened by variables other than just visitor numbers – e.g. the type of visitors and visitor behaviour (Washburne, 1982). Following the shift from the focus on numbers to the focus on objectives, broader alternative planning and management frameworks were developed, including the Recreation Opportunity Spectrum (Clark & Stankey, 1979), LAC (McCool, 1994) and Visitor Impact Management (Graefe, Kuss & Vaske, 1990).

As such, particularly in the 1980s and the 1990s, the concept of numerical carrying capacity was widely questioned (Buckley, 1999; Dhondt, 1988; Lindberg, McCool & Stankey, 1997; Washburne, 1982). Besides the aforementioned limitations, carrying capacities are far from being universal constants (Seidl & Tisdell, 1998). The numerical approach leads to misguided simplicity, implying that objective criteria exist and that these are transferable from destination to destination (Lindberg, McCool & Stankey, 1997). Therefore, the traditional concept of carrying capacity is inappropriate (McCool & Lime, 2001), if not meaningless (Buckley, 1999), since it "carries a number of assumptions that are unsupported in the real world" (McCool & Lime, 2001, p. 372). It is a concept that should be avoided (Lindberg, McCool & Stankey, 1997).

3.2 Carrying Capacity in Tourism Management: Focus on (Numerical) Social Carrying Capacity

Due to growing concerns about the environmental and social impacts of steadily increasing tourism development, calls for establishing carrying capacities for tourism emerged and, in the 1980s, 1990s, and at the beginning of the new millennium, the carrying capacity of a destination became one of the central areas of tourism research (Butler, 1980, 1996; Coccossis, Mexa, Collovini, Parpairis & Konstandoglou, 2002; De Ruyck, Soares & McLachlan, 1997; Getz, 1983; O'Reilly, 1986; Saveriades, 2000; Swarbrooke, 1999; UNWTO, 1983). In other words, the shortcomings of the concept were ignored and carrying capacity entered the tourism literature at a time when its limitations were widely acknowledged within recreation settings (Wall, 2019).

Increasingly, though, attempts to convey the complexity of carrying capacity have arisen (Getz, 1983; O'Reilly, 1986; Swarbrooke, 1999). Contrary to the original notion, more attention has been dedicated to its social component, reflecting the perspectives of host communities. Having observed that host communities have a certain psychological tolerance threshold beyond which their perceptions of tourism become negative and might result in unwelcoming behaviour towards visitors (Doxey, 1975; Pizam, 1978), social/socio-cultural carrying capacity – in simple terms, the level of tolerance of host communities to the presence and behaviour of visitors (O'Reilly, 1986; Saveriades, 2000) – has been delineated (Saveriades, 2000; Swarbrooke, 1999). Social carrying capacity has been measured through local communities' perceptions of tourism impacts. Identifying social carrying capacity thresholds has therefore been particularly problematic since it relies entirely on subjective value judgements. Consequently, attempts to determine social carrying capacity in terms of visitor numbers (De Ruyck, Soares & McLachlan, 1997; Saveriades, 2000) have not been without criticism (McCool & Lime, 2001).

Until the second half of the 2010s, the research interest in carrying capacity slightly decreased. However, the growing resistance of local communities to the increasing touristification of their living space – a challenge labelled as overtourism – has led to a renewed interest in the topic, particularly in Europe (Mihalic, 2020; Namberger, Jackisch, Schmude & Karl, 2019; Postma, Koens & Papp, 2020; Tokarchuk, Barr & Cozzio, 2021). The concept of (social) carrying capacity has been applied in the pursuit of framing and defining overtourism, often in quantitative terms (Tokarchuk, Gabriele & Maurer, 2021) and in complex urban destinations where its application has been fraught with difficulties (Namberger, Jackisch, Schmude & Karl, 2019). As summarised by Wall (2020), "concern with and application of carrying capacity has migrated across continents, from natural to urban settings, and has re-emerged as 'overtourism'" (p. 213).

4 Overtourism

The term overtourism describes a complex socio-psychological phenomenon that is simplistically approached as a tipping point of tourism development beyond which adverse impacts occur (Milano, Cheer & Novelli, 2019). The phenomenon is considered to be the ultimate consequence of the growth-oriented mindset of governments, marketing organisations, and providers of tourism services combined with the absence of a long-term view among decision-makers (at all levels) (Dodds & Butler, 2019; Goodwin, 2019). Overtourism has been mostly associated with urban settings (Peeters et al., 2018). However, it may also arise in rural areas, coastal and island environments, or natural and cultural heritage sites. Apart from locations such as Barcelona, Venice and Amsterdam, signs of the phenomenon have been reported from destinations as varied as rural Bavaria (Hockenos, 2020), the Isle of Skye, Bali, Reykjavík and Santorini (Milano, Cheer & Novelli, 2018; Smith, 2018), and the Glacier, Rocky and Yosemite National Parks (Girma, 2021). Therefore, overtourism represents a global challenge that is not specific to a particular type of destination. Overtourism is a temporally and spatially delineated phenomenon that mostly occurs seasonally and is predominantly limited to a few sub-areas of a tourism destination (Peeters et al., 2018). An overview of the definitions of overtourism that have been introduced in recent years is provided in the following section.

4.1 Beyond the Threshold? An Overview of the Definitions of Overtourism

Kirstges (2020) defines overtourism as "the temporary overcrowding of a tourist destination by too many tourists" (p. 103), thus approaching the phenomenon from a quantitative perspective. According to Goodwin (2019), "overtourism describes destinations where hosts or guests, locals or visitors, feel that there are too many visitors and that the quality of life in the area or the quality of the experience has deteriorated unacceptably" (p. 110). This definition implies the subjective nature of overtourism that arises from individual evaluation of the perceived negative impacts of tourism, either from the perspective of local inhabitants or visitors. Gössling, McCabe and Chen (2020) posit the phenomenon as "a psychological reaction of residents to tourist pressure, in which place–person interrelationships are affected and damaged, triggering different types of emotional and behavioural responses" (p. 3). The above definitions by Goodwin (2019) and Gössling, McCabe and Chen (2020) highlight the importance of local inhabitants in discussions on overtourism. Accordingly, overtourism is explicitly connected with the perspective of local inhabitants and their subjective perceptions of tourism impacts.

Peeters et al. (2018), Postma, Koens and Papp (2020) and Benner (2020) draw on carrying capacity when defining overtourism. Acknowledging the limitations of the traditional concept of carrying capacity and the complex nature of overtourism, the authors apply the concept to overtourism only in a broader, not (solely) quantitative sense. Peeters et al. (2018) address the multidimensionality of carrying capacity (O'Reilly, 1986) and refer to overtourism as "the situation in which the impact of tourism, at certain times and in certain locations, exceeds physical, ecological, social, economic, psychological, and/or political capacity thresholds" (p. 19). Postma, Koens and Papp (2020) underline that each capacity threshold varies from destination to destination. A similar point of view is held by Benner (2020), describing overtourism as a scenario where a destination's tolerance for tourism is exceeded. A destination's tolerance is argued to be a relative, subjective and destination-specific term and overtourism is understood as a phenomenon whose complete picture cannot be captured by a quantitative view alone – qualitative factors such as visitor behaviour also need to be considered (Benner, 2020).

4.2 Overcrowding as a Synonym for Overtourism? The Distinction of the Terms

Although the qualitative essence of overtourism has been suggested in the above definitions, the quantitative component in terms of overcrowding remains an inherent part. Nevertheless, while overcrowding may be one of the factors that led to the subsequent occurrence of overtourism, it does not automatically imply the occurrence of overtourism per se (Butler, 2019; Koens, Postma & Papp, 2018; Wall, 2019). Research has shown that destinations with high objective tourism criteria such as tourism density and intensity may not face the state of overtourism if local communities' perceptions of tourism impacts are favourable. South Tyrol is a concrete example of a destination whose inhabitants consider tourism to be a particularly important part of the local economy (Eurac Research, 2021). On the contrary, in the Bavarian destination of Tölzer Land, increasing resentment of local inhabitants towards same-day visitors (from Munich) manifested by public protests against tourism has been observed (Vecchiato, 2020) – not least due to tourism-induced traffic issues, litter in the landscape, and a low added value for the destination as such.

These examples distinctly point to the insufficiency of objective tourism criteria to interpret overtourism (Kagermeier & Erdmenger, 2019a) and demonstrate that overcrowding in the sense of high density is not an appropriate term to describe overtourism accurately. The qualitative, social component of overtourism in terms of the subjectively perceived impacts of tourism on the quality of life and well-being of host communities is essential. Whether overcrowding evolves into overtourism thus ultimately depends on destination-specific acceptance of tourism by local inhabitants (Herntrei, Pillmayer, Scherle & Nikitsin, 2022).

Consequently, approaches striving for quantification of overtourism – either by calculating social carrying capacity as a threshold value for the starting point of overtourism (Tokarchuk, Gabriele & Maurer, 2021) or by developing sets of quantifiable indicators of overtourism, whether in the form of early warning tools (McKinsey & Company & WTTC, 2017) or checklists (Peeters et al., 2018) – have been fraught with criticism (Arlt, 2018). As implied in the previous paragraphs, local communities' subjective judgement is inherent in the process of perceiving and evaluating the impacts of tourism, making it infeasible to assess overtourism in terms of any carrying capacity and/or overcrowding per se. Assessing overtourism as a destination-specific situation where a threshold in terms of local communities' tolerance to tourism is exceeded and their perceptions of tourism impacts change from positive to negative is conceptually agreeable. However, such a threshold cannot be expressed numerically since overtourism is essentially a qualitative and highly subjective phenomenon (Arlt, 2018; Gössling, McCabe & Chen, 2020).

4.3 Benefits Below Costs? Overtourism as a Result of the Unsatisfactory Tourism Impact Weighing Process from the Perspective of Local Communities

Overtourism is based on the perceptions of tourism impacts from the perspective of local communities (Gössling, McCabe & Chen, 2020). Given the complexity and, at times, the ambiguity of tourism impact studies (Herntrei, 2019), Ap's (1992) Social Exchange Process Model is considered a fundamental contribution to theory development in the field of local communities' perceptions of tourism (Getz, 1994). Expressed in simple terms, the model suggests that local inhabitants evaluate tourism development in the destination according to the benefits that tourism brings to their economic, social and psychological needs – thus according to the extent to which tourism contributes to an increase in their quality of life. Accordingly, local inhabitants are supposed to be supportive of tourism only if the perceived positive effects of tourism (benefits) outweigh its perceived negative effects (costs). Applying the Social Exchange Process Model to the debate on overtourism, the phenomenon can be defined as a situation in which tourism has a negative impact on quality of life in the perception of local communities because the local communities perceive the effects of tourism as more negative than positive.

5 Conclusion and Outlook

The discussion on tourism is influenced by the way the destination is perceived. Conversely, the way the destination is perceived is influenced by the current social, political and economic spirit of the time. The advent of overtourism has brought a renewed interest in the perceptions of local communities compared to previous decades in which the focus was on the perceptions of visitors and their experiences (Becken & Simmons, 2019). Following this shift of attention, local communities are increasingly placed at the centre alongside the developments and needs of the market.

Regardless of how tourism is planned, developed and organised in a destination, the role of local communities in the tourism industry has been repeatedly recognised as crucial (Ap, 1992; DTV & DIFT, 2022; Eckert, Zacher, Pechlaner, Namberger & Schmude, 2019; Haywood, 1988; Herntrei, 2014, 2019; Herntrei, Pillmayer, Scherle & Nikitsin, 2022; Murphy, 1983). The culture and hospitality of local inhabitants are an integral part of the tourism product, attracting visitors to the destination itself (Murphy, 1983). Local inhabitants thus act in various roles as hosts and co-creators of visitor experiences and are powerful ambassadors of a destination's brand (Herntrei, Pillmayer, Scherle & Nikitsin, 2022). Consequently, a successful destination needs not only satisfied visitors who feel welcomed but also satisfied local inhabitants who feel comfortable with the local development of tourism (DTV & DIFT, 2022). However, in many sought-after destinations, the opposite seems to be true. Insisting on growth-focused tourism strategies and planning models centred on the needs of visitors and markets – as has been the case for decades – does not seem to be an advisable option against the background of growing dissatisfaction. As destinations have become to be seen primarily as living spaces rather than competitive units, the ultimate objective of destination development needs to be the quality of life rather than the economic objective of competitiveness (Herntrei, 2014).

From this perspective, the concept of carrying capacity is obsolete since it fits neither into the contemporary discussions on overtourism nor into destination development per se (see Figure 2.1). Carrying capacity implies the search for the maximum number of visitors an area can absorb, drawing on the question of how

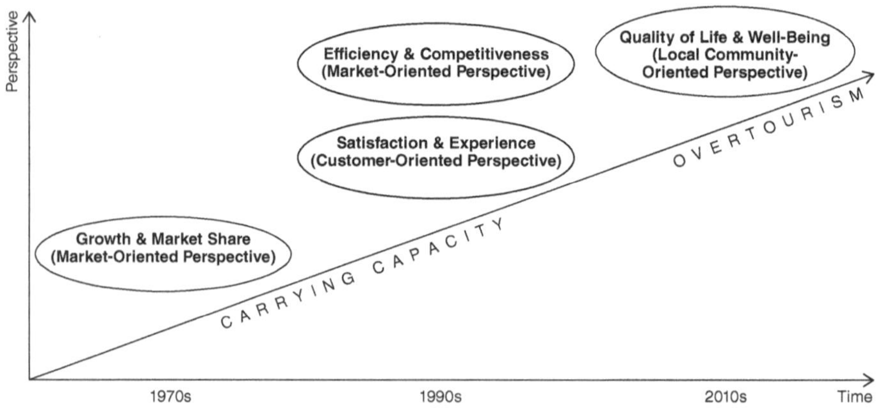

Figure 2.1 Carrying capacity and overtourism in the course of time: A question of changing perspective

Source: Own Illustration

much pressure can be put on a destination before its limits are exceeded. Indeed, that notion contradicts the shift of attention outlined above. Rather, the question of how tourism can contribute to the quality of life of local inhabitants has become relevant and this question will determine the destination development of the future – development based on qualitative growth and social responsibility.

Since the tourism industry has a strong impact on the living space of local communities and thus on their quality of life, the legitimisation of tourism is necessary. Otherwise, tourism will be replaced by other profitable forms of land use. As such, the quality of life of local communities needs to become a key success indicator of a thriving tourism industry of the future. Furthermore, the tourism industry must extend decision-making beyond the policy and business spheres and consider the long-term interests of host communities on which it rests. Therefore, destination development and management models based on (implemented!) participatory planning approaches are required, enabling local communities to benefit from tourism and shape the nature of future destinations – destinations that will be attractive to visitors and liveable for local communities.

References

Ap, J. (1992). Residents' perceptions on tourism impacts. *Annals of Tourism Research, 19*(4), 665–690. https://doi.org/10.1016/0160-7383(92)90060-3.

Arlt, W. (2018). Overtourism als Weckruf für die Tourismusindustrie und die Tourismuswissenschaft [Overtourism as a wake-up call for the tourism industry and tourism science]. *Forum Wohnen und Stadtentwicklung, 10*(2), 63–66.

Bauer, A., Gardini, M. A., & Skock, A. (2020). Overtourism im Spannungsverhältnis zwischen Akzeptanz und Aversion [Overtourism in the tension between acceptance and aversion]. *Zeitschrift für Tourismuswissenschaft, 12*(1), 88–114. https://doi.org/10.1515/tw-2020-0014.

Becken, S., & Simmons, D. (2019). Stakeholder management: Different interests and different actions. In R. Dodds & R. W. Butler (Eds.), *Overtourism: Issues, realities and solutions* (pp. 234–249). Berlin, Boston: De Gruyter.

Benner, M. (2020). The decline of tourist destinations: An evolutionary perspective on overtourism. *Sustainability, 12*(9), 3653.

Boissevain, J. (1979). The impact of tourism on a dependent island: Gozo, Malta. *Annals of Tourism Research, 6*(1), 76–90. https://doi.org/10.1016/0160-7383(79)90096-3.

Boissevain, J. (Ed.). (1996). *Coping with tourists: European reactions to mass tourism.* Providence, Oxford: Berghahn Books.

Boughton, J. M. (2022). Globalisation and the silent revolution of the 1980s. International Monetary Fund. www.imf.org/external/pubs/ft/fandd/2002/03/bought.htm.

Brougham, J. E., & Butler, R. W. (1981). A segmentation analysis of resident attitudes to the social impact of tourism. *Annals of Tourism Research, 8*(4), 569–590. https://doi.org/10.1016/0160-7383(81)90042-6.

Buckley, R. (1999). An ecological perspective on carrying capacity. *Annals of Tourism Research, 26*(3), 705–708.

Burgen, S. (2018, June 25). "Tourists go home, refugees welcome": Why Barcelona chose migrants over visitors. *The Guardian.* www.theguardian.com/cities/2018/jun/25/tourists-go-home-refugees-welcome-why-barcelona-chose-migrants-over-visitors.

Butler, R. W. (1974). The social implications of tourist developments. *Annals of Tourism Research, 2*(2), 100–111. https://doi.org/10.1016/0160-7383(74)90025-5.

Butler, R. W. (1980). The concept of a tourist area cycle of evolution: Implications for management of resources. *Canadian Geographer, 24*(1), 5–12. https://doi.org/10.1111/j.1541-0064.1980.tb00970.x.

Butler, R. W. (1996). The concept of carrying capacity for tourism destinations: Dead or merely buried? *Progress in Tourism and Hospitality Research, 2*(3–4), 283–293.

Butler, R. W. (2019). Overtourism and the tourism area life cycle. In R. Dodds & R. W. Butler (Eds.), *Overtourism: Issues, realities and solutions* (pp. 76–89). Berlin, Boston: De Gruyter.

Carey, D. I. (1993). Development based on carrying capacity: A strategy for environmental protection. *Global Environmental Change, 3*(2), 140–148.

Carvalho, F. L., Guerreiro, M., & Matos, N. (2020). Overtourism: A systematic review of literature. In C. Ribeiro de Almeida, A. Quintano, M. Simancas, R. Huete, & Z. Breda (Eds.), *Handbook of Research on the Impacts, Challenges, and Policy Responses to Overtourism* (pp. 12–36). IGI Global.

Clark, R. N., & Stankey, G. H. (1979). *The recreation opportunity spectrum: A framework for planning, management, and research* (Vol. 98). US Department of Agriculture, Forest Service, Pacific Northwest Forest and Range Experiment Station.

Coccossis, H., Mexa, A., Collovini, A., Parpairis, A., & Konstandoglou, M. (2002). *Defining, measuring and evaluating carrying capacity in European tourism destinations*. Environmental Planning Laboratory, Athens.

Cracolici, M. F., Nijkamp, P., & Rietveld, P. (2006). Assessment of tourism competitiveness by analysing destination efficiency. https://papers.ssrn.com/sol3/papers.cfm?abstract_id=942729.

Davis, H. D. (1967). Investing in tourism. *Finance and Development, 4*(1), 1–8.

De Ruyck, M. C., Soares, A. G., & McLachlan, A. (1997). Social carrying capacity as a management tool for sandy beaches. *Journal of Coastal Research*, 822–830.

Dhondt, A. A. (1988). Carrying capacity: A confusing concept. *Acta Oecologica, 9*(4), 337–346.

Dodds, R., & Butler, R. W. (Eds.). (2019). *Overtourism: Issues, realities and solutions*. Berlin, Boston: De Gruyter.

Dogan, H. Z. (1989). Forms of adjustment: Sociocultural impacts of tourism. *Annals of Tourism Research, 16*(2), 216–236. https://doi.org/10.1016/0160-7383(89)90069-8.

Doxey, G. V. (1975). A causation theory of visitor-resident irritants: Methodology and research inferences. In *Travel and Tourism Research Association's Sixth Annual Conference Proceedings* (pp. 195–198). San Diego, CA.

Dredge, D. (2017, September 13). "Overtourism" old wine in new bottles? LinkedIn. www.linkedin.com/pulse/overtourism-old-wine-new-bottles-dianne-dredge.

Eckert, C., Zacher, D., Pechlaner, H., Namberger, P., & Schmude, J. (2019). Strategies and measures directed towards overtourism: A perspective of European DMOs. *International Journal of Tourism Cities, 5*(4), 639–655.

Eurac Research. (2021, February 15). Zwischen Skepsis und Solidarität – die Wahrnehmung des Tourismus in Südtirol [Between scepticism and solidarity – The perception of tourism in South Tyrol]. www.eurac.edu/de/press/wahrnehmung-des-tourismus-in-suedtirol-in-zeiten-der-pandemie.

Forster, J. (1964). The sociological consequences of tourism. *International Journal of Comparative Sociology, 5*, 217.

Frissell, S. S., & Stankey, G. H. (1972). *Wilderness environmental quality: Search for social and ecological harmony*. Society of American Foresters, Hot Springs, Arkansas.

German Tourism Association & German Institute for Tourism Research. (2022). Tourismus im Einklang mit den Einheimischen vor Ort möglich machen. Maßnahmen zur Förderung der Tourismusakzeptanz [Enabling tourism in harmony with local inhabitants. Measures to promote tourism acceptance]. www.deutschertourismusverband.de/fileadmin/Mediendatenbank/Bilder/Impulse/LIFT_Wissen_Tourismusakzeptanz_Broschuere.pdf.

Getz, D. (1983). Capacity to absorb tourism: Concepts and implications for strategic planning. *Annals of Tourism Research, 10*(2), 239–263.

Getz, D. (1994). Residents' attitudes towards tourism: A longitudinal study in Spey Valley, Scotland. *Tourism Management, 15*(4), 247–258. https://doi.org/10.1016/0261-5177(94)90041-8.

Girma, L. L. (2021, July 28). US national parks boom is great but overtourism solutions needed urgently, say senators. *Skift*. https://skift.com/2021/07/28/u-s-national-parks-boom-is-great-but-overtourism-solutions-needed-urgently-say-senators/.

Goodwin, H. (2019). Overtourism: Causes, symptoms and treatment. *Tourismus Wissen – Quarterly*, 110–114. https://responsibletourismpartnership.org/wp-content/uploads/2019/06/TWG16-Goodwin.pdf.

Goodwin, H. (2021). City destinations, overtourism and governance. *International Journal of Tourism Cities, 7*(4), 916–921. https://doi.org/10.1108/IJTC-02-2021-0024.

Gössling, S., McCabe, S., & Chen, N. C. (2020). A socio-psychological conceptualisation of overtourism. *Annals of Tourism Research, 84*. https://doi.org/10.1016/j.annals.2020.102976.

Graefe, A. R., Kuss, F. R., & Vaske, J. J. (1990). *Visitor impact management: The planning framework*, Vol II. National Parks and Conservation Association.

Haywood, K. M. (1988). Responsible and responsive tourism planning in the community. *Tourism Management, 9*(2), 105–118. https://doi.org/10.1016/0261-5177(88)90020-9.

Herntrei, M. (2014). *Wettbewerbsfähigkeit von Tourismusdestinationen: Bürgerbeteiligung als Erfolgsfaktor?* [Competitiveness of tourism destinations: Citizen participation as a success factor?]. Wiesbaden: Springer Gabler.

Herntrei, M. (2019). Tourist go home! Beobachtungen zu Overtourism und einer sinkenden Tourismusakzeptanz in Europa. Welche Folgen ergeben sich für Wissenschaft und Praxis? [Tourist go home! Observations on overtourism and decreasing tourism acceptance in Europe. What implications arise for science and practice?]. In H. Pechlaner (Ed.), *Destination und Lebensraum* (pp. 107–123). Springer Gabler.

Herntrei, M., Pillmayer, M., Scherle, N., & Nikitsin, V. (2022). Nachhaltige Destinationsentwicklung im Freistaat Bayern [Sustainable destination development in the Free State of Bavaria]. https://bzt.bayern/wp-content/uploads/2022/11/Projektbericht_Nachhaltige-Destinationsentwicklung-in-Bayern.pdf.

Hockenos, P. (2020, November 23). Bavaria's hard lesson learned: Local tourism isn't necessarily sustainable. *Energy Transition*. https://energytransition.org/2020/11/23503/.

Jafari, J. (2005). Bridging out, nesting afield: Powering a new platform. *Journal of Tourism Studies, 16*(2), 1–5.

Jamal, T. B., & Getz, D. (1995). Collaboration theory and community tourism planning. *Annals of Tourism Research, 22*(1), 186–204. https://doi.org/10.1016/0160-7383(94)00067-3.

Jungk, R. (1984). Wieviel Touristen pro Hektar Strand? [How many tourists per hectare of beach?]. *geo, 10*, 154–156.

Kagermeier, A., & Erdmenger, E. (2019a). Das Phänomen Overtourism: Erkundungen am Eisberg unterhalb der Wasseroberfläche [The phenomenon of overtourism: Explorations on the iceberg below the water's surface]. In J. Reif & B. Eisenstein (Eds.), *Tourismus und Gesellschaft: Kontakte-Konflikte-Konzepte* (pp. 97–110). Berlin: Erich Schmidt Verlag.

Kagermeier, A., & Erdmenger, E. (2019b). Overtourismus: Ein Beitrag für eine sozialwissenschaftlich basierte Fundierung und Differenzierung der Diskussion [Overtourism: A contribution to a social science-based foundation and differentiation of the discussion]. *Zeitschrift für Tourismuswissenschaft, 11*(1), 65–98. https://doi.org/10.1515/tw-2019-0005.

Keogh, B. (1990). Public participation in community tourism planning. *Annals of Tourism Research, 17*(3), 449–465. https://doi.org/10.1016/0160-7383(90)90009-G.

King, B., Pizam, A., & Milman, A. (1993). Social impacts of tourism: Host perceptions. *Annals of Tourism Research, 20*(4), 650–665. https://doi.org/10.1016/0160-7383(93)90089-L.

Kirstges, T. (2020). *Tourismus in der Kritik: Klimaschädigender Overtourism statt sauberer Industrie?* [Tourism under criticism: Climate-damaging overtourism instead of clean industry?]. UVK Verlag.

Koens, K., Postma, A., & Papp, B. (2018). Is overtourism overused? Understanding the impact of tourism in a city context. *Sustainability, 10*(12), 4384. https://doi.org/10.3390/su10124384.

Krippendorf, J. (1975). *Die Landschaftsfresser: Tourismus und Erholungslandschaft, Verderben oder Segen?* [The landscape eaters: Tourism and recreational landscape, spoilage or blessing?]. Hallwag

Lindberg, K., McCool, S. F., & Stankey, G. (1997). Rethinking carrying capacity. *Annals of Tourism Research, 24*(2), 461–465. https://doi.org/10.1016/S0160-7383(97)80018-7.

McCool, S. F. (1994). Planning for sustainable nature-dependent tourism development: The limits of acceptable change system. *Tourism Recreation Research, 19*(2), 51–55.

McCool, S. F., & Lime, D. W. (2001). Tourism carrying capacity: Tempting fantasy or useful reality? *Journal of Sustainable Tourism, 9*(5), 372–388. http://dx.doi.org/10.1080/09669580108667409.

McIntosh, R. W. (1964). First UN conference held on international travel and tourism. *Cornell Hotel and Restaurant Administration Quarterly, 5*(2), 49–52.

McKinsey & Company & World Travel & Tourism Council (WTTC). (2017). Coping with success: Managing overcrowding in tourism destinations. www.mckinsey.com/~/media/mckinsey/industries/travel%20logistics%20and%20infrastructure/our%20insights/coping%20with%20success%20managing%20overcrowding%20in%20tourism%20destinations/coping-with-success-managing-overcrowding-in-tourism-destinations.pdf.

Meadows, D. H., Meadows, D. L., Randers, J., & Behrens III, W. W. (1972). *The limits to growth: A report to The Club of Rome*. Universe Books.

Mihalic, T. (2020). Conceptualising overtourism: A sustainability approach. *Annals of Tourism Research, 84*, 103025. https://doi.org/10.1016/j.annals.2020.103025.

Milano, C., Cheer, J. M., & Novelli, M. (2018, July 18). Overtourism: A growing global problem. *The Conversation*. https://theconversation.com/overtourism-a-growing-global-problem-100029.

Milano, C., Cheer, J. M., & Novelli, M. (Eds.). (2019). *Overtourism: Excesses, discontents and measures in travel and tourism*. CABI Publishing.

Milano, C., Novelli, M., & Cheer, J. M. (2019). Overtourism and tourismphobia: A journey through four decades of tourism development, planning and local concerns. *Tourism Planning & Development, 16*(4), 353–357.

Murphy, P. E. (1981). Community attitudes to tourism: A comparative analysis. *International Journal of Tourism Management, 2*(3), 189–195. https://doi.org/10.1016/0143-2516(81)90005-0.

Murphy, P. E. (1983). Tourism as a community industry – An ecological model of tourism development. *Tourism Management, 4*(3), 180–193. https://doi.org/10.1016/0261-5177(83)90062-6.

Namberger, P., Jackisch, S., Schmude, J., & Karl, M. (2019). Overcrowding, overtourism and local level disturbance: How much can Munich handle? *Tourism Planning & Development, 16*(4), 452–472.

Novy, J., & Colomb, C. (Eds.). (2017). *Protest and resistance in the tourist city*. Routledge.

Novy, J., & Colomb, C. (2019). Urban tourism as a source of contention and social mobilisations: A critical review. *Tourism Planning & Development, 16*(4), 358–375. https://doi.org/10.1080/21568316.2019.1577293.

O'Reilly, A. M. (1986). Tourism carrying capacity: Concept and issues. *Tourism Management, 7*(4), 254–258. https://doi.org/10.1016/0261-5177(86)90035-X.

Pechlaner, H. (2003). *Tourismus-Destinationen im Wettbewerb* [Tourism destinations in competition]. Deutscher Universitätsverlag.

Pechlaner, H., Innerhofer, E., & Erschbamer, G. (Eds.). (2019). *Overtourism: Tourism management and solutions*. Routledge.

Peeters, P., Gössling, S., Klijs, J., Milano, C., Novelli, M., Dijkmans, C., Eijgelaar, E., Hartman, S., Heslinga, J., Isaac, R., Mitas, O., Moretti, S., Nawijn, J., Papp, B., & Postma, A. (2018). Research for TRAN Committee – Overtourism: Impact and possible policy responses. European Parliament, Policy Department for Structural and Cohesion Policies. www.europarl.europa.eu/RegData/etudes/STUD/2018/629184/IPOL_STU(2018)629184_EN.pdf.

Perdue, R. R., Long, P. T., & Allen, L. (1990). Resident support for tourism development. *Annals of Tourism Research, 17*(4), 586–599. https://doi.org/10.1016/0160-7383(90)90029-Q.

Pizam, A. (1978). Tourist impacts: The social costs to the destination community as perceived by its residents. *Journal of Travel Research, 16*(4), 8–12. https://doi.org/10.1177/004728757801600402.

Porter, M. E. (1979). How competitive forces shape strategy. *Harvard Business Review, 57*(2), 137–145.

Porter, M. E. (1980). *Competitive strategy: Techniques for analysing industries and competitors*. Free Press.

Postma, A., Koens, K., & Papp, B. (2020). Overtourism: Carrying capacity revisited. In J. A. Oskam (Ed.), *The overtourism debate* (pp. 229–249). Emerald Publishing.

Rosenow, J. E., & Pulsipher, G. L. (1979). *Tourism, the good, the bad, and the ugly*. Century Three Press.

Saarinen, J. (2006). Traditions of sustainability in tourism studies. *Annals of Tourism Research, 33*(4), 1121–1140.

Saveriades, A. (2000). Establishing the social tourism carrying capacity for the tourist resorts of the east coast of the Republic of Cyprus. *Tourism Management, 21*(2), 147–156.

Seidl, I., & Tisdell, C. A. (1998). Carrying capacity reconsidered: From Malthus' population theory to cultural carrying capacity. *Ecological Economics, 31*(3), 395–408.

Simmons, D. G. (1994). Community participation in tourism planning. *Tourism Management, 15*(2), 98–108. https://doi.org/10.1016/0261-5177(94)90003-5.

Smith, O. (2018, June 4). Is Greece on the brink of an overtourism crisis? *The Telegraph*. www.telegraph.co.uk/travel/destinations/europe/greece/articles/greece-overtourism-santorini/.

Swarbrooke, J. (1999). *Sustainable tourism management*. CABI.

Timothy, D. J. (1999). Participatory planning: A view of tourism in Indonesia. *Annals of Tourism Research, 26*(2), 371–391. https://doi.org/10.1016/S0160-7383(98)00104-2.

Tokarchuk, O., Barr, J. C., & Cozzio, C. (2021). Estimating destination carrying capacity: The big data approach. Travel and Tourism Research Association: Advancing Tourism Research Globally. https://scholarworks.umass.edu/ttra/2021/research_papers/51.

Tokarchuk, O., Gabriele, R., & Maurer, O. (2021). Estimating tourism social carrying capacity. *Annals of Tourism Research, 86*. https://doi.org/10.1016/j.annals.2020.102971.

Tosun, C. (1999). Towards a typology of community participation in the tourism development process. *Anatolia, 10*(2), 113–134. https://doi.org/10.1080/13032917.1999.9686975.

Turner, L., & Ash, J. (1975). *The golden hordes: International tourism and the pleasure periphery*. Constable.

Vecchiato, A. (2020, December 27). Ärger mit dem “Overtourism” [Trouble with “Overtourism”]. *Süddeutsche Zeitung*. www.sueddeutsche.de/muenchen/wolfratshausen/2020-in-bad-toelz-wolfratshausen-aerger-mit-dem-overtourism-1.5158209

Wagar, J. A. (1964). The carrying capacity of wild lands for recreation. *Forest Science*, 10. https://doi.org/10.1093/forestscience/10.s2.a0001.

Wagar, J. A. (1974). Recreational carrying capacity reconsidered. *Journal of Forestry, 72*(5), 274–278.

Wall, G. (2019). Perspectives on the environment and overtourism. In R. Dodds & R. W. Butler (Eds.), *Overtourism: Issues, realities and solutions* (pp. 27–44). De Gruyter.

Wall, G. (2020). From carrying capacity to overtourism: A perspective article. *Tourism Review, 75*(1), 212–215. https://doi.org/10.1108/TR-08-2019-0356

Washburne, R. F. (1982). Wilderness recreational carrying capacity: Are numbers necessary? *Journal of Forestry, 80*(11), 726–728.

Wood, M. E. (1991). Global solutions: An ecotourism society. In T. Whelan (Ed.), *Nature tourism: Managing for the environment*. Island Press.

World Tourism Organisation (UNWTO). (1983). *Risk of saturation or tourist carrying capacity overload in holiday destinations*. UNWTO.

World Tourism Organisation (UNWTO). (2018). “Overtourism”? Understanding and managing urban tourism growth beyond perceptions: Executive summary. UNWTO. https://doi.org/10.18111/9789284420070.

World Tourism Organisation (UNWTO). (2019). *International tourism highlights, 2019 edition*. UNWTO. https://doi.org/10.18111/9789284421152.

Young, G. (1973). *Tourism: Blessing or blight?* Penguin.

Zeit Online. (2017, September 24). Die überbuchte Insel [The overbooked island]. www.zeit.de/entdecken/reisen/2017-09/palma-de-mallorca-protest-gegen-touristen?utm_referrer=https%3A%2F%2Fwww.google.fr%2F

3 Destinations During and After the Lockdown

Evidence from Venice, Italy

Jan van der Borg

1 Introduction

Sustainable tourism development has become an essential ingredient of virtually all destination management strategies, whether they are dealing with nature- or with urban-based tourism. The overtourism debate has clearly contributed hugely to this shift in the focus of destination management: from tourism as means to boost economic development to tourism as a key instrument to enhance the wellbeing of the local population and local entrepreneurs.

Surprisingly, the COVID-19 pandemic has not fundamentally changed the overtourism debate and the consequences it has for destination management. In fact, it seems that the question of how to use public tourism spaces and public facilities of destinations more intelligently, a use that adequately caters to the needs of locals, of local firms and of visitors, has probably become more pressing than ever before.

This chapter aims to investigate the relationship between sustainable tourism development and the way a destination management strategy ought to be designed, using the iconic case of Venice, Italy, as a continuous point of reference. Attention will be paid to the consequences of the lockdown, particularly severe in Italy, for the local economy and society, and to the impact this lockdown had (or not) on the strategy of Venice to finally make a start with dealing with the unsustainability of tourism development.

At the moment of writing, the world has almost entirely shed the COVID-19 pandemic. Almost all restrictions have been alleviated; this is especially true for travel restrictions. Tourism has globally returned to business as usual and the first figures that are emerging are fuelling the expectation that 2023 will abundantly surpass the record year 2019.

Tourism has been hurt more than any other economic and social activities by the pandemic. The United Nations World Tourism Organization (UNWTO, 2021) estimated a total decline of about 60 to 80 per cent in international tourist arrivals in 2020, with strong knock-on effects on global tourism-generated gross domestic product and jobs. This is true for the global tourism industry but is even more true for a country like Italy and a destination like Venice, which possess economies that are particularly dependant on tourism. In fact, when Italy and Venice went into a very strict lockdown on March 12, 2020, visitors to Venice abruptly disappeared, and with the visitors vaporised much of tourism's impact.

DOI: 10.4324/9781003365815-4

Many academics, I among them, very much hoped that this disruption would have created a unique moment of reflection that convinced tourists, the tourism industry and policy makers to embrace a greener and more sustainable tourism business model. Questions concerning how visitors should use public tourism spaces and public facilities in an intelligent way, also because of rules on social distancing enacted during the pandemic, continued to be pertinent, notwithstanding the dramatic fall in tourism consumption in 2020 and 2021. Indeed, the idea that the pandemic would leave some permanent marks on the behaviour of tourists and travellers appears to have vanished altogether.

To answer the question as to what extent public space and facilities can be used by tourism and under what conditions, a thorough analysis of the impacts (i.e. the collective benefits and costs) of tourism for the various sectors and for the different stakeholders that together form the destination is propaedeutic. As Van der Borg (2022a) has argued, only when all these sectors and all these stakeholders involved in the sector benefit from tourism development can the use of the tourism assets and the facilities the destination offers become (Pareto) optimal. In practice, since the tourism system is effectively full of trade-offs, given its implicit complexity, and per definition very dynamic nature, striving for sustainable tourism development becomes a balancing act to optimise the use of public space and public facilities in the long term (see for instance Stoffelen and Ioannides, 2022).

In the next section, I will use some ideas relating to Raworth's (2017) innovative way to describe the global economy as a doughnut to gain a grasp of the trade-offs and dynamism that tend to shape tourism destinations. I will concentrate mainly on the economic dimension of sustainable tourism. My first aim is to explain what can be seen as sustainable tourism development and what not, what role the different impacts related to tourism development play in all this, and whether the concept of the Tourist Carrying Capacity (TCC) is indeed useful to design policies that can make and, even more urgently, keep tourism development sustainable (Bertocchi et al., 2020). This analysis allows me to revisit the case of the North Italian city of Venice, which I have been writing about since the end of the 80s, in the third section.

2 Unsustainable Tourism Development: The Economic Dimension of Under- and Overtourism

2.1 The Persistent Unsustainability of Tourism

As Van der Borg (2022b) argued, it was only after World War II that tourism as an economic and societal activity grew and transformed both continuously and rapidly. In many places, especially in industrialised countries, the average income per person, the number of paid vacation days, and private car ownership increased dramatically, and tourism changed from an activity that was exclusive to small numbers of very wealthy people into a mundane phenomenon for the masses. With the growth of mass tourism, a corresponding business model emerged. This model was based on the mere replication of formulas that seemed to work elsewhere and by searching for economies of scale and price-based competitiveness, thus focusing on quantity rather than on quality. It is this mass-tourism-oriented business model that continues to dominate to this day.

The first criticisms aimed at this model appeared already a few decades ago (see for example Young, 1973; Krippendorf, 1986), especially with respect to tourism's effects on places such as coastal communities. By the 1990s, several authors had argued that additional types of destinations, including cities (see for example Van der Borg, 1992 or Costa, Gotti and Van der Borg, 1996), suffered to varying degrees from an excessive touristic pressure which was related to the mass touristic business model they had been embracing. Today, many destinations like Venice (see for instance Visentin and Bertocchi, 2019), the case that will be treated in this chapter, and Barcelona (see for example Russo and Scarnato, 2017) are plagued by what has been frequently called "overtourism". Indeed, overtourism has now become a trendy research topic for many academics (Koens, Postma and Papp, 2018). However, despite admonitions concerning the dangers of overtourism and the need to better regulate the industry (see for example Fletcher et al., 2019; Hall et al., 2020; Higgins-Desbiolles, 2021), the different stakeholders that have traditionally been involved in tourism's development processes, especially policy makers and tourism entrepreneurs, still very much foster the traditional business model, stressing economic objectives rather than the other facets of life in destinations.

As I have often written (see for example Van der Borg, 2022a, and Van der Borg, 2022b), unsustainability is, unfortunately, very much the inevitable consequence of the very nature of the core of the tourism product. First, a destination's primary tourism assets (or its attractions) are uniquely linked to a specific geographic context, which means that they are not reproducible and, hence, from an economist's perspective, extremely scarce. Moreover, most of them are public or common goods, which means that leaving their use simply to the market forces generally does not lead to an optimal use of these assets. The combination of extreme scarcity and the impossibility of being able to count on the market as far as the optimal allocation of tourism assets is concerned lies at the heart of the persistent unsustainability of tourism development. I will explain this further in the next section.

2.2 *Theoretical Foundations of the Issue of Overtourism*

In this section, I focus on the theoretical foundations of the process that tends almost inevitably to push successful destinations towards overtourism, which is the most frequent and recognisable outcome of their non-optimal use. I should stress, first of all, that the structure of the (macro) tourism product is incredibly complex and of a composite nature. A vacation or a day trip is, in fact, a composition of an infinite number of micro touristic products. These products range from cultural and natural attractions to accommodation facilities; from catering services to entertainment; from intermediation to transportation. In turn, various types of organisations and businesses produce or manage them. Some of these organisations are public entities, while many are private. Some are huge, multinational firms, while others are very small family firms. This fragmentation of the sector complicates decision-making and meeting the interests of all involved stakeholders (Hartman, in Stoffelen and Ioannidis, 2022).

A distinction that is frequently made to simplify this complexity is that which exists between primary and secondary tourism products (Van den Berg et al., 1995). While the first category encompasses all the goods and services that attract people to a place, the secondary tourism product is made up of all the goods and services that allow visitors to enjoy the primary tourism product. Secondary tourism products are an essential ingredient of the macro tourism product. However, it is the primary tourism products that, because of their uniqueness and close relationship to the places within which they are located, are the heart and soul of any tourism system. This is especially true for heritage cities, a particular type of tourism system that offers a package of unique cultural assets to those who decide to visit them. Moreover, since the primary tourism product is per se a central ingredient of the reputation or the brand of a destination, the decision to travel to a place is often based on the perception people have of the attractions that they expect to visit once they have arrived at their destination.

Not only are many primary tourism products extremely unique but they also belong to a category of goods that economists call public or common goods. These resources are non-exclusive and non-excludable. Obviously, this holds true for many natural resources (such as beaches, forests, lakes, wildlife and so on), but also, perhaps more frequently, for many cultural-historic resources (such as churches, palaces, gardens and town squares). Some cities of art, including Venice – the subject of this case study – are monuments in their totality. Examples of secondary products are accommodation, catering services, and shopping. Secondary tourism products are reproducible (in fact, the barriers to entry are often rather low for tourism firms) and are subject to market forces.

This gives rise to the intrinsic risk that destinations fall victim to what Hardin called the "tragedy of the commons" (Hardin, 1968). In Hardin's article in *Science*, he describes a pasture that is "open to all" (Hardin, 1968, p. 1244). He asks us to imagine the grazing of animals on a common ground. Individuals who aim to increase their wealth are pushed to add to their flocks. Yet, every animal added to the total helps to degrade the commons marginally but significantly. Although the degradation for each additional animal is small relative to the gain in wealth for the owner, if all owners follow this pattern, the commons will ultimately be destroyed (De Young, 1999). If all actors have an implicit drive to pursue their individual interests, each owner continues to add animals to their flock:

> Therein is the tragedy. Each man is locked into a system that compels him to increase his herd without limit – in a world that is limited. Ruin is the destination toward which all men rush, each pursuing his own interest in a society that believes in the freedom of the commons.
>
> (Hardin, 1968, p. 1243)

Hardin's insight was far from new, but it was he who fleshed out that the concept of the tragedy of the commons applies in principle to environmental problems at large. In many such cases the problem relates to the fact that perfectly rational individual behaviour causes long-term damage to the *common good*, to *others* and, eventually, to *oneself.*

In the case of tourist destinations, the problems tend to be even more complicated. Not only are local inhabitants, local tourism firms and individual visitors competing for the unlimited use of the destination's amenities, but also (non-local) tourism firms are directly or indirectly using these resources without paying a fee that expresses the intrinsic value these amenities possess. Effectively, while heritage, due to its uniqueness and lack of reproducibility, is a scarce good in an absolute way, the fee – especially what the tourism industry is paying for using the heritage – often equals zero or is, at any rate, not in line with its use-value or with the costs linked to produce and conserve it. The absence of a market (or pricing) mechanism as an implicit and automatic instrument of regulating the use of heritage means that tourism development eventually tends to become unsustainable: visitors are not paying (enough) for using the cultural-historic assets to compensate for the collective costs they are generating.

Indeed, as clearly is the case for Venice, neither the tourism industry nor the visitors will ever perceive the scarcity of these assets and, consequently, the demand for them tends to be infinitely large. Once total effective tourism demand has reached the destination's capacity to absorb these visitors, negative externalities, such as wear-and-tear, congestion, pollution and neighbourhood gentrification, will rapidly emerge, rendering the destination unattractive for inhabitants, commuters and, eventually, even for visitors themselves (Van der Borg, 2017).

Overtourism is, of course, not the only form of non-optimal allocation of tourism assets. Underutilisation, which reflects the opposite direction towards which the market inefficiency described earlier might move, is socially and economically undesirable, especially in places where tourism features as a key economic revitalisation strategy and where alternative development trajectories have not been explored or are hard to implement.

What also seems to be counterproductive is not to recognise the fact that mismanagement is the principal cause of unsustainable destination development. Destinations and academics continue to look for culprits. Tourists are ignorant, uninformed, not mindful enough or of the "wrong type". Or they simply blame the low-cost airlines, the cruise industry, online travel agencies such as Airbnb or Booking, for just being greedy, for free riding on tourism assets and for exploiting their personnel. And a simplistic, incorrect diagnosis will probably lead to tourism policies that fail to address the true causes of unsustainability.

2.3 Undertourism and Overtourism in the Doughnut Destination

It is easy to see that the misallocation of tourism assets, either in the form of overtourism or undertourism, is compatible with Raworth's (2017) vision of the economy, which she depicts as a doughnut, with an inner and upper boundary. I contend that the economy of a tourism system can also be represented in the same manner.

The tourism doughnut has two boundaries: an inner one and an outer one. The inner boundary of the metaphorical doughnut is the social boundary, which designates the threshold that should be crossed if people are to collectively experience (existential) benefits from tourism. The outer boundary seems to correspond to the "critical range of elements of capacity" that Butler introduced decades ago in the context of his Tourism Area Life Cycle (or TALC) model (Butler, 1980), which are defined as the limits that a tourism destination can endure before impacts become unsustainable. Or, following O'Neill et al. (2018), pursuing a doughnut-type development for a tourism destination means using tourism for "meeting the people's basic needs" (e.g. satisfying the social, inner boundary mentioned earlier), but not so intensively as to "transgress planetary boundaries" (e.g. satisfying the ecological, outer boundary, that contains various thresholds that are linked to the UN sustainability goals).

In this doughnut model of the destination, the level and type of development of the economy that remains stuck below the social boundary is unable to satisfy the basic needs of the people. Raworth (2017) calls this a "shortfall", a situation that resembles the one prevailing in Butler's initial stages of tourism development, where tourism assets are (still) underutilised. Some authors have labelled this phase in the destination's development undertourism (Barač Miftarević, 2023), and state that, in comparison to the attention that is paid to overtourism, undertourism is not yet studied enough. Overshooting the ecological boundary has similarities with that of overshooting the TCC, a concept that I will explicitly refer to in the next section. In fact, exceeding the carrying capacity involves the emergence of environmental, economic and societal costs that are incompatible with the destination's sustainability. Overtourism is, therefore, reconcilable with Raworth's overshooting of the outer boundary of the destination doughnut. Obviously, as argued already in Van der Borg (1991), unsustainability of tourism development is not just matter of numbers. In fact, initially, most visitors will be day tourists and their socioeconomic footprint will be modest. And when a destination accelerates towards Butler's "critical range of elements of capacity", slow tourism will often be replaced by superfast tourism, which contributes less to the destination than the number of negative externalities it tends to generate.

Therefore, representing a tourism system using the doughnut metaphor clearly helps to better understand the two dilemmas of tourism development and to place them in an omni comprehensive framework of sustainability. Namely, the dilemmas are the need to develop tourism to such a level and quality that: (1) the needs of the stakeholders are served; and (2) that both the impoverishing quality of the visitors and the overshooting of the TCC are avoided so that the different negative impacts do not get out of hand. Considering the concept's intuitive metaphorical usage not to mention its alignment with influential concepts for discussing tourism impacts like the TALC and TCC, several cities that are also looking for a sustainable tourism development path, like Amsterdam (see for example Van den Bosch, 2020), have embraced doughnut thinking to completely revise their urban development strategy, including their tourism policy.

In the next section, I provide an analysis of a concrete example of the "overtourism" family of misallocating tourism assets by investigating Venice. I will illustrate this famous Italian destination's struggle with tourism development and will try to see whether understanding the underlying mechanisms may help us to find ways to escape the mass touristic business model that has always been based on growth and on the sheer number of visitors. It will become evident that the pandemic has done nothing to catapult Venice to a more sustainable trajectory of tourism development. On the contrary, it might even seem that Venice has never been further away from embracing a business model that is compatible with the needs of the Venetians.

3 Understanding Venice Helps Us Understand Unsustainable Tourism Development

3.1 Tourism in Venice Before the Pandemic: An Increasingly Difficult Relationship

Venice is an urban tourism destination *par excellence* that has been presented as an iconic example of overtourism since the early 80s (see for example CoSES, 1979; Van der Borg, 1991). It is a UNESCO world heritage site and boasts numerous monuments and museums. Images of the city can be found on the cover of promotional materials of virtually every major tour operator (Van der Borg, 1994) and in numerous commercial websites that deal with city tourism. Moreover, many international movies are filmed in the the city, and a series of popular crime books use Venice as their stage. Even before the legendary Pink Floyd concert in 1989, the San Marco square had hosted important musical events. Some of its most iconic buildings have been reconstructed in full scale to add to the flavour of casinos, exhibition areas and theme parks in places like Macau and Las Vegas. Venice's uniqueness, the incredibly strong brand it possesses, and its continuous media exposure have turned it into a magnet for visitors from all over the world.

Consequently, the numbers of visitors the city received annually before the COVID-19 pandemic struck were truly incredible, especially when compared to heritage cities of a similar size: unofficial, unpublished estimates for 2019 spoke of approximately 30 million visitors, of which only 4.3 million arrivals were overnight tourists, who stayed in Venice for slightly more than two nights on average, generating almost 10 million bed nights in official tourism accommodation. However, of the two mega segments that form the Venetian tourism market, day tourism has grown much faster than that of overnight tourism. This was mainly because the local government decided to curb the number of hotels in the 1990s by virtually blocking the possibility to change the designation of real estate from residential into touristic. Even back in the 1990s, the continued expansion of the number of visitors raised the awareness among inhabitants, academics and some policy makers that the number of people who visited Venice was already incompatible with the city's economic and social needs.

Although there have been ups and downs in recent decades, the trend was decisively headed in an upward direction in the period ending with the pandemic in

March 2020. Most of the changes in Venice's tourism market were triggered by global societal and economic transformations; only a few of them were induced by changing local circumstances, let alone changes in tourism policy.

In fact, as early as 2015, Van der Borg described the changes that have characterised the development of tourism in Venice before the pandemic in detail. The following seem to stand out:

- The emergence and the subsequent boom of *low-cost airlines* that have allowed much more people to engage in city trips. This has not only positively influenced the number of people visiting urban destinations and the City of Venice but has also made shorter stays in short haul destinations much cheaper with respect to the total costs of the trip. Moreover, in the case of Venice, it has allowed people to come from further away, thus rapidly eroding the dominance of the neighbouring Austrian, Swiss and German markets that was so characteristic of the mostly car-based tourism of the 1980s.
- The increasing diversification of the *supply of accommodation* was driven by the law that was implemented to facilitate the organisation of the *Grande Giubileo 2000* by legalising Bed and Breakfast establishments (B&Bs) to cheaply provide accommodations to the pilgrims attracted by this religious mega event. This boosted the number of B&Bs, a situation that has now been accentuated through the appearance of dedicated portals such as Airbnb and Couchsurfing that allow entire homes or parts of them to be rented out to visitors. The emergence of cheaper forms of accommodation has, in turn, slowed down the growth of the number of day tourists, freeing up capacity for an expansion of overnight stays, but it has also created substantial additional tensions in the Venetian housing market. The City of Venice had already timidly started to discuss ways of curbing short rentals before the pandemic, looking, however, very much to the national government to introduce laws and regulations at the national level.
- The widespread diffusion of *the internet* and of *smart phones and tablets* as indispensable instruments for tourists to inform themselves, reserve tourism products and share their experience with others allows policy makers to intercept tourists using new information and communication technologies, for example when implementing visitor management strategies. Venice has recently invested millions of euros in smart information and telecommunication technologies to dress up what is now known as the "Smart Control Room", which will be discussed in detail below.
- A pre-pandemic increase in the purchasing power of households in Asia, Africa, Central Europe and South America has made tourism in Venice even more global and has boosted the number of arrivals in the city. In fact, tourists from these parts of the world have gained much importance in the total number of visitors to Venice, and their contribution to total tourist expenditures had grown considerably in the decade leading up to the pandemic. With the pandemic, especially the long-haul visits disappeared quickly, and tourists from America and Asia are gradually returning only now. Last but not least, due to the war in Ukraine, the Russian market has lost most of its pre-pandemic importance.

- The increasing popularity of *cruise tourism*. Many operators have found Venice an attractive port of call, and cruise ships have been growing bigger and bigger. This increasing popularity has led to a situation where the industry's impact on Venice's relationship with tourism at large has truly become disturbing. Since cruise tourism was one of the most important and probably emblematic reasons prompting UNESCO to consider putting Venice on the endangered heritage list (Petric et al., 2020), the City of Venice has urged the Port Authorities to consider an alternative location for the cruise terminal. This new terminal is located a few kilometres outside the perimeters of the historical centre and compels the gigantic cruise ships to avoid the canal in front of St. Mark's square and take the "petrol tankers" canal instead.

These structural changes in holidaying have boosted the number of visitors to Venice and have undoubtedly changed the profile of the visitor to Venice. Consequently, since different types of visitors possess varying economic and logistical patterns of behaviour, it seems plausible to presume that the visitor mix has also influenced the overall impact tourism has on Venice. Additionally, this means that the identification of adequate and differentiated tourism policies that account for the impact of various visitor segments is of the utmost importance. Most importantly, the case of Venice shows us that a distinction must be made between overnight tourism and day visitors when attempting to understand the causes of overtourism and developing policies to target it.

In effect, according to a survey performed by the University of Venice on behalf of the City of Venice in 2012 (Van der Borg, 2017), day tourists spent much less than overnight tourists. In fact, day tourists were spending 124 euros per visit, for the entire travel company (e.g. the "leader" of the travel company was interviewed and for simplicity asked to answer for everybody under his or hers responsibility). This translated into an average expenditure of 40 euros per person per day, only a quarter of the amount that is daily spent by a visitor staying overnight. Apart from the qualitative differences in the profiles between overnight and day tourists described earlier, this figure alone illustrates that the economic footprint of the former type of visitors is four times that of the latter type. As I will show below, however, a positive economic impact alone is not the only input needed to design an innovative tourism development strategy.

3.2 How Much Is Too Much in Venice?

The terms undertourism and overtourism are both intrinsically subjective concepts, that is, they always relate to some ideal situation. This was made clear in the previous section when the relationship between the concepts and Raworth's inner and outer boundaries was described. Violating the outer boundary gives rise to important tensions between the different stakeholders in the destination that are sometimes even pursuing opposite goals. In cities like Amsterdam, Barcelona, Berlin, Dubrovnik and Venice, tourism has conquered a top position on the political agendas.

A clear confirmation of the tensions that continue to exist between Venice's tourism development process and the necessities of the city's inhabitants and non-touristic economic activities has been given through the tourist carrying capacity model that Costa and van der Borg proposed back in 1988 and that has been implemented in more sophisticated forms and updated by various others (Canestrelli and Costa, 1991; Bertocchi et al., 2020).

One may argue that understanding urban tourism impacts from a sustainability lens using the concept of TCC is both *controversial* and *intuitive*.

Controversial, because the discussion as to whether TCC is a useful concept for tourism management has not reached a moment of convergence. Bertocchi et al. (2020) wrote that there is a certain "difficulty of determining a maximum visitor number, given the fact that destinations respond to various thresholds of the capacity exist, each linked with a different dimension of sustainability". Seidl and Tisdell (1999) and Saarinen (2006) argue similarly, by reflecting that "the weakness of the concept stays in the use of values and perceptions on which it is based" (Bertocchi et al., 2020, p. 3). Conversely, Watson and Kopachevsky (1996) correctly state that the lack of practical tools to implement TCC does not justify dismissing the TCC concept altogether. Papageorgiou and Brotherton (1999) claim that in any case TCC continues to be a useful concept for environmental management issues since it fosters the deepening of relationships between the environment and human activities. Moreover, Mexa and Coccossis (2004) also stress the importance of TCC as a valuable concept for the planning and management of sustainable tourism. In short, the last word has not yet been said about the utility of the TCC concept for evaluating tourism impacts.

Intuitive, because it is undeniable that any economic and social activity that makes an intensive use of (public or private) space, (public or private) natural or cultural assets, or (public or private) facilities and infrastructures is necessarily bound to the capacity of this space, these natural or cultural assets or facilities and infrastructures. Tourism does not constitute an exception to this issue. The COVID-19 pandemic has made it crystal clear that the optimal allocation of tourism assets is of the utmost importance for the sustainable development of tourism. Therefore, it seems obvious that there is some kind of threshold in destinations that recalls the "critical range of elements of capacity" of Butler (1980) or the "outer boundary of the doughnut" of Raworth (2017).

My research has shown that the second characteristic of the TCC concept is often the dominating one. In fact, in 1988, when the TCC model was applied for the first time, the overall TCC of the historical centre of Venice was supposed to be around 10 million visitors per year, of which 45% were overnight tourists and 55% were day tourists. In 2018, the TCC rose to 17 million visitors a year, equally distributed among overnight tourists and day tourists. This updated TCC using data from 2018 has been obtained by implementing the same model, but by accounting for several novel factors: the investments made in local public transport; the improvements in the management system of solid waste; and, evidently, the expansion of the number of beds in tourist accommodation, including B&Bs. In other words, through these investments the outer boundary of the destination's "doughnut" has widened somewhat (Bertocchi et al., 2020).

Notwithstanding that the capacity to absorb visitors has increased substantially in Venice over time, the optimal number of 17 million visitors is still well below the 30 million visitors who arrived in Venice in 2019. Moreover, Bertocchi et al. (2020) confirmed once more that the problem was not only of a quantitative but also of a qualitative character; in fact, the *de facto* composition of the visitor flow (20% overnight tourists versus 80% day tourists) is very distant from the optimal one (50% overnight tourists versus 50% day tourists) the model had calculated.

After two years of pandemic, it not only seems that already in 2022 the number of visitors to Venice was close to the 30 million visitors of the record year 2019, but also that in 2023 a new record will be established.

3.3 Impacts of Tourism, the Tourist Carrying Capacity and Sustainable Destination Management in Venice

The quantitative and qualitative mismatch between tourism demand and supply in Venice, as Bertocchi et al. (2020) have illustrated by using the carrying capacity model, and the excessive pressure that the host community perceives in their daily lives, as expressed by the ratio of overnight stays (9 million) to the total number of inhabitants (less than 50,000), are the reason to assume that overall costs that tourism generates are decisively greater than its collective benefits. Moreover, and perhaps even more importantly, those who do pick up the bill are in many cases not at all, or only very indirectly, benefitting from tourism.

Obviously, this is not surprising for those who have been studying Venice since the first signs of excessive tourism pressure had already appeared in the late 1970s (CoSES, 1979). Nevertheless, the idea that not only nature-based destinations, but also urban destinations might be subject to excessive tourism pressure is still less widespread than expected (see Van der Borg 2022a for an overview of urban tourism research and the role sustainability has been playing in its development). However, with a growing number of cities in Europe that share exactly these same symptoms of overtourism, the impression that they are turning into Venice lookalikes is strong. And given the radical changes that have characterised urban tourism since the 1990s that were discussed before, the iconic case of Venice continues to show how strongly an analysis of the impact of tourism is intertwined with the TCC and, hence, with sustainable destination management.

The Venetian overtourism problems appeared to have vanished almost instantly in 2020. Italy was one of the first countries of the world to take draconian measures to stop the pandemic from killing people. Consequently, the COVID-19 crisis almost fatally hit the Venetian tourism industry. From being one of the most important sectors of the economy during the first days of March 2020, tourism went abruptly to an almost total state of inactivity. The crisis thus caused Venice to shed almost 90% of its visitors and experience a massive decline in income turnover and jobs. Only a small part of this loss was briefly reversed during the relaxation of travel restrictions in Europe during the summer of 2020.

Against this backdrop, it has been perfectly understandable that the travellers and tourism industry everywhere, but in places like Venice in particular, have been pestering governments as well as regional and local administrations with pressing demands to return to normal as soon as possible. However, instead of simply picking up where the destinations and the tourism industry were in January 2020, the disruptive moment without precedents should have been used to design a more conscious, sustainable and safer form of tourism (Benjamin et al., 2020; Cheer, 2020; Ioannides & Gyimóthy, 2020; Nepal, 2020).

Now that the pandemic seems to have definitively left the world, and Italy and Venice and the tourism market seem to be on the development course that the UNWTO formulated in 2015 (a doubling of the number of international travellers before 2030), this will probably prove to be an incredible lost opportunity. In fact, in April 2023, the total number of inhabitants of the historical centre of Venice dropped to its all-time low of 49,365, while the total number of beds in all forms of tourism accommodation has reached 48,596. Soon, on an average night in the high season there will be more tourists sleeping in Venice than Venetians, notwithstanding the symbolic measures that have been taken before and during the pandemic and that were mentioned above.

This gentrification process, a clear illustration that the tensions between tourism and the local population and non-touristic activities have become unbearable, has induced the local administration to redesign its tourism development policy more vigorously. Two new visitor management instruments are supposed to make tourism smarter and more sustainable.

The first is the Smart Control Room mentioned previously. This Smart Control Room is a room that has principally been designed to address security and safety issues, by monitoring situations of crowding in specific locations in real time through a system of interconnected security and web cams and by monitoring the information that is provided by mostly public providers of local services, such as the company that manages local public transportation and parking lots, the museums, and so forth. The resulting dashboard should make interventions in cases of congestion and of accidents more effective. The data that are collected daily are being stored and should provide the local police with time series with which forecasts of future situations of crowding can be made. Moreover, the Smart Control Room is supposed to be a basic condition of a system that allows the City of Venice to make a reservation of the visit mandatory for all types of visitors.

The second is a new tourism tax that is not levied on bed nights but on arrivals in the historical centre. The idea is that it is not correct to tax only the 20% of visitors that already contribute most to Venice's economy, and to exempt the 80% of the visitors that contribute heavily to the collective costs that tourism generates but just a little to its collective benefits. Moreover, this tax might also become a disincentive for all those visitors that come to Venice without a clear motivation and whose user value is rather low.

After years of discussion and an infinite number of press communiqués announcing the introduction of an entrance fee for all those who are willing to visit Venice, the Smart Control Room still makes absolutely no difference, and the introduction of the "landing fee" has been postponed numerous times. Both new measures, although in theory at least going in the right direction, suffer from the same lack of vision about Venice and tourism the local government seems to possess. In fact, without a clear idea what indicators you should be using before intervening, what the critical values of these indicators are, and what interventions should be taken when one or more of these critical values have been reached, the "Smart" Control Room remains a "Dumb" one. The same is true for the landing fee. It is not clear what the central objective of the fee is: simply raising money, curbing free riding or discouraging day tourists to come to Venice from their homes or their holiday destination? Moreover, and these issues are clearly related to the previous one, it is not clear which visitors should be paying what, and who is going to be excluded from paying the tax (people living in the Veneto Region or those coming from Italy)? Without developing a clear vision about what future lies ahead for Venice and what tourism's position should be in such a vision, the implementation of theoretically valid policies is useless.

Obviously, and as has been suggested already numerous times here, the two years of the pandemic were an ideal opportunity to formulate such a vision, translate this vision into a few concrete pilot projects, and make sure the policies based on these pilot projects were well in place before the tourism tsunami started to hit Venice once more. Not having done so in a timely fashion makes it increasingly difficult to implement a coherent destination management strategy that brings the interest of using Venice as a destination and that of Venice as a place to live, study and work closer together.

4 Closing Remarks

The intention in this paper has been to investigate sustainable tourism development in theory and practice and its implications for destination management. The city of Venice, an iconic example of overtourism, has been used to illustrate the various dimensions of these concepts, both before and after the pandemic. Central in this investigation is the declination of Raworth's idea of a doughnut economy, which has been called the doughnut destination, to cover both forms of unsustainable tourism – that is, under- and overtourism – and to look into the second form more closely by studying Venice.

To synthesise, I have argued that understanding urban tourism impacts from a sustainability lens using the concept of TCC is both controversial and intuitive. This dominant, second consideration suggests that it might be an effective tool to design and to fine-tune destination management strategies.

Given that overtourism occurs and is most noticeable when the number of arrivals exceeds the threshold above which tourism's impacts cannot be absorbed, we can argue that the TCC describes one type of unsustainability of tourism: the *overutilisation* of space, of assets and of facilities. In particular, the phenomenon of overtourism equates to the appearance of all sorts of negative externalities that determine that the community suffers from tourism rather than benefitting from it. These externalities include, among others, the rising costs of living; the loss of inhabitants and those firms that are not serving tourism demand, overcrowding, wear-and-tear, pollution, and the loss of local identity (Costa et al., 1996).

This article, which has drawn heavily on my own extensive past investigations into the role of tourism in the historical centre of Venice, has made some of the above-mentioned mechanisms and relations more concrete using this iconic case of excessive tourism development. In fact, evidence that the collective costs have surpassed the collective benefits has been collected by various authors (CoSES, 1979; Page, 1995; Russo, 2002; Quinn, 2007; Van der Borg, 1991) and has even been recognised by organisations that until recently embraced a boosterism approach to tourism development, like the United Nations World Tourism Organisation (UNWTO (2019) and the World Tourism and Travel Council (WTTC and McKinsey (2017) and WTTC (2020)).

Despite the widespread critique of Venice as one of the worst cases of overtourism around the globe, the city is in fact an exception in that its officials should have understood by now where the outer boundary lies and what the optimal visitor mix looks like, with respect to the capacity of its cultural assets and facilities. Bertocchi et al. (2020) have shown that in 2018 this optimum for Venice was approximately 17 million visitors as opposed to the 30 million people that actually visited the city that year. The optimal visitor mix consists of 50% of overnight tourists and 50% of day tourists, while day tourists dominate (80% of the total number of visitors). The year 2023 will probably confirm this structural analysis. The symptoms of Venice's violation of the carrying capacity have been well documented: a monocultural development that might turn a city into a theme park, massive overcrowding, an unstoppable process of gentrification that is enhanced by the unbearable cost of living and that results in a loss of authenticity and of identity. Moreover, the national and local administrations face the challenge of identifying funding sources to pay almost double the amount of money needed to keep the city clean, safe and relatively well conserved, compared to what would be expected for the city without visitors.

Obviously, the COVID-19 pandemic has only temporally dampened the urgency to identify ways of overcoming the overtourism problem in Venice. In fact, given the virtual annihilation of global tourism during the pandemic, cities like Venice, which have witnessed a huge downturn in arrivals and in economic benefits, appear to have downgraded their attempts to strengthen their sustainability agendas by heavily focusing on measures to overcome the massive problems resulting from the pandemic. National governments embraced a relaunch strategy based on financial support, often without any strings attached.

However, all the suggestions that were formulated to fight overtourism prior to the pandemic should not have been abandoned. On the contrary, they should become a fundamental starting point for a profound discussion on how to "build tourism back better". This disruptive moment should have been used to question the pre-COVID mindset that was dominating the tourism sector that "more is always better than less"; that more income and jobs always justify the sacrifices that were being made to generate them. In other words, as soon as the global tourism sector sees its growth pick up again a new business model for the entire tourism sector is urgently needed.

A new, innovative business model must be found quickly to make a clean break with the obsolete mass tourism business model, and help make travellers more conscious and mindful, tourism firms more responsible and environmentally and socially engaged, and, hence, tourism destinations more sustainable, safer and more resilient to future disruptions. This business model should clearly place quality over quantity, and the wellbeing of the local population and of local firms over the success of the global tourism industry. Various options that any destination that suffers from overtourism like Venice has to adopt policies to sustain such a strategy were discussed in this paper.

Whether you think a destination resembles a "doughnut" or not, it is obvious that tourism is not a means to an end, but in principle an important tool aimed primarily at fulfilling the needs of the local population and entrepreneurs. Therefore, tourism should not be demonised but managed carefully. And if destination management fails, under- or overtourism are likely to appear or persist.

Venice is teaching us that the momentum that the pandemic has offered the tourism industry has largely been wasted and that the pre-pandemic discourse about unsustainable tourism development and overtourism has not lost any of its usefulness.

References

Barač Miftarević, S. (2023). Undertourism vs. overtourism: A systematic literature review. *Tourism: An International Interdisciplinary Journal*, *71*(1), 178–192.

Benjamin, S., Dillette, A., & Alderman, D. H. (2020). "We can't return to normal": Committing to tourism equity in the post-pandemic age. *Tourism Geographies*, *22*(3), 476–483.

Bertocchi, D., Camatti, N., Giove S., & Van der Borg, J. (2020). Venice and overtourism: Simulating sustainable development scenarios through a tourism carrying capacity model. *Sustainability*, *12*(2), 512.

Butler, R. W. (1980). The concept of a tourist area cycle evolution: Implications for the management of resources. *Canadian Geographer*, *24*(1), 5–12.

Canestrelli, E., & Costa, P. (1991). Tourist carrying capacity: A fuzzy approach. *Annals of Tourism Research*, *18*(2), 295–311.

Cheer, J. (2020). Human flourishing, tourism transformation and COVID-19: A conceptual touchstone. *Tourism Geographies*, *22*(3), 514–524.

CoSES. (1979). *Turismo a Venezia*. COSES Publicazioni.

Costa, P., Gotti, C., & Van der Borg, J. (1996). Tourism in European heritage cities. *Annals of Tourism Research*, *23*(2), 306–321.

Costa, P., & van der Borg, J. (1988). Un modello lineare per la programmazione del turismo [A linear model for tourism planning]. *CoSES Informazioni*, *18*(32/33), 21–26.

De Young, R. (1999). Tragedy of the commons. In D. E. Alexander & R. W. Fairbridges (Eds.), *Encyclopaedia of environmental science*. Kluwer Academic Publishers.

Fletcher, R., Mas, I. M., Blanco-Romero, A., & Blázquez-Salom, M. (2019). Tourism and degrowth: An emerging agenda for research and praxis. *Journal of Sustainable Tourism*, *27*(12), 1745–1763.

Hall, C. M., Lundmark, L., & Zhang, J. J. (Eds.). (2020). *Degrowth and tourism*. Routledge.

Hardin, G. (1968). The tragedy of the commons. *Science*, *162*, 1243–1248.

Higgins-Desbiolles, F. (2021). The "war over tourism": Challenges to sustainable tourism in the tourism academy after Covid-19. *Journal of Sustainable Tourism*, *29*(4), 551–569.

Ioannides, D., & Gyimóthy, S. (2020). The COVID-19 crisis as an opportunity for escaping the unsustainable global tourism path. *Tourism Geographies*, *22*(3), 624–632.

Koens, K., Postma, A., & Papp, B. (2018). Is overtourism overused? Understanding the impact of tourism in a city context. *Sustainability, 10*(12), 4384.

Krippendorf, J. (1986). *Die Ferienmenschen* [The vacation people]. Dtv Deutscher Taschenbuch.

Mexa, A., & Coccossis, H. (2004). Tourism carrying capacity: A theoretical overview. In H. Coccossis & A. Mexa (Eds.), *The challenge of tourism carrying capacity assessment: Theory and practice*. Ashgate.

Nepal, S. K. (2020). Adventure travel and tourism after COVID-19 – business as usual or opportunity to reset? *Tourism Geographies, 22*(3), 646–650.

O'Neill, D.W., Fanning, A.L., Lamb, W.F., & Steinberger, J.K. (2018). A good life for all within planetary boundaries. *Nature Sustainability, 1*(2), 88–95. https://doi.org/10.1038/s41893-018-0021-4

Page, S. (1995). *Urban Tourism*. Routledge.

Papageorgiou, K., & Brotherton, I. (1999). A management planning framework based on ecological, perceptual and economic carrying capacity: The case study of Vikos-Aoos National Park, Greece. *Journal of Environmental Management, 56*(4), 271–284.

Petric, J. L., Hell, M., & Van der Borg, J. (2020). Process orientation of the world heritage city management system. *Journal of Cultural Heritage 46*, 259–267.

Quinn, B. (2007). Performing tourism. *Annals of Tourism Research, 34*(2), 458–476.

Raworth, K. (2017). *Doughnut economics: Seven ways to think like a 21st-century economist*. Random House Business.

Russo, A. (2002). *The sustainable development of heritage cities and their regions: Analysis, policy, governance*. Thela Thesis, Amsterdam.

Russo, A., & Scarnato, A. (2017). "Barcelona in common": A new urban regime for the 21st-century tourist city? *Journal of Urban Affairs, 40*, 1–20.

Saarinen, J. (2006). Traditions of sustainability in tourism studies. *Annals of Tourism Research, 33*(4), 1121–1140

Salerno, M-S., & Russo, A. P. (2020). Venice as a *short-term city*. Between global trends and local lock-ins. *Journal of Sustainable Tourism, 30*(5), 1040–1059.

Seidl, I., & Tisdell, C. A. (1999). Carrying capacity reconsidered: From Malthus' population theory to cultural carrying capacity. *Ecological Economics*, 31, 395–408

Stoffelen, A. & Ioannides, D. (Eds.). (2022) *Handbook of tourism impacts: Social and environmental perspectives*. Edward Elgar Publishing.

UNWTO. (2019). *"Overtourism"? Understanding and managing urban tourism growth beyond perceptions*, Vol. 2: Case Studies. UNWTO.

UNWTO. (2021). *2020: Worst year in tourism history with 1 billion fewer international arrivals*.www.unwto.org/news/2020-worst-year-in-tourism-history-with-1-billion-fewer-international-arrivals

Van den Berg, L., Van der Borg, J., & Van der Meer, J. (1995). *Urban tourism: Performance and strategies in eight European cities*. Aldershot.

Van den Bosch, H. (2020). Humane by choice: Smart by default – 39 building blocks for cities of the future. *IET Smart Cities*. https://doi.org/10.1049/iet-smc.2020.0030.

Van der Borg, J. (1991). *Tourism and urban development*. Thesis Publishers.

Van der Borg, J. (1992). Tourism and urban development: The case of Venice, Italy. *Tourism Recreation Research, 17*(2), 46–56.

Van der Borg, J. (1994). The demand for city trips in Europe: Tour operators' catalogues. *Tourism Management, 15*(1), 66–69.

Van der Borg, J. (2015). *Toerisme en Erfgoed: Zijn er Grenzen aan Toeristische Ontwikkeling? Lessen voor de XXI eeuw* [Tourism and Heritage: Are there Limits to Tourism Development? Lessons for the XXI century]. Leuven University Press.

Van der Borg, J. (2017). Sustainable tourism in Venice: What lessons for other fragile cities on water. In S. Caroli and S. Soriani (Eds.), *Fragile and Resilient Cities on Water*. Scholars Publishing.

Van der Borg, J. (Ed.). (2022a). *A research agenda for urban tourism*. Edward Elgar Publishing.

Van der Borg, J. (2022b). The role of the impacts of tourism on destinations in determining the tourism-carrying capacity: Evidence from Venice, Italy. In A. Stoffelen & D. Ioannides (Eds.), Handbook of tourism impacts: Social and environmental perspectives. Edward Elgar Publishing.

Visentin, F., & Bertocchi, D. (2019). Venice: An analysis of tourism excesses in an overtourism icon. In C. Milano, J. M. Cheer & M. Novelli (Eds.), *Overtourism: Excesses, discontents and measures in travel and tourism*. Cap Intl.

Watson, G. L., & Kopachevsky, J. P. (1996). Tourist carrying capacity: A critical look at the discursive dimension. *Progress in Tourism and Hospitality Research*, *2*, 169–179.

WTTC & McKinsey. (2017). *Coping with success: Managing overcrowding in tourism destinations*. World Travel and Tourism Council, McKinsey Company.

WTTC. (2020). *The importance of travel and tourism in 2019: Economic impacts report 2020*. WTTC.

Young, G. (1973). *Tourism: Blessing or blight?* Pelican.

Excursus

Overtourism in Austria Using the Example of Hallstatt

Peter Zellmann

The romantic village of Hallstatt, situated on the shores of Lake Hallstatt in the Salzkammergut region, is extremely well suited for a case study on the subject of overtourism in Austria. Declared a World Heritage Site in 1997, it has become a major tourist attraction. However, this idyllic scene is deceptive. With an exponential increase in day visitors, the village of just under 800 inhabitants currently has to cope with up to 10,000 tourists a day from all over the world. Excluding the pandemic period, over one million guests visit Hallstatt every year. The narrow, romantic alleys soon become terribly crowded. Apart from taking photographs, the visitors also fly drones over the properties, make a corresponding amount of noise, and leave behind piles of waste.

The majority of these tourists come from Asia. They plan to travel through Europe within only a few days, including Hallstatt in the Salzkammergut as a fixed holiday destination. At the entrance to the village, you can rent traditional costumes for 22 euros (per hour!) in order to take home authentic souvenir photographs. Moreover, so-called "Hallstatt air" filled in containers can be purchased and taken home.

The "Citizens for Hallstatt" association states that in 2010 only 3,440 coaches came to Hallstatt (www.bfhallstatt.at). Since then, the number has increased by up to 30% per year. Of course, only those buses that have paid parking fees can be counted. In 2019, before the pandemic restrictions, there were over 21,000 buses whose passengers stormed the otherwise idyllic little village of Hallstatt. At the same time, 211,400 cars were counted in the public car parks. According to the association's website, "the noise level of the buses parked with their engines running is unmistakable and the exhaust emissions are enormous. The lakeside is filled with buses" (Lassner, 2022a).

The mayor of Hallstatt, Alexander Scheutz, outlines both sides of the coin. On the one hand, Hallstatt's attractiveness has become a curse. On the other hand, the increasing number of visitors provides added value and employment for the people of the region. Who would have thought it possible "that the operating turnover of sanitary facilities exceeds the municipal income from property tax by 1.5 times" (A. Scheutz, personal communication, June 19, 2023).

"Some tourists believe that Hallstatt is merely a scenery, a kind of Disneyland, and that those who work here live in tower blocks behind the mountains.

DOI: 10.4324/9781003365815-5

Sometimes they are genuinely surprised when we tell them that we are living right in the middle of it" (Kazim, 2018), as the newspaper *Spiegel* also quoted the mayor.

While the majority of day visitors come from Asia, there are also visitors from Germany, Austria and the rest of Europe. One of the main reasons for its popularity in Asia is the South Korean TV soap opera "Spring Waltz", which was partly filmed in Salzburg and Hallstatt (Lassner, 2022b). This led to a replica of Hallstatt in China, which further drives the numbers of Asian visitors who want to visit the original on their European trip.

It probably applies to all places confronted with overtourism that the residents are the ones who suffer most from this development. Hosts become victims insofar as people curse the tourist crowds at the regulars' table. A considerable number of locals leave their hometown despite possible financial benefits, because quality of life is more important to them than standard of living.

There are no patent remedies for mitigating tourist crowds in villages or towns. The residents as hosts have to decide, as the local conditions vary considerably. It is the task of local politicians to involve them in solution concepts by means of well-moderated decision-making processes. Approaches include restricting coach slots, parking space management and providing information for visitors, whose quality of experience is also at stake. The mayors concerned usually refuse to charge entrance fees in order to prevent their tourist attraction from becoming a museum.

References

Kazim, H. (2018). Warum so viele Chinesen nach Hallstatt kommen [Why so many Chinese come to Hallstatt]. www.spiegel.de/reise/europa/hallstatt-in-oesterreich-warum-kommen-so-viele-touristen-aus-china-a-1233291.html

Lassner, J. (2022a). Hallstatt: Ein malerisches Dorf wird vom Massentourismus überrannt [Hallstatt: picturesque village is overrun by mass tourism]. https://globusliebe.com/hallstatt-massentourismus/

Lassner, J. (2022b). Salzburg im Winter: Die schönsten Orte in der Weihnachtszeit [Salzburg during winter: The most beautiful places in the Christmas time]. https://globusliebe.com/salzburg-im-winter/

Excursus

From Graffiti to Regulations: Addressing Overtourism in Palma de Mallorca

Hans Müller

"Mr. Müller, where are your overbooked guests going to sleep tonight?" That was the question the editor of a major German tabloid abruptly asked me when he called me on a Saturday morning at the NUR TOURISTIC SERVICE office in Palma. It was high season, and the year was 1994. Mallorca was completely overcrowded. All tour operators were struggling with overbookings because hoteliers were overwhelmed by reservations and couldn't handle the situation anymore. Another major tour operator had even docked a cruise ship in the Bay of Palma de Mallorca to accommodate at least some of the guests rejected by hotels. Antalya was not an option as a holiday destination that year due to sporadic terrorist attacks, so everyone flocked to Spain, especially Mallorca.

This little anecdote describes the first intense experience I had with "overtourism" – even though this term didn't exist back then. It was simply high season, and the island was bursting at the seams.

In the past 30 years, I have experienced similar phases in other destinations as well, but I want to focus on Mallorca here. The term overtourism gained popularity in 2016 and 2017 when specific groups in the population wanted to draw attention to the situation using graffiti on facades of emblematic buildings in the old town of Palma. I remember slogans like "Tourists go home" and "Tourists no, refugees welcome".

Obviously, these circles wanted to show the guests that they were no longer welcome. It was never possible to determine who was specifically responsible for defacing the old town. I never believed that the residents of the city centre would deface their own walls. However, there have always been groups, including political ones, that opposed tourism and capitalism. It was a time when communist and separatist parties were forming and restructuring in Spain and the Balearic Islands.

German media, in particular, took up the issue intensively, as it guaranteed high ratings, clicks and circulation. Almost every television station sent film crews to publish original statements from supposedly affected individuals and corresponding images. These were easy to find because, unlike in the past phases of full occupancy, there were two additional factors that had significantly contributed to the island becoming even more crowded than in previous years.

Firstly, a new phenomenon had gained momentum: spending vacations not only in hotels but also in close proximity to the hosts. Airbnb had risen to power and

DOI: 10.4324/9781003365815-6

popularised this type of vacation, leading to the emergence of more platforms following suit. Every year, more property owners jumped on the bandwagon to participate in this boom. As a result, approximately 250,000 beds in these more or less legal vacation rentals were available on Mallorca. This roughly corresponds to the number of beds in all the legal hotels and aparthotels on the island.

This number also corresponded to the statistics of the airport company AENA, which determined the so-called "floating population". After all, AENA knows how many passengers arrive and depart. In those years, the peak of this floating population always occurred around 12 August. It amounted to approximately 510,000 guests being accommodated on the island simultaneously. This was about twice as many guests as traditionally stayed in legal hotel accommodations. The number of official hotel beds had only increased by a few percentage points since the 1980s. There were no spectacular new construction projects of hotel complexes on Mallorca. Thus, over the decades, the infrastructure had been able to adapt well to the "legal" guest numbers. However, it was not prepared for a doubling of the number of guests accommodated on the island due to the phenomenon of vacation rentals, which had established itself during the same period through the ongoing development of various booking platforms.

Previously, tourists stayed in the well-known tourist areas of Playa de Palma, Magaluf, Cala Ratjada or other famous or infamous areas. Now they populated former quiet neighbourhoods in Palma, Inca or idyllic villages nearby. This inevitably led to conflicts because guests, especially those geared towards partying, had different sleep patterns and noise levels than their local neighbours. Municipalities also struggled to handle the mountains of garbage. Tourists and locals shop differently, and tourists produce significantly more waste (whenever they stay in private apartments or houses instead of hotels). As a result, overflowing garbage containers became a common sight because the municipalities were unable to adjust their waste disposal cycle to the sudden population growth. The streets were also significantly busier because unlike package tourists traditionally transported to their hotels by transfer or shuttle buses, individual vacation rental tenants also booked rental cars. There were about 100,000 rental cars on the island, putting a strain on traffic flow and parking situations.

Secondly, another form of vacationing had become popular among the masses: vacationing on a cruise ship. While just a few years ago, cruise ships had a capacity of several hundred passengers, nowadays this genre is almost extinct. Today, we are talking about several thousand beds on modern cruise ships. Since Palma is one of the most beautiful cities in Europe, hardly any ship in the Mediterranean skips a stop here. In 2012, 985,000 cruise ship passengers visited Palma for a brief stopover. In 2019, this number had already more than doubled to over 2.2 million. During peak times, I have counted up to ten cruise ships in the harbour on some days. The result was that, in addition to hotel and vacation rental guests, thousands of people crowded through Jaime III Avenue or in front of the cathedral.

During the tourist standstill imposed due to COVID, everything came to a halt, and the opponents of tourism fell silent because they suddenly had other concerns. However, the numbers have now returned to pre-COVID levels, and the debate on overtourism is gaining momentum once again. Cruise ship tourism is booming like never before, and people are booking even more individually, resulting in an intensification of vacation rentals. Another significant consequence of this is an unnatural increase in rental prices. Many property owners prefer to rent to tourists at significantly higher prices than a local long-term tenant would be willing to pay.

This also means that employees of traditional hotels, some of whom come from the mainland to Mallorca and need to rent an apartment for the season, can no longer find suitable accommodation at an affordable price. The exorbitant increase in vacation rentals, therefore, harms the existing system and society in multiple ways. Graffiti that I recently found on an old building sums it up: "AirBNB raises my rent – Tourists Go Home" it said in big black letters.

There have been some initial political decisions to regulate the influx somewhat. No licences are being issued for additional tourist beds, either for new hotels or for vacation rentals. In addition, hefty fines for illegal rentals have been announced. The relevant booking platforms must provide a licence number for each legal property, and if they fail to do so, they will be fined. The number of cruise ships allowed to dock simultaneously in the port of Palma has also been restricted. Now, it remains to be seen whether these measures, along with inspections to identify illegal vacation rental providers, will be sufficient to tackle the problem.

Regardless, an improvement in infrastructure, transportation routes, parking situations and public transportation networks would be desirable. Hopefully, the new government after the elections will handle logistics better than its predecessors, who exacerbated the problem by thinning out some traffic arteries and eliminating parking spaces in the city. The new mayor of Palma de Mallorca is an architect by profession and not a professional politician. He has the ability and the desire to restore Palma de Mallorca to a city with the highest quality of life. Therefore, at least regarding the situation in the city, I am very optimistic about the future.

Part II

Towards a New Sustainability

4 Beyond Sun and Sand

How the Algarve Region is Pioneering Innovative Sustainable Practices in Tourism

Paulo Martins and Antonia Correia

1 Introduction

Sustainability is an essential concept in our modern world that requires examining current practices and implementing creative and sustainable solutions to foster significant societal change. Sustainable practices have become increasingly important, with organisations focusing on reducing their environmental impact and promoting social responsibility. It has been suggested in recent research by Vizcaino-Suarez and Diaz-Carrion (2018) that the tourism industry is closely interconnected and shaped by power dynamics among nations, organisations and social groups. As a result of these power dynamics, the tourism sector must continuously evolve through innovative and creative ideas, especially regarding sustainability and innovation.

Sustainability in tourism primarily focuses on environmental preservation efforts, which involve the effective management of resources to enhance the economic and social aspects of the industry (Aslan & Rahman, 2018). As a result, Gössling et al. (2012) argue that sustainable tourism practices should promote biodiversity while protecting natural resources by reducing tourism's environmental impact and supporting conservation efforts. Gökalp and Gökmen's (2019) study recognises that natural resources and economic sustainability are interrelated, and that small and medium-sized business can significantly contribute to sustainable tourism. Thus, energy-efficient smart city planning has become vital to the sustainable green tourism industry and contributes to economic sustainability (Lu et al., 2021). In addition, Dredge and Jenkins (2007) suggest that sustainable tourism practices should balance the needs of tourists and residents in urban areas by establishing policies and regulations that encourage sustainable development, creating a link between sustainable economies, infrastructure and planning. Therefore, it is crucial to assess beach quality and understand tourist priorities to advance sustainable coastal tourism, such as in the Algarve region (Correia & Crouch, 2003; Lukoseviciute & Panagopoulos, 2021).

The Algarve region of Portugal has gained worldwide recognition for its magnificent beaches, picturesque natural parks, and significant historical landmarks, establishing itself as a top-rated tourist destination worldwide (Correia & Crouch, 2003). As a result, the region is vital to the country's tourism industry's sustainability.

DOI: 10.4324/9781003365815-8

Sustainability is one of the top priorities of a region born in the sixties with a model of tourism development where more is better. Several actions have been undertaken to mitigate past mistakes to recover the Algarve and ensure a sustainable destination. Sustainability is also one of the most researched topics in the Algarve, covering several perspectives. Bienvenido-Huertas et al. (2020) suggest that most Algarve hotels adopt sustainable practices, but energy and water deserve more research. Farinha et al. (2019) frame the sustainability indicators in tourism to monitor sustainable practices. Furthermore, Carvalho and Fernandes (2023) assessed the region's current state of sustainable tourism, suggesting that seasonality, limited public transportation and limited local involvement inhibit the region's sustainable development. The University of Algarve (2021) released a sustainability report covering all Sustainable Development Goals of Agenda 2030 to promote and stimulate sustainable practices aligned with the Sustainable Development Goals (United Nations, 2015).

Therefore, the region of Algarve in Portugal has made noteworthy endeavours by incorporating sustainable and innovative approaches into the tourism industry, encompassing diverse aspects such as natural resources use and management, economy, urban planning, safety and monitoring. This book chapter aims to highlight some of the best sustainable practices of the region, considering a sustainable model based on three axes. By sharing those best practices, we aim to inspire other regions to adopt those practices and divulge their best ones, as sustainability is also learning by doing.

2 The Concept of Sustainability

Sustainability is a complex and multi-faceted concept that involves three interdependent and tripartite pillars: the *environment*, the *economy* and *social* equity. These pillars are crucial for achieving sustainable development, which is development that meets the present needs while safeguarding the ability of future generations to meet their own needs (Murphy, 2012). According to the Aalborg Charter, policymakers must integrate environmental protection with people's fundamental social needs, including healthcare, employment and housing programmes (ESCTC, 1994). This integration ensures that sustainability is not just focused on environmental protection but also prioritises the well-being of society. Also, the European Union's Sustainable Development Strategy emphasises the integration of economic, social and environmental considerations to ensure coherence and mutual reinforcement (Council of European Union, 2006). This integration is critical to ensuring that economic growth is sustainable, socially responsible and environmentally friendly. Therefore, policymakers, researchers and stakeholders must balance environmental protection, economic growth and social fairness to achieve sustainable development. In this section, we examine the importance of sustainability from a three-way perspective:

Environment: A key pillar of sustainability, the environment consists of all natural resources and systems that sustain life on earth. Ecosystems, biodiversity and natural resources such as air, water and soil are all vital to achieving environmental sustainability. It is possible to incorporate tourism with the environment using environmentally appropriate planning and practices (Wu et al., 2021). While Harris et al. (2012) emphasise that the environment and tourism are separate concepts, they can both promote sustainable development at various levels, from businesses to national policies. As a result, businesses and governments are crucial to promoting environmental sustainability (Yasmeen et al., 2020). Farinha et al. (2019) emphasise that companies should engage with stakeholders and seek expert consultation to identify sustainable indicators that can assist in tracking and promoting sustainable development in the Algarve region to decrease their carbon footprint. The authors reinforce that policymakers, stakeholders and companies can collaborate to balance economic and environmental conservation, ensuring the preservation of the region's natural resources for future generations.

Economy: A system's economy is the second pillar of sustainability, including the production, distribution and consumption of goods and services. It is crucial to create an economic system that supports sustainable development, promotes social equity and minimises environmental impact (Collins, 2001). This is particularly relevant in the context of tourism, which influences the economy and the environment. Pulido-Fernández et al. (2015) suggest that sustainable tourism can play a role in reducing resource costs and creating market differentiation. However, sustainability measures can reduce profitability and hinder competitiveness. This perception has posed a substantial challenge in promoting sustainable economic growth and tourism yield (Northcote & Macbeth, 2006). The Algarve region has successfully addressed this by adopting sustainable business practices, investing in infrastructure and implementing policies to support sustainable economic growth. By doing so, the region has promoted economic sustainability while minimising the impact on the environment and helping to create a balance between economic growth and environmental protection, which is crucial for achieving sustainable development (Bienvenido-Huertas et al., 2020). Therefore, sustainable economic growth is an essential component of sustainability. The relationship between sustainability, economy and tourism is complex and requires careful consideration of various factors. The Algarve region's successful adoption of sustainable business practices and policies is an excellent example of how economic sustainability can be achieved while protecting the environment.

Social: Social sustainability is the third pillar and a fundamental principle of sustainability that focuses on promoting the well-being of people and communities by ensuring access to critical essentials, including healthcare, education and employment, while also advocating for fairness and equal opportunities in society. According to the research of Roca-Puig (2019), reciprocity and trust between companies and society are essential to social sustainability. Santos and Moreira (2022) suggest that stakeholder engagement and communication are important for improving the social sustainability and overall sustainability performance of companies in the Algarve, as tourism plays a significant role in the region's social development. This highlights the importance of tourism and corporate social responsibility in tourism as positive factors for improving the lives of residents. As a result, Franzoni (2015) emphasises tourism's importance for sustainable economic growth and employment.

Given the significance of the three interdependent pillars of sustainability and the role of tourism in the Algarve in fostering sustainable and innovative practices across various areas – such as natural resources use and management, economy, urban planning – seafront redevelopment, safety, social, and monitoring – the proposed best practices address all three components of the triad, and the examples cited may pertain to more than one of these pillars.

3 Natural Resources Use and Management

Although the Algarve region is widely recognised for its attractive beaches, white sand, pleasant climate and vibrant culture, it faces significant challenges in the sustainable management of its natural resources, particularly water management. Given the Mediterranean climate and dependence on tourism, there is an urgent need to adopt sustainable practices such as water conservation, wastewater treatment and efficient irrigation systems to reduce the environmental impact while ensuring the availability of water for local communities and the tourism industry (Rodrigues & Antunes, 2021). Sustainable water management practices are essential to maintaining the quality of water resources and reducing the tourism industry's carbon footprint.

Golf, as highlighted by Videira et al. (2006), is an industry that effectively addresses its impact on the environment and sustainability performance indicators by prioritising waste and water management practices. Sustainable practices in golf course management include using recycled water, reducing chemical inputs and adopting energy-efficient technologies (Gonzalez-Perez et al., 2023).

In this vein of resource management, **urban water consumption** is also an example of good practice, and includes water metering, leak detection and conservation campaigns. Such practices reduce water usage, ensure equitable distribution and promote sustainable urban development (Moreira da Silva et al., 2022; Smith, 2022).

In addition to water management, **sustainable agricultural practices** are essential for the Algarve's natural resource sustainability. Agrotourism is a promising way to promote sustainable agricultural practices while providing economic opportunities for local communities. The **cultivation of the Algarve carob**, also known as "black gold", is a prime example of a sustainable agricultural practice that can help promote the region's natural resource sustainability. The Algarve carob is a versatile crop used in various industries, including food, cosmetics and biofuels. Furthermore, the carob tree is well adapted to the Algarve's climate and requires little water, making it a sustainable crop for the region (Correia & Pestana, 2018). Born and developed in Algarve, "Carob World" and "Gran Carob" are two notable examples of innovation and sustainability in the carob industry. Carob World is a cooperative that promotes sustainable and responsible production of carob products, offering high-quality items such as carob powder, flour and syrup while promoting sustainable agricultural practices. Gran Carob, on the other hand, has developed a range of innovative and sustainable carob products, including chips and snacks.

Adopting sustainable practices in these areas ensures a prosperous future while preserving the region's natural resources for future generations.

4 Economy and Social Equity

Tourism is an economic and social phenomenon due to its ability to stimulate regional development (Franzidis, 2018). The Algarve economy heavily relies on the tourism industry, which provides employment opportunities and generates income for its people. Managing tourism sustainably could positively impact the environment, society and economy. Therefore, implementing sustainable tourism practices, such as the Region Intelligent Algarve (RIA) project that encompasses SMART tourism, is crucial for the region's economic growth while minimising negative environmental and societal impacts. The main objective of this project is to enhance the partner entities' capabilities in achieving a **Smart Region Algarve** through collaboration with the region's stakeholders and focusing on regional competitiveness and innovation in the digital economy. The project also establishes a governance model led by the partnership to ensure its viability and sustainability, with an action plan for implementing the RIA Platform, Smart Tourism Destination and Smart Mobility solutions. The project also aims to create institutional cooperation dynamics and share best practices while identifying and disseminating national and international best practices and lessons learned that can be replicated in the region. The Algarve Regional Coordination and Development Commission (CCDR Algarve) coordinates the RIA project with the University of Algarve, the Algarve Tourism Region (RTA) and the Algarve Intermunicipal Community (AMAL). Therefore, adopting sustainable tourism practices can boost the significance of such initiatives, enabling the region to appeal to travellers who prioritise eco-friendly and socially responsible travel. This can enhance the region's reputation and lead to an influx of tourists.

The region's heavy reliance on tourism makes it susceptible to **seasonality's** negative impacts. Algarve is known for its seasonality, with a significant influx of tourists during the summer months and a decline in the off-season. This seasonal fluctuation can have negative economic impacts, as many businesses rely on tourism for their livelihoods, leading to potential job losses and economic instability in the off-season. As a result, promoting year-round tourism, so that local businesses and communities can benefit from tourism revenue throughout the year rather than only during peak tourist season, is vital for the region's sustainability. This can help create additional stable and sustainable economic growth for the region, ultimately leading to job creation, economic stability, and the preservation of the region's unique culture. However, tourism also poses some environmental challenges, such as desertification, which can significantly impact agriculture, biodiversity and other industries, leading to negative implications for tourism. Therefore, it is essential to implement initiatives like **"Project 365 Algarve"**, launched by RTA, which promotes sustainable land use practices and responsible tourism, minimising negative impacts on the environment and ensuring the long-term sustainability of the region's economy. The initiative seeks to diversify the region's tourism offerings throughout the year, encouraging visitors to explore Algarve's natural, cultural and gastronomic attractions during the off-season. Therefore, this project strives to establish a more robust, forward-thinking and sustainable tourism sector by promoting eco-friendly practices and environmental responsibility. The objective is to reduce the adverse effects of tourism on the environment while safeguarding the region's natural resources and enhancing the cultural experience of visitors (Correia et al., 2022).

To reinforce the connection between social sustainability, tourism and employment, the study "*O Capital Humano na Hotelaria e Empreendimentos Turísticos do Algarve*" (Human Capital in Tourism and Hospitality Companies in the Algarve) conducted by Associação dos Hotéis e Empreendimentos Turísticos do Algarve (AHETA), Knowledge to Innovate Professions in Tourism (KIPT) Co-Lab, and the University of Algarve (2022), highlighted that the human resource needs of surveyed companies are expected to increase from current levels by 45%, ranging from 4,484 to 7,906, by the end of 2023. These findings emphasize the shortage of the labour market in the Algarve, and the relevance of increased research and action plans in this direction. The main activity of KIPT Colab is to promote and value professions in tourism and hospitality, to sustain tourism competitiveness and sustainability. Therefore, it is vital to understand that sustainable practices considering environmental and economic growth can only be realised by deepening our understanding of social reality. Thus, businesses and governments must invest in social infrastructure, foster inclusive economic growth, and establish policies that support social equality. By promoting social sustainability, we can build strong and resilient communities and achieve sustainable development that fulfils present needs while preserving future generations' ability to meet them.

In summary, the region's economy and sustainability are intermingled. Therefore, it is essential to implement innovative initiatives to promote sustainable tourism practices and environmental stewardship to create a more diverse, innovative and sustainable future for the Algarve economy.

5 Urban Planning – Seafront Redevelopment

In Algarve, seafront redevelopment plays a significant role in urban planning, especially as the seafront is a major attraction for tourists. However, to ensure the sustainability of these projects, it is essential to consider environmental, social and economic factors. Sustainable practices, such as using eco-friendly construction materials and preserving natural habitats, should be prioritised. General urban planning should also be integrated into these projects, considering traffic flow, pedestrian access and public transportation. Community engagement and participation are critical for success, and measures like beach cleanups, waste reduction and sustainable transportation options should be implemented to minimise negative impacts. Tourism is another crucial consideration, with seafront areas as major tourist destinations. Designing attractive public spaces, providing amenities and ensuring waterfront accessibility are critical factors in meeting tourists' needs.

However, balancing tourism demands with preserving the local culture and heritage is equally important. One example of a sustainable seafront redevelopment project with walkways is the **Praia de Quarteira project in Algarve**, completed in 2019. This project was designed to provide visitors with a more enjoyable and sustainable experience while preserving the natural environment and promoting local economic development (Freitas & Dias, 2019).

The seafront redevelopment project in Quarteira involved a series of sustainable initiatives to enhance the visitor experience while minimising negative impacts on the environment and local community. One of the project's main features was the construction of a wooden walkway that spans the entire seafront, providing a safe and accessible route for pedestrians and cyclists. The walkway was made from sustainably sourced wood and was designed to blend in harmoniously with the natural surroundings.

In addition to the walkway, the project included the installation of solar-powered lighting strategically placed to minimise light effluence and avoid disturbing the natural habitats of local fauna. To manage waste, recycling bins were installed along the walkway, and a waste management plan was implemented, which included regular cleaning and garbage collection.

Another critical aspect of the project was its focus on promoting local economic development by creating public spaces for events and festivals, supporting small businesses and showcasing the local cultural heritage. These initiatives aimed to enhance the visitor experience and benefit the local community. As a result, the Praia de Quarteira seafront redevelopment project is an excellent example of how innovation, sustainability and urban planning can be combined to create a more attractive, sustainable and enjoyable destination for tourists and locals. According to Baloch et al. (2023) by prioritising environmental, social and economic factors, seafront development balances meeting visitors' needs and preserving the natural environment and local community. As a result, the new seafront walkway has become a popular and sustainable destination that provides an enjoyable experience for everyone.

6 Safety

The COVID-19 pandemic significantly impacted tourism in the Algarve, as in many other destinations worldwide. Safety and sustainability have become even more critical concerns for the tourism industry in the region.

To ensure the safety of travellers during the COVID-19 pandemic, several measures were implemented in the tourism industry in Algarve and Portugal. These measures included mandatory mask-wearing in public spaces, social distancing guidelines, frequent disinfection of high-touch surfaces, and capacity restrictions in indoor spaces. Additionally, the tourism industry has adopted new safety protocols such as contactless check-ins and increased use of technology to reduce person-to-person interactions.

Turismo de Portugal launched the **safe and clean seal** to promote sustainability in the sector in 2020. This seal is awarded to tourism businesses and destinations implementing rigorous health and safety protocols such as increased sanitation and hygiene measures, physical distancing and staff training on COVID-19 prevention.

For several reasons, the safe and clean seal is an important safety measure in the tourism industry in Algarve and Portugal. Firstly, it reassures travellers that tourism businesses and destinations are taking their health and safety seriously, which can increase their confidence and encourage them to travel again. This is particularly important in the current climate, where travellers may be hesitant to travel due to the risks posed by COVID-19 (Orîndaru et al., 2021).

Secondly, by implementing rigorous health and safety protocols, tourism businesses and destinations can demonstrate their commitment to protecting the health and well-being of residents and workers. This is essential for building trust between the tourism industry and local communities. As a result, local communities are more likely to support tourism development in their area, which can have positive economic impacts.

Thirdly, the safe and clean seal can help to standardise health and safety protocols across the tourism industry in Algarve and Portugal. Thus, it ensures that all businesses and destinations follow best practices and can help prevent the spread of COVID-19 and other infectious diseases.

However, it is essential to note that the safe and clean seal does not guarantee safety, and travellers should still take precautions to protect themselves and others while traveling in Algarve and Portugal. This includes following local guidelines and regulations, practicing good hygiene, wearing masks and practicing physical distancing. Furthermore, its implementation reassures travellers, helps build trust between the tourism industry and local communities, and standardises health and safety protocols (Santos & Moreira, 2021). Despite this, travellers must remain cautious and take measures to ensure their safety and the safety of those around them. By doing so, not only can they help to reduce the spread of COVID-19, but they can also contribute to the sustainable growth of the regional and national tourism industry.

7 Monitoring

As one of the world's major industries, tourism significantly benefits from an international tourism observatory that gathers and analyses tourism-related data. By providing information, the observatory facilitates informed decision-making for sustainable tourism development, encourages collaboration and networking among countries, organisations and stakeholders, and drives innovation and research to address emerging challenges and opportunities (Ignarra et al., 2014). To support these efforts, the United Nations World Tourism Organization (UNWTO, 2021) launched the **International Network of Sustainable Tourism Observatories (INSTO)**, a global network of observatories that brings together stakeholders across the globe to share knowledge, best practices and innovations related to sustainable tourism.

One of the strengths of INSTO is its focus on data-driven decision-making, as the observatories collect and analyse data on a range of sustainability indicators such as energy efficiency, waste management and biodiversity conservation.

This data can help inform policy decisions, guide actions toward more sustainable practices, and measure progress toward sustainability goals. Observatories use a similar methodology to ensure that data is comparable across different destinations, which can facilitate benchmarking and knowledge sharing. In addition to data collection and analysis, the international network plays a significant role in capacity building by providing training and technical assistance to destinations to help them develop and implement sustainable tourism policies and practices. Moreover, according to UNWTO (2021), the INSTO's collaborative approach is another strength, as networking brings together a broader range of stakeholders, such as governments, non-governmental organisations, tourism industry professionals and academic institutions, to share knowledge and best practices. This collaboration can foster innovation, facilitate knowledge sharing and promote the development of more effective and sustainable policies and tourism practices.

Using a data-driven and collaborative approach, this network can inform policy decisions and guide actions toward sustainable tourism practices. Thus, such a network promotes innovation, facilitates knowledge sharing and advances sustainable tourism development worldwide.

8 Conclusions

In summary, sustainable innovation in tourism has brought about positive changes in the Algarve region's natural resources, economy, urban planning, safety and monitoring. Implementing sustainable practices has reduced environmental degradation and preserved the region's natural resources, including water, energy and biodiversity. By adopting sustainable tourism practices, the Algarve region has protected and preserved its natural assets, which benefits the environment and attracts tourists who seek eco-friendly destinations. Furthermore, sustainable tourism practices have helped the region's economic expansion by creating employment, amplifying local businesses and increasing tourist outflows. Such initiatives have proved to be a significant driving force for the sustainable development of the tourism industry in the region. Incorporating sustainable practices into their operations can result in financial gains for businesses and enhance their reputation by showcasing their dedication to social and environmental responsibility. This can eventually translate into improved brand awareness and customer loyalty, increasing brand recognition and loyalty.

Sustainable tourism practices are also linked to urban planning, which has led to eco-friendly infrastructure development, such as coastal development. These developments made it possible to reduce carbon emissions and improve the overall environmental quality in the region. Moreover, by investing in sustainable infrastructure, the Algarve region has demonstrated its commitment to building a sustainable and safer future for its residents and visitors. Initiatives such as the safe and clean seal have played a vital role in elevating Portugal and Algarve's standing as a trustworthy tourist destination, catering to the demands of safety-conscious travellers. Initiatives such as an international observatory provide a platform for sharing knowledge, best practices and experiences, enabling identifying common challenges and solutions. By working together, stakeholders and policymakers can promote and monitor sustainable tourism development and ensure the long-term viability of the tourism industry while safeguarding the environment and well-being of local communities. Adopting sustainable tourism practices has brought benefits to the Algarve region. However, it is crucial to continue implementing sustainable practices and to ensure that the region's natural resources and cultural heritage are protected for future generations. By doing so, Algarve can continue to attract tourists, create new job opportunities and enhance its reputation as a sustainable and responsible tourism destination.

References

AHETA – Associação Dos Hotéis E Empreendimentos Turísticos Do Algarve, KIPT Inovação e Turismo Laboratório Colaborativo & University of Algarve. (2022, October 26). *O Capital Humano na Hotelaria e Empreendimentos Turísticos do Algarve* [Human Capital in tourism and hospitality companies in the Algarve]. Mais Algarve. https://maisalgarve.pt/2022/10/26/aheta-o-capital-humano-na-hotelaria-e-empreendimentos-turisticos-do-algarve/

Aslan, E., & Rahman, M. (2018). Sustainable tourism: A comprehensive literature review on frameworks and applications. *Annals of Tourism Research, 73*, 137–154. doi:10.1016/j.annals.2018.09.005

Baloch, Q. B., Shah, S. N., Iqbal, N., Ahmed, M., & Hussain, M. (2023). Impact of tourism development upon environmental sustainability: A suggested framework for sustainable ecotourism. *Environmental Science and Pollution Research, 30*(5), 5917–5930. https://doi.org/10.1007/s11356-022-22496-w

Bienvenido-Huertas, D., Farinha, F., Oliveira, M. J., Silva, E. M. J., & Lança, R. (2020). Challenge for planning by using cluster methodology: The case study of the Algarve region. *Sustainability, 12*(4), 1536. https://doi.org/10.3390/su12041536

Carvalho, F. L., & Fernandes, S. C. (2023). Sustainable tourism and an analysis of opportunities for and challenges to researchers and professionals. In G. Altinay & A. Paraskevas (Eds.), *Handbook of research on sustainable tourism and hotel operations in global hypercompetition*. https://doi.org/10.4018/978-1-6684-4645-4.ch025

Collins, A. (2001). Thinking economically about sustainable tourism. *Annals of Tourism Research, 28*(3), 809–811. doi:10.1016/S0160–7383(00)00032-3

Correia, A., & Crouch, G. I. (2003). Tourist perceptions of and motivations for visiting the Algarve, Portugal. *Tourism Analysis, 8*(2–4), 165–169. doi:10.3727/108354203774076670

Correia, A., Lopes, A., Portugal, J., & Santos, M. (2022). A esperança de trazer cultura para o Algarve [The hope to bring culture to the Algarve]. In A. Correia (Ed.), *Turismo Algarve, Segredos por Revelar*. Escolar Editora.

Correia, P. J., & Pestana, M. (2018). Exploratory analysis of the productivity of carob tree (Ceratonia siliqua) orchards conducted under dry-farming conditions. *Sustainability, 10*(7), 2250. https://doi.org/10.3390/su10072250

Council of European Union. (2006). *Renewed European Union Sustainable Development Strategy*. European Council.

Dredge, D., & Jenkins, J. (2007). Sustainable tourism planning and development in New Zealand. *Tourism Management, 28*(2), 475–486. doi:10.1016/j.tourman.2006.04.013

European Sustainable Cities and Towns Charter (ESCTC). (1994). *The Aalborg Charter*. http://ec.europa.eu/environment/urban/pdf/aalborg_charter.pdf

Farinha, F., Oliveira, M. J., Silva, E. M. J., Lança, R., Pinheiro, M. D., & Miguel, C. (2019). Selection process of sustainable indicators for the Algarve region – OBSERVE project. *Sustainability, 11*(2), 444. doi:10.3390/su11020444

Franzidis, A. (2018). An examination of a social tourism business in Granada, Nicaragua. *Tourism Review, 73*(2), 214–227. doi:10.1108/TR-08–2017–0143

Franzoni, S. (2015). Measuring the sustainability performance of the tourism sector. *Tourism Management Perspectives, 16*, 22–27. https://doi.org/10.1016/j.tmp. 2015.05.007

Freitas, J. G., & Dias, J. A. (2019). Governance and management of coastal zones. Algarve (Portugal): A historical view of the impacts of seaside tourism. *Global Environment, 12*(2), 375–403. https://doi.org/10.3197/ge.2019.120208

Gökalp, F., & Gökmen, B. (2019). Sustainable tourism development: A case study of North Cyprus. *Journal of Hospitality and Tourism Management, 40*, 52–62. doi:10.1016/j.jhtm.2019.04.002

Gonzalez-Perez, D. M., Martín Martín, J. M., Guaita Martínez, J. M., & Morales Pachón, A. (2023). Analyzing the real size of the tourism industry on the basis of an assessment of water consumption patterns. *Journal of Business Research, 157*, 113601. https://doi.org/10.1016/j.jbusres.2021.09.006

Gössling, S., Hall, C. M., Ekström, F., Engeset, A. B., & Aall, C. (2012). Transition management: A tool for implementing sustainable tourism scenarios? *Journal of Sustainable Tourism, 20*(6), 899–916. https://doi.org/10.1080/09669582.2012.699062

Harris, R., Williams, P., & Griffin, T. (2012). *Sustainable Tourism: A Global Perspective*. Routledge.

Ignarra, M., Giaoutzi, A., & Pungetti, M. (2014). The importance of an international network observatory for the implementation of sustainable tourism in the Mediterranean. *Journal of Sustainable Tourism, 22*(3), 473–492.

Lu, C. W., Huang, J. C., Chen, C., Shu, M. H., Hsu, C. W., & Bapu, B. R. T. (2021). An energy-efficient smart city for sustainable green tourism industry. *Sustainable Energy Technologies and Assessments, 47*, 101494. https://doi.org/10.1016/j.seta.2021.101494

Lukoseviciute, G., & Panagopoulos, T. (2021). Management priorities from tourists' perspectives and beach quality assessment as tools to support sustainable coastal tourism. *Ocean & Coastal Management, 208*, 105646. https://doi.org/10.1016/j.ocecoaman.2021.105646

Moreira da Silva, M., Resende, F. C., Freitas, B., Aníbal, J., Martins, A., & Duarte, A. (2022). Urban wastewater reuse for citrus irrigation in Algarve, Portugal – Environmental benefits and carbon fluxes. *Sustainability, 14*(17), 10715. https://doi.org/10.3390/su141710715

Murphy, K. (2012). The social pillar of sustainable development: A literature review and framework for policy analysis. *Journal of Sustainable Development, 5*(6), 15–29. https://doi.org/10.1080/15487733.2012.11908081

Northcote, J., & Macbeth, J. (2006). Conceptualising yield: Sustainable tourism management. *Annals of Tourism Research, 33*(1), 199–220. doi:10.1016/j.annals.2005.10.012

Orîndaru, A., Popescu, M-F., Alexoaei, A. P., Căescu, S-C., Florescu, M. S., & Orzan, A-O. (2021). Tourism in a post-COVID-19 era: Sustainable strategies for industry's recovery. *Sustainability, 13*(2), 1–19. doi:10.3390/su13020356

Pulido-Fernández, J. I., Andrades-Caldito, L., & Sánchez-Rivero, M. (2015). Is sustainable tourism an obstacle to the economic performance of the tourism industry? Evidence from an international empirical study. *Journal of Sustainable Tourism, 23*(1), 47–64. doi:10.1080/09669582.2014.909447

Roca-Puig, V. (2019). The circular path of social sustainability: An empirical analysis. *Journal of Cleaner Production, 212*, 916–924. doi:10.1016/j.jclepro.2018.12.078

Rodrigues, M., & Antunes, C. (2021). Best management practices for the transition to a water-sensitive city in the South of Portugal. *Sustainability, 13*(5), 2983. https://doi.org/10.3390/su13052983

Santos, E., & Moreira, J. (2022). Social sustainability of water and waste management companies in Portugal. *Sustainability, 14*(1), 221. https://doi.org/10.3390/su14010221

Santos, N., & Moreira, C. O. (2021). Uncertainty and expectations in Portugal's tourism activities. Impacts of COVID-19. *Research in Globalization, 3*, 100071. https://doi.org/10.1016/j.resglo.2021.100071

Smith, J. (2022). Sustainable water management in urban tourism: Balancing consumption and conservation. *Annals of Tourism, 49(2)*, 145–160. doi:10.1016/j.annals.2022.02.001

United Nations. (2015). *Sustainable Development Goals*. www.un.org/sustainabledevelopment/sustainable-development-goals/

University of Algarve. (2021). *Sustainability report: Sustainable Development Goals*. www.ualg.pt/en/content/sustainability-report

Videira, N., Correia, A., Alves, I., Ramires, C., Subtil, R., & Martins, V. (2006). Environmental and economic tools to support sustainable golf tourism: The Algarve experience, Portugal. *Tourism and Hospitality Research, 6*(3), 204–217. https://doi.org/10.1057/palgrave.thr.6050013

Vizcaino-Suarez, L., & Diaz-Carrion, I. (2018). Gender in tourism research: Perspectives from Latin America. *Tourism Review, 73*(3), 311–325. doi:10.1108/TR-02–2018–0028

World Tourism Organization. (2021). *International Tourism Highlights, 2020 Edition*. UNWTO. https://doi.org/10.18111/9789284422456.

Wu, J. S., Barbrook-Johnson, P., & Font, X. (2021). Participatory complexity in tourism policy: Understanding sustainability programmes with participatory systems mapping. *Annals of Tourism Research, 90*, 103269. https://doi.org/10.1016/j.annals.2021.103269

Yasmeen, H., Wang, Y., Zameer, H., & Ismail, H. (2020). Modeling the role of government, firm, and civil society for environmental sustainability. In S. Ali & S. Y. Ali (Eds.), *Developing eco-cities through policy, planning, and innovation: can it really work?* IGI Global. doi:10.4018/978-1-7998-0441-3.ch003.

5 Transformative Urban Projects as Stimulators for the Quality of Stay

Elina Störmann, Hannes Thees and Daniel Zacher

1 Residents Reconquer the City Spaces

Cities are the crystallisation point of megatrends and express our living habits and needs. Recent trends and popular buzzwords are smartness, vertical densification, co-living, healing architecture, 15-minutes-city and urban manufacturing. Those examples illustrate and trigger urban life transformation, including housing and mobility, leisure and the way we work (Sodiq et al., 2019). In addition, urban planners initiate and push the necessary sustainability transformation (Tonne et al., 2021). Together with the local community, they develop projects that respond to the residents' changing needs and the need for sustainability – whether sustainable housing, lively neighbourhoods, co-working or the re-planning and construction of a (new) city district and traffic zones.

Although cities are growing globally, Europe witnessed an urban escape, as frequently stated; however, the borders between rural and urban are blurring (Wu, Long, Zhao & Hui, 2022). Regionality is introduced in trends such as urban farming and co-living, redefining the urban community and responding to emerging lifestyles. How is sustainability recognised in this transformation?

The city is not the devil of sustainable development, reduced to pollution, crowding and land grabbing. As a creative centre of a pluralistic society and hotspot of globalisation, it can drive innovation for sustainability and resource-efficient modes of living (Hanna & Comín, 2021). A central turn in urban planning has been the recognition of the relevance of the smallest urban unit, whether complexes of buildings, districts or flagship projects. In those transformed or modern districts, central components such as work, housing and leisure are combined and prioritised differently than during the previous urbanisation phase: The residents seem to reconquer the streets with the design of calmer and more sustainable habitats in contrast to hectic and pollution-intensive traffic and industries (Thees, Zacher & Eckert, 2020).

Struggling with diverging interests is natural in a high-density system such as a city. However, attempts to balance particular claims or use cases have mixed success. A combination of interests and expression of an attractive space is tourism, which brings, in its early understanding, foreigners to the city. Although destination marketing aims at integrating tourists into the specific local context, it often

DOI: 10.4324/9781003365815-9

remains a marketing phrase to convince the residents to support tourism development because of mutual interests – or at least to promote a peaceful co-existence (Kantsperger, Thees & Eckert, 2019). Despite resulting in overtourism and overcrowding, destination marketing for tourists is often given precedence over urban planning for residents, disregarding sustainability. Developing more sustainable and resilient urban futures requires focusing on residents' quality of life as the owners of particular spaces rather than attraction-seeking for foreigners (Zacher & Gavriljuk, 2021).

Against this background, this chapter does not intend to separate "owners" and "foreigners", but rather changes the perspectives and assumes that resident-centred planning and innovation will also boost the quality stay for tourists (Del Chiappa, Atzeni & Gallarza, 2019). We assume that transformative projects, lively districts, flair and atmosphere are key for experiencing a particular urban space (Volgger, 2019), rather than traditional urban planning that leads to overcrowded attractions, marketing superlatives and chasing inauthentic experiences. In this context, the quality of stay in cities hereafter roughly refers to the experience and satisfaction of individuals or communities while visiting or residing in a particular public space (Section 4). Therefore, this chapter asks: How do transformative urban projects stimulate a change in the quality of stay in city spaces?

We analyse planned or realised projects of transformational scope and quality in cities and their effects. The urban transformation represents the interplay of infrastructural re-configuration, design and quality of urban spaces from a resident's perspective, as described above. Such transformation promotes a stronger focus on sustainability, city and tourism interfaces, eudemonic experiences, alternative points of interest and intensive interaction. Consequently, this means promoting a new era of urban planning with the transformation impetus from the COVID-19 pandemic and a new understanding of stakeholder management and local governance.

To support the guiding question, the subsequent sections ask:

- How do urban transformation projects work? (Section 2)
- What projects are planned/implemented? What is their character (objectives and visions)? (Section 3)
- How do transformative projects change urban districts? Which factors are crucial for the quality of stay? (Section 4)

2 The Theory of Change in Sustainable Urban Planning

Transforming urban spaces is a complex process that leads to various changes on different city levels. In literature, different theories and approaches seem to be suitable: e.g. sustainable urban development, theories of change, sustainable change, system change, theory of the city, community development and others. This contribution focuses on the Theory of Change (ToC), which "[…] is an ongoing process of reflection to explore the change and how it happens" (Rengarajan & Sivasubramaniyan, 2020, p. 280). ToC is often used to develop a holistic logic planning model for programmes and initiatives (Vogel, 2012). Besides, it is suitable as a theoretical foundation because it highlights the needed stages in a transformation process – as it is being implemented in many disciplines – including strategic planning and sustainable urban development to promote sustainable and inclusive growth, enhancing quality of life and reducing inequity (Ibrahim et al., 2017).

The ToC emerged in the mid of 90s from social change theory and evaluation theory. The purpose was mainly to provide new ways for analysing and evaluating political and social programmes and initiatives (Stein & Valters, 2012). In relation to evaluation theory, ToC is often used as a model to evaluate social and community initiatives and programmes (Connell & Klem, 2000; Vogel, 2012). Various definitions of the ToC exist. They all depend on the field in which the theory is applied. Nevertheless, all definitions have something in common: they describe ToC as a theory to analyse why and how specific actions could lead to a specific change (Ibrahim et al., 2017). Most theoretical frameworks can be traced back to evaluation theory and authors such as Huey Chen, Peter Rossi, Michael Quinn Patton and Carol Weiss (Rengarajan & Sivasubramaniyan, 2020). Weiss (1995) characterised the ToC as a theory that gives insights into how and why initiatives work. She used the approach to explain how activities or actions can lead to long-term change. More recently, Stein and Valters (2012, p. 4) stated that despite the varieties of views and interpretations, the "ToC is most often defined in terms of the connection between activities and outcomes, with the articulation of this connection the key component of the ToC process".

According to Connell and Kubisch (1998) and Connell and Klem (2000), the ToC is characterised by seven key elements:

1. **Chain of outcomes**: The pathway of initiatives is characterised by strategies and their outcomes. Outcomes could be *early*, *intermediate* or *long-term*. Nevertheless, a strategy has to keep the achievement of these outcomes in mind.
2. **Backward planning**: This is indicated as "moving backward", starting from the long-term goals and outcomes and considering the early and intermediate outcomes as a prerequisite to achieving those long-term goals.
3. **Standards of quality**: The approach depends on the quality and considers how *plausible*, *doable*, *testable* and *meaningful* the ToC is.
4. **Change as a process**: The theory considers how much change – for which stakeholders, in what setting and when – is expected to occur.
5. **Resource planning**: Available and potential resources must be always considered when thinking about the different outcomes and strategies to achieve them.
6. **Stakeholder management**: Different and multiple stakeholders can be addressed.
7. **Evaluation and adaptability**: The pathway of the initiative could also change as it is tested and evaluated over the course of an initiative.

The ToC is increasingly attracting interest in different fields – also in urban transformation and urban change planning (Mitlin et al., 2019; Rengarajan & Sivasubramaniyan, 2020). Urban transformation is a radical change in urban (sub-)systems in terms of fulfilling local needs, urban functions or implications towards more resilience and sustainability (Hölscher & Frantzeskaki, 2021). Radical change is not only addressing the urban (sub-systems); it also implies changes in practices (e.g. mobility behaviour), changes in cultures and values, and changes in dominant structures (e.g. infrastructure) (Ernst et al., 2016). Thus, urban spaces can benefit from a ToC approach, given the complex and interconnected nature of many urban challenges and recognising the need for a systemic change. With the aid of ToC, it is possible to understand how problems arise and to identify the key stakeholders, resources and activities necessary to achieve the desired change (Oberlack et al., 2019; Stein & Valters, 2012).

Current directions and approaches to transformation processes in urban spaces often aim to create an enhanced quality of life for all residents and refer to integrated and sustainable urban development (Mikelsone et al., 2021), which can be achieved – among other ways – by creating resource-efficient, low-emission and climate-resilient cities, where the needs and desires of the residents are the focus of the transformation projects (Kabisch et al., 2017).

Strategies to achieve urban transformation considering ToC may include:

1. The densification of urban spaces. This is driven by the reutilisation of brownfield sites, an increase in the height of buildings, and the fundamental re-densification of neighbourhoods. Innovative neighbourhood development concepts aim to transform entire urban districts into liveable environments. Vertical densification, on the one hand, saves space and, on the other hand, promotes sustainability because it enables better utilisation of the existing infrastructure (McFarlane, 2020).
2. Innovation, technology and digitalisation, which can also help to make urban spaces more efficient and sustainable. In this regard, smart city concepts in particular are gaining greater attention. For example, intelligent traffic systems can help relieve traffic congestion and reduce emissions simultaneously (Keshavarzi et al., 2021).
3. Cultural diversity and creating vibrant, inclusive and open neighbourhoods, which are special in transforming urban spaces (Tretter et al., 2014). Creative urban development concepts rely on the creative industries as a driver for urban development. For example, cultural events, art installations and creative businesses can help improve a city's image, attractiveness and competitiveness (Störmann & Lill, 2022).
4. Participatory methods to ensure that the perspectives and needs of all relevant stakeholders are taken into account (Zohar et al., 2022). Participation in urban spaces means involving all residents in the transformation process and considering their needs and ideas. This involves the broad participation of different stakeholders as well as transparent and open communication between all actors (Jupp, 2008).

Further strategies are compact city, urban regeneration, functional mix, no land take, green city and high density (Cortinovis et al., 2019). Thus, transformation in urban spaces may involve multiple stakeholders – including government agencies, local communities, civil society organisations and private sector actors – and may require the integration of multiple strategies and interventions over a long period of time. Overall, ToC helps these actors reflect and be aware of the whole transformation process (Deutsch et al., 2021; McLellan, 2021).

3 Classifying Transformative Urban Projects

Cities are focal points of economic activity and social attraction (UNESCO & World Bank, 2021). However, they are subject to growing pressure for change due to overarching societal trends and increased awareness of social and environmental sustainability (Sodiq et al., 2019, sec. 1; cf. SDGs). European cities in particular are often structures that have evolved over centuries and bear witness to historical eras. Therefore, urban planners cannot address the pressure for sustainable change with fundamental new planning processes. Rather, transformation occurs selectively or in specific urban micro spaces, whereby these projects have a pilot character for the long-term and overarching transformation of urban structures.

3.1 Methodological Remarks

This chapter introduces our case study method and develops a classification of urban projects. Therefore, our investigations are based on case study research. Using multiple case studies provides a rich and in-depth understanding of the phenomena, allowing for a comprehensive analysis and classification of the project characteristics. When comparing different case studies, it is important to identify similarities and differences among the transformative urban projects. This comparative analysis can provide valuable insights into the factors contributing to successful outcomes, as well as the challenges and limitations faced in different contexts (Yin, 2009).

As a first step, desk research led to a collection of around 40 projects that are spread all over the world. As illustrated in Figure 5.1, the urban projects cover a range of topics: Modern mobility concepts in the scope of transformational projects (e.g. "Superblocks" in Barcelona, "Supergrätzl" in Vienna) aim to reduce traffic in certain neighbourhoods and open up space for pedestrians, cyclists and public transport instead. Innovative urban projects are also partly the result of converting former infrastructure into attractive public spaces (e.g. "High Line" in New York City; Seoul's Cheonggye Stream) and projects are often created in former industrial sites (e.g. "De Ceuvel" in Amsterdam; Berlin Tegel Airport) to establish sustainable and innovative urban neighbourhoods (Ersoy & van Bueren, 2020). Many of the projects observed intend long-term transformation of infrastructure and place. However, they also exhibit a high degree of design experimentation, including innovative participation methods and a conscious monitoring of changes, as typically seen in real-world laboratory projects.

To further narrow the collection, we decided on the criteria the urban projects should meet. The overall intention of this sampling is to represent a diverse range

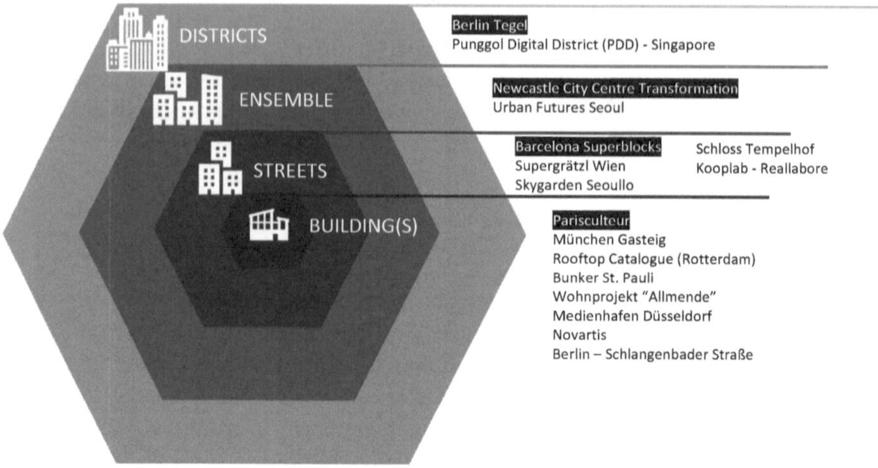

Figure 5.1 Observed scales in urban transformation projects

Source: Own elaboration

of transformative urban projects. This diversity helps in capturing the complexities and variations inherent in different projects. In addition, different geographical locations, scales and project types should be included to avoid bias and generalisation.

Criteria for case selection:

1. **City Size**: Cities bigger than 250,000 inhabitants.
2. **Transformation/re-configuration**: Intention to transform existing urban spaces with a certain degree of innovation to align with future challenges. No brownfield project included.
3. **Impacting the quality of stay**: To create visible change and increase attractiveness for the local population (and visitors).
4. **Integrated urban development**: Projects that take into account at least two aspects of urban development, such as housing, new work, green spaces, infrastructure and mobility.
5. **Public accessibility**: Re-configured spaces that represent a public and accessible space with a high recreational and leisure function.

Applying those criteria led to the selection of 15 relevant transformative projects. Here it became clear that the different projects can be attributed to different spatial scales. The smallest are individual buildings or building complexes. The spatially largest projects are located at the level of independent city districts with a complete range of living, working and leisure environments.

3.2 Selected Cases

Out of a variety of projects that evolved in the first screening and following the criteria, one example from each scale was selected to provide further details and present insights into the urban transformation.

Berlin – Tegel

On the grounds and in the bordering areas of the former Berlin Tegel Airport, a completely new district with mixed-use functions has been developing since the airport's closure. The state of Berlin founded a state-owned limited company – Tegel Projekt GmbH – to develop this quarter (see Figure 5.2). This is responsible for planning building construction and utility infrastructure, as well as project communication and marketing of the area. A substantive focus of land development is the formation of the *Urban Tech Republic*, a tech and creative quarter intend to attract technology companies and start-ups focusing on urban development solutions. In this way, Berlin is creating a centre for technology-oriented innovations in urban spaces. The aim is that great diversity and a modern start-up environment with up to 1,000 individual companies should emerge. The Berliner Hochschule für Technik will open up a new location onsite. With this new educational home for 2,500 students, close cooperation with the local tech companies and start-up environment is expected.

The second pillar of the project is the construction of the *Schumacher Quarter*, which will provide residential space for up to 10,000 people. About half of the flats are subsidised by Berlin housing associations and there will be a legal rent control. This is intended to provide much-needed affordable housing for people with lower and middle incomes in Berlin and to create a social mix in the district from the outset. One focus is on sustainable mobility and the public space is enriched by two large parks. By the time the quarter is scheduled to be occupied in 2027, all major construction and development activities will have been completed, so Tegel will be a fully planned model project from the beginning. The neighbourhood should offer all activities for daily needs. A Neighbourhood Society will systematically promote participation and involvement in the new neighbourhood, with participation formats for the conscious appropriation of public space by residents. In addition, the neighbourhood is seen as an experimental space, as the innovative urban technologies from the Urban Tech Republic provide a testing environment for identifying social space innovations in modern urban neighbourhoods. Thus,

Figure 5.2 Berlin Tegel

Source: Berlin Tourismus and Kongress GmbH (n.d.)

urban transformation is to be achieved through the technological results of the researching companies and the socio-spatial model character. In the project, the future of urban development is intended to be made tangible for residents as well as for visitors to Berlin, and leading local politicians attribute image-shaping effects to an international target group showing Berlin as a centre for innovative urban development.

Newcastle City Centre Transformation

With the central aim of creating climate-resilient spaces, the City of Newcastle is undertaking functional and physical changes to eight different inner-city squares and streets. Based on the strategic objective of economic regeneration and an overall increase in the attractiveness of the city centre for locals and tourists, the plans for all eight transformation zones have been shaped through citizen participation processes. The single projects are implemented against the backdrop of a consciously perceived process transforming Newcastle from a rather unattractive industrial city to an attractive residential city and to a city that is welcoming for business and leisure tourism (see Figure 5.3).

Saville Row, as one example, is being transformed into a new kind of garden street that increases the dwelling quality. Garden streets are areas where land sealing is reduced or reversed, and the proportion of vegetation and green places is increased. This positively addresses various resilience factors, such as reducing heat islands or the risk of local flooding (Vannieuwenhuyze, 2020). Newcastle is openly

Figure 5.3 Northumberland Street of the Newcastle City Centre transformation
Source: Newcastle City Council (2023)

addressing the issue that city centres are affected by an irreversible change in the retail structure (Gardner & Sheppard, 2012). Where occasions for shopping are disappearing, alternative uses need to be created. The aim of the various sub-projects is to create attractive places to stay that invite people to visit and meet in the community. Grey Street, as another example project, will be rebuilt in 2023 and 2024 in several phases. The event areas will be expanded so that outdoor events can take place there. Overall, the projects aim for transparent communication of the implementation steps, which also refers to the temporary burdens caused by the conversion (such as limited accessibility, noise pollution, etc.).

Barcelona Superblocks

Barcelona has evaluated its negative effects of urban densification, such as noise, air pollution, and a lack of space for exchange and encounter. The *Superblocks* are pursuing an urban planning approach that mitigates these effects and turns them into the opposite. To this end, several neighbourhoods have been transformed since 2016 by expanding and greening public spaces, enforcing traffic speed reduction, or making local shopping more attractive. The Superblocks' primary goal is to make the heat-affected city more climate-resilient through decentralised interventions. The central planning instrument is the avoidance of individual motorised vehicles through traffic bans in broad areas of the city (see Figure 5.4). The decision for the project was made politically top-down and certain development goals were non-negotiable. To increase acceptance, a citizen monitoring committee

Figure 5.4 Barcelona Superblocks

Source: Barcelona.de Tourist Info & Distribution (n.d.)

composed of supporters and critics of the transformation is being organised. In addition, when starting a new Superblock, temporary concepts are implemented first, which are then evaluated in terms of their ability to implement functional solutions in the long term.

The Barcelona Superblocks initiative has attracted a lot of attention (e.g. ADFC, 2020), and tourist offerings such as the Architectural Walks are already available. These contribute to bring socially more sustainable tourism to neighbourhood areas of Barcelona, distributing it more evenly across the city. Politically, far-reaching decisions were made by pushing through controversially discussed topics such as the banning of cars. To balance this, the initiative promotes decentralised local production and shopping areas that make the use of cars in everyday life less necessary, along with sustainable transport options, including an extensive network of bicycle lanes. The combination of several Superblocks and associated projects in Barcelona has now resulted in a network of transformed streets and squares spanning the entire urban space, increasing the overall quality of stay for the population and mobility for pedestrians and cyclists. The model is one of the earliest and most far-reaching transformation projects in Europe and has had a noticeable positive impact on the quality of life in the city. Due to the development processes, a behavioural change in the use of the city quarters among the population and visitors to the city is intended. The urban planning approach meets the zeitgeist of a conscious increase in the attractiveness of everyday life in the city centre. In principle, it is transferable to many other cities and already has some imitators, for example, Vienna.

Parisculteur

The case of *Parisculteurs* can be understood as a thematically focused programme. The project started in 2016 and is looking for potential areas in Paris for urban farming (see Figure 5.5). Urban planners set a target of 100 hectares for urban farming employing a city declaration of intent. Today, 30 hectares are already implemented, mainly in publicly owned land, educational institutions or non-profit organisations.

On the one hand, the project aims to green sealed surfaces and thus create cooler and more pleasant places to stay in the urban area. On the other hand, it pursues education on sustainable development through local food production and the individual experience of cultivating the land, which is usually impossible for metropolitan inhabitants. The associated partners and individual projects are brought together via an online networking platform. A central communication strategy is aimed at landowners, established or future urban farmers, and the general public in order to raise awareness of sustainable food production and to provide training programmes for cultivation. In Parisculteurs, private persons can take responsibility for public space. Compared to other urban transformation projects, the change in the urban structure takes place in a less visible and more decentralised way. However, Parisculteur can be seen as a good example of how urban transformation can also be tested and experienced on a smaller scale.

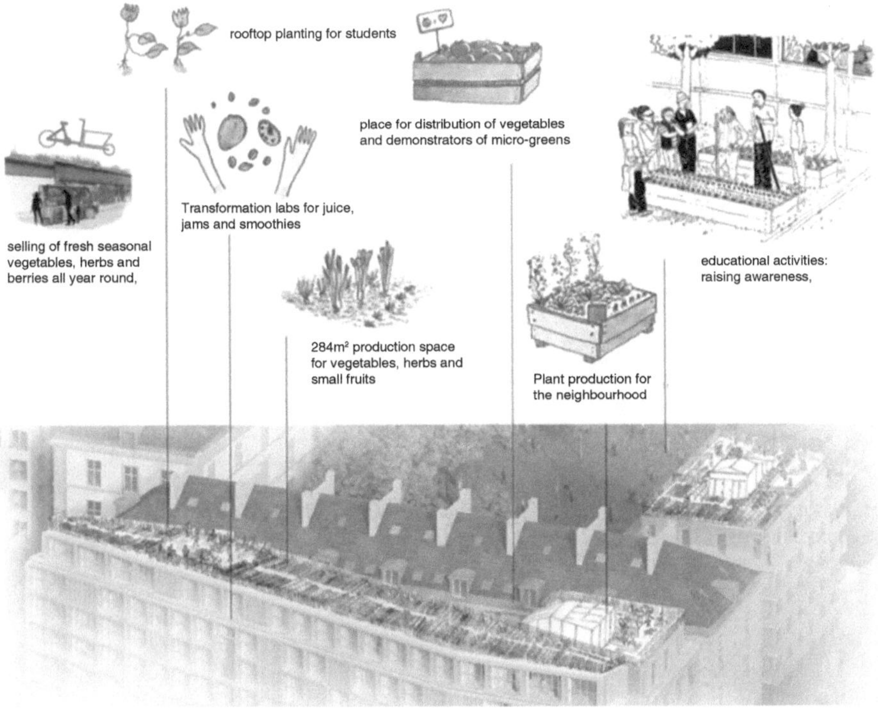

Figure 5.5 Parisculteurs

Source: Parisculteurs (n.d.), translated

3.3 Pattern Shaping Characteristics

Starting from the four projects presented in detail above, overarching characteristics for urban transformation projects can be derived. This classification covers the central phases of project conception, management and implementation. The cases draw heterogeneous approaches regarding spatial-structural preconditions, strategic approach, innovation paradigm, research orientation, political steering and project communication. Each criterion is contrasted with two exemplary characteristics that can also be derived from the project examples (see Table 5.1), whereby mixed forms of these characteristics can be observed in reality.

4 The Quality of Stay in Transformative Urban Projects

Drawing conclusions on the transformative value of urban projects means discussing their particular qualities. Besides describing technical (architectural, infrastructure) or ecological (water consumption, energy supply) factors, we focus on a quality that directly relates to all visitors and residents: the quality of stay. This represents the desired outcome of the social dimension for sustainable urban development. In this context, the quality of stay in cities hereafter roughly refers to

Table 5.1 Characteristics and project examples

Characteristic	*Contrasting expression*	*Example / Explanation*
Structural transformation claim	Improvement of functional quality vs. functional transformation	Change in the use of **Newcastle Saville Row** vs. airport site in residential district at **Berlin Tegel**
Spatial concentration of transformation project	High spatial concentration vs. decentralised/multicentralised change	**Berlin Tegel:** entire transformation vs. **Parisculteur:** Network of urban gardens
Type of innovation	Technological innovation (construction and infrastructure oriented) vs. social innovation (process-oriented)	**Urban Tech Republic** (at Tegel) creates new types of residential and work buildings vs. creation of new meeting spaces through closure to road traffic in **Barcelona**
Interest of the accompanying research	Applied interest (implementation) vs. research-oriented interest (experiment)	**Barcelona Superblocks** evaluation of behavioural change vs. **Urban Tech Republic:** Creating completely new structures for testing new forms of co-working and co-living
Political Governance	Bottom up: by creating incentive structures vs. top down: through central guidelines and coordination	All projects have some bottom-up approaches (e.g. Barcelona's Citizen Monitoring Committee) but are initiated top-down to a certain extent (e.g. city declaration in Paris)
Socio-political ambitions	Passive: creation of preconditions through building infrastructure vs. active: targeted promotion of social mix through participation projects	**Berlin Tegel's** quota for social housing vs. promoting outdoor events in **Newcastle**
Financing	Public money and non-profit character vs. private investors	**Barcelona Superblocks** consumption-free recreation areas vs. selling office space in Industry Park (**Tegel**)
Communication Strategy	Internally oriented (promoting participation and acceptance) vs. externally oriented (visualisation and marketing)	All projects pursue both aspects in a twin strategy with different weighting.

Source: Own elaboration.

the experience and satisfaction individuals or communities have while visiting or residing in a particular public space (based on Netzwerk Innenstadt NRW, 2015). It encompasses various factors contributing to its residents' and visitors' liveability, comfort and well-being. For urban planning, the quality of stay includes areas for rest, movement, green spaces or markets. In this regard, it is a central goal of urban planning to increase the quality of stay in different kinds of districts to achieve a balance/harmony between various forms of usage.

There is also close conceptual proximity visible in those cases to city hospitality, a multi-stakeholder approach to co-create places, provide a feeling of welcome and recognise the guest–host encounters, leading to satisfaction, delight, engagement and thus, experience (Wiegerink & Huizing, 2022). The examples presented may not have tourism development as their core strategy, but they indicate a long-term commitment to securing the image and quality of life in cities, which, in turn, contributes to the ongoing development of the city as a tourist destination. Additionally, these projects are not implemented in the current tourist hotspots of the cities but rather in everyday neighbourhoods, which helps counteract potential tourist overcrowding. The quality of encounter and the quality of experience need to be recognised further to illustrate the overlapping of city and tourism management.

The transformative potentials and intermediate steps results of the various transformative urban projects studied are analysed against the background of their quality of stay. In this way, a change of perspective is adopted and the longer-term impact of the projects on the population is considered. From the various starting points and cases, a triad of quality factors emerges that can be considered as socially perceived results of successful urban transformation projects:

1. **Encounter**: The projects aim in different ways to facilitate and promote encounters between residents and visitors within the neighbourhoods. This is being pursued in Newcastle in particular, where the transformation is intended to make places more attractive for guests and locals alike. Entering the public space should become attractive. In most cases, this is achieved by creating dedicated meeting spaces or developing new open spaces through the conversion of existing areas or structural alterations. In some projects, however, encounter can also be understood in the socio-demographic sense, for instance through the promotion of local retail in Barcelona. A social mix is to take place through the targeted development of different housing concepts, the promotion of social housing and the integrated approach to residential and commercial areas such as in the Schumacher Quarter at Tegel. An additional contribution is to be made by the settlement of start-ups or cultural workers, which will enable frequency and offers for the local area. Since the aim is to achieve a basic participation of all sections of the population, recreational areas without compulsory consumption also play an important role. Encounter is also promoted by the projects contributing to improving urban ecology (especially air quality) and safety in public spaces.
2. **Atmosphere**: Another quality of urban transformation spaces is the atmosphere that accompanies their initiation and development. This is characterised by an adequate appropriation of the created meeting spaces by the population and visitors (see Newcastle's Grey Street Event Areas). In the respective individual projects, a planned intention can become more or less clear. While some projects have relatively clear objectives through specifically designated spaces for festivals and events, other areas are deliberately designed so that appropriation

takes place in a bottom-up process that is open to results (e.g. neighbourhood society at Tegel). The latter holds the potential for social innovations and the consideration of the specific needs of the different stakeholders, which usually cannot be excluded in the planning process.

3. **Purpose**: An indirect quality of stay can be seen in the enabling of a purpose that includes the positive impact of the encounter opportunities and atmosphere, such as a stronger social togetherness. The purpose is expressed both in the internal and external communication of the projects analysed. Besides communication, urban transformation projects are often subject to the normative claim of contributing to the community and social development (e.g. climate protection in Parisculteur). The promotion of encounters and exchange go beyond simply providing entertainment and fun in an eventized experience, adding a deeper meaning. Ultimately, encounters in urban space are also a mirror for the situation of the self and the other. They are thus – consciously or unconsciously – always also processes of understanding and negotiation about socio-political issues intended to prevent subgroups from drifting apart (e.g. Tegel's "Neighborhood Society"). Densified urban spaces are also always spaces for creativity and innovation, whereby they can be regarded as pioneering places where sustainable social interaction and experimentation should occur.

The three quality factors described above build on each other but are also mutually dependent. For example, questions about the purpose can require a new orientation

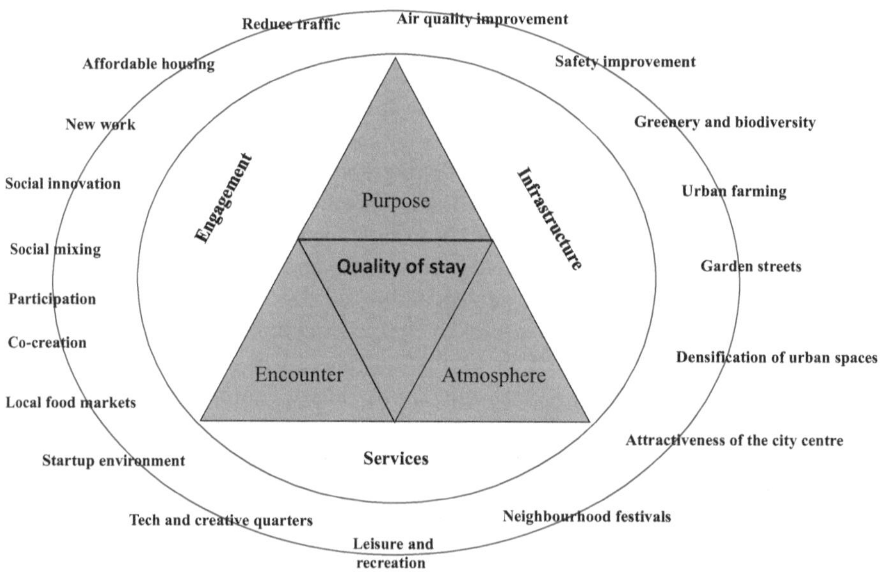

Figure 5.6 The triad of quality factors with regard to the quality of stay in transformative urban projects

Source: Newcastle City Council (2023)

and adaptation of the infrastructural development. Commercial offers, services and entrepreneurship can contribute to specialisation in certain areas, which can help professionally shape the district's transformation. Local production systems and a culture of local engagement are required here that successively reduce the use of public funds. Ultimately, the voluntary commitment of residents and visitors to co-creatively shape their urban space is an important contribution to successful transformation. Figure 5.6 illustrates aspects that complement the triad of quality factors.

For example, purpose and encounter can be associated with a certain level of engagement. Encounter and atmosphere, on the other hand, require special services. Services in this sense are not always consumption-oriented, but relate to well-being of residents and guests. Finally, infrastructure is necessary to meet the basic requirements of atmosphere and purpose. The outer framework illustrates selected examples of how this may be applied in practice.

5 Critical Thoughts on the Transformative Power of Urban Projects

The contribution has shown that the transformative power of urban projects analysed is on the right track, achieving change while taking a long-term perspective on achieving sustainable impacts on society. In this context, the ToC is particularly well suited as a theoretical frame of reference. For example, cause–effect relationships can be better understood but flexibility and adaptability also have a special importance. Projects can change, so continuous evaluation and adaptation is necessary to cope with changing conditions and challenges. However, a critical look at challenges is always necessary. Based on the seven key elements of the ToC, the following critical points can be highlighted in the context of this contribution:

1. **Chain of outcomes**: The goals of transformative projects might sometimes be overambitious. Nevertheless, it is important to justify and explain even futuristic projects and their particular values for local transformation. This requires a high sensitivity and openness in communication.
2. **Backward planning**: Starting from the long-term goals and outcomes is essential for planning transformative projects. Some of the projects analysed explain and communicate the relevant steps in a plausible way. However, urban transformation requires the integration of multiple projects and interdependencies: how should the city as a whole look in the future?
3. **Standards of quality**: Change needs to be meaningful and doable. Some urban mega projects have a scope that is hardly manageable, while small-sized projects (e.g. Parisculteur) can have a significant impact through many entities. A political frame and the necessary demand should be taken into account when defining the scope of projects.
4. **Change as a process**: Project managers should assess how change will occur for specific stakeholders. Most of the projects presented include a certain

participation. However, this participation needs to be conducted in an open and honest way to define the reasonability for the stakeholders and give enough time to adapt to change. Regarding the cities as tourist destinations, there needs to be a broader understanding of attractions that value quieter, more integrated experiences as a draw for visitors. This way, these projects can not only transform urban development, but they can also act as pioneering projects for a new era of "touristic" offers with a much wider portfolio of travel motivations.

5. **Resource planning**: The accuracy of resource planning is often unsatisfactory. Against the background of a sustainability transformation in cities, a clear focus on the ecological aspects helps to improve land sealing through the generation of green space and irrigation systems. Many examples show the benefits of vertical compression in cities.
6. **Stakeholder management**: Participation in urban projects should therefore include architecture, urban planners, politics, residents, mobility experts, guests and even artists and gardeners to meet the complex targets of urban transformation.
7. **Evaluation and adaptability**: Whether long-term mega projects or small-scale impulses, monitoring of projects in cities is insufficient and needs to cover the sustainability dimensions in the long run. Such projects can learn from evaluation methods that are used and experiences that are gained in real-world laboratory projects.

References

ADFC. (2020). Innovative Radverkehrslösungen auf Deutschland übertragen. InnoRAD-Factsheet 4/6 [Transferring innovative cycling solutions to Germany. Inno RAD factsheet 4/6]. www.adfc.de/fileadmin/user_upload/Expertenbereich/Politik_und_Verwaltung/Download/adfc_innorad_superblocks_web.pdf

Barcelona.de Tourist Info & Distribution. (n.d.). Barcelona wird super dank Superblocks! Weniger Verkehr, mehr Grün, mehr Lebensqualität durch das neue Stadtentwicklungsprojekt [Barcelona will be super thanks to Superblocks! Less traffic, more green, more quality of life thanks to the new urban development project]. www.barcelona.de/de/barcelona-superblocks.html

Berlin Tourismus & Kongress GmbH. (n.d.). Urban tech republic – Berlin TXL. www.visitberlin.de/de/urban-tech-republic-berlin-txl

Connell, J., & Klem, A. (2000). You can get there from here: Using a Theory of Change approach to plan urban education reform. *Journal of Educational and Psychological Consultation, 11*(1), 93–120. https://doi.org/10.1207/s1532768Xjepc1101_06

Connell, J., & Kubisch, A. C. (1998). Applying a Theory of Change approach to the evaluation of comprehensive community initiatives: Progress, prospects, and problems. *New Approaches to Evaluating Community Initiatives, 2*(15–44), 1–16.

Cortinovis, C., Haase, D., Zanon, B., & Geneletti, D. (2019). Is urban spatial development on the right track? Comparing strategies and trends in the European Union. *Landscape and Urban Planning, 181*, 22–37.

Del Chiappa, G., Atzeni, M., & Gallarza, M. (2019). Collaborative policy making and stakeholder engagement: A resident-based perspective. In N. Kozak & M. Kozak (Eds.), *Tourist destination management: Instruments, products, and case studies*. Springer.

Deutsch, L., Belcher, B., Claus, R., & Hoffmann, S. (2021). Leading inter- and transdisciplinary research: Lessons from applying theories of change to a strategic research program. *Environmental Science & Policy, 120*, 29–41. https://doi.org/10.1016/j.envsci.2021.02.009

Ernst, L., de Graaf-Van Dinther, R. E., Peek, G. J., & Loorbach, D. A. (2016). Sustainable urban transformation and sustainability transitions; conceptual framework and case study. *Journal of Cleaner Production, 112*, 2988–2999. https://doi.org/10.1016/j.jclepro.2015.10.136

Ersoy, A., & van Bueren, E. (2020). Challenges of urban living labs towards the future of local innovation. *Urban Planning, 5*(4), 89–100. https://doi.org/10.17645/up.v5i4.3226

Gardner, C., & Sheppard, J. (2012). *Consuming passion (RLE retailing and distribution): The rise of retail culture*. Routledge.

Hanna, E., & Comín, F. A. (2021). Urban green infrastructure and sustainable development: A review. *Sustainability, 13*(20), 11498. doi:10.3390/su132011498

Hölscher, K., & Frantzeskaki, N. (2021). Perspectives on urban transformation research: Transformations in, of, and by cities. *Urban Transformations, 3*(1). https://doi.org/10.1186/s42854-021-00019-z.

Ibrahim, M., El-Zaart, A., & Adams, C. (2017). Theory of Change for the transformation towards smart sustainable cities. In *2017 Sensors Networks Smart and Emerging Technologies (SENSET)*. IEEE. https://doi.org/10.1109/SENSET.2017.8125067

Jupp, E. (2008). The feeling of participation: Everyday spaces and urban change. *Geoforum, 39*(1), 331–343. https://doi.org/10.1016/j.geoforum.2007.07.007

Kabisch, S., Koch, F., Gawel, E., Haase, A., Knapp, S., Krellenberg, K., Nivala, J., & Zehnsdorf, A. (Eds.). (2017). *Future city: Vol. 10. Urban transformations: Sustainable urban development through resource efficiency, quality of life and resilience* (1st ed.). Springer International Publishing.

Kantsperger, M., Thees, H., & Eckert, C. (2019). Local participation in tourism development – Roles of non-tourism related residents of the Alpine Destination Bad Reichenhall. *Sustainability, 11*(24), 6947. doi:10.3390/su11246947

Keshavarzi, G., Yildirim, Y., & Arefi, M. (2021). Does scale matter? An overview of the "smart cities" literature. *Sustainable Cities and Society, 74*, 103151. https://doi.org/10.1016/j.scs.2021.103151.

McFarlane, C. (2020). De/re-densification. *City, 24*(1–2), 314–324. https://doi.org/10.1080/13604813.2020.1739911.

McLellan, T. (2021). Impact, Theory of Change, and the horizons of scientific practice. *Social Studies of Science, 51*(1), 100–120. doi: https://doi.org/10.1177/0306312720950830

Mikelsone, E., Atstaja, D., Koval, V., Uvarova, I., Mavlutova, I., & Kuzmina, J. (2021). Exploring sustainable urban transformation concepts for economic development. *Studies of Applied Economics, 39*(5). https://doi.org/10.25115/eea.v39i5.5209

Mitlin, D., Bennett, J., Horn, P., King, S., Makau, J., & Masimba Nyama, G. (2019). Knowledge matters: The potential contribution of the co-production of research to urban transformation. *SSRN Electronic Journal*. https://doi.org/10.2139/ssrn.3470133.

Netzwerk Innenstadt NRW. (2015). Magazin Innenstadt. Aufenthalts – und Gestaltqualität [Magazine Downtown. Quality of stay and design]. www.innenstadt-nrw.de/fileadmin/user_upload/Service/Veroeffentlichungen/Magazin_Innenstadt/201501_Gestaltqualita%CC%88t/Magazin_Innenstadt_0115_web.pdf

Newcastle City Council (2023). Improvement works to Northumberland Street now confirmed. www.newcastle.gov.uk/citylife-news/community/improvement-works-northumberland-street-now-confirmed

Oberlack, C., Breu, T., Giger, M., Harari, N., Herweg, K., Mathez-Stiefel, S-L., Messerli, P., Moser, S., Ott, C., Providoli, I., Tribaldos, T., Zimmermann, A., & Schneider, F. (2019). Theories of change in sustainability science: Understanding how change happens. *GAIA – Ecological Perspectives for Science and Society*, *28*(2), 106–111. https://doi.org/10.14512/gaia.28.2.8

Parisculteurs. (n.d.). *Parisculteurs 2: une seconde saison prometteuse* [A promising second season]. www.parisculteurs.paris/fr/actualites/1535-parisculteurs-2-une-seconde-saison-prometteuse.html

Rengarajan, V., & Sivasubramaniyan, K. (2020). Theories of change in the process of rural transformation: A refined way forward. *International Journal of Research – Granthaalayah*, *8*(7), 279–297. https://doi.org/10.29121/granthaalayah.v8.i7.2020.727

Sodiq, A., Baloch, A. A., Khan, S. A., Sezer, N., Mahmoud, S., Jama, M., & Abdelaal, A. (2019). Towards modern sustainable cities: Review of sustainability principles and trends. *Journal of Cleaner Production*, *227*, 972–1001.

Stein, D., & Valters, C. (2012). *Understanding Theory of Change in international development*. The Justice and Security Research Programme (JSRP).

Störmann, E., & Lill, E-M. (2022). Resilience in regional development: Culture and creativity as a driving force to strengthening resilience? In H. Pechlaner, N. Olbrich, J. Philipp, & H. Thees (Eds.), *Towards an ecosystem of hospitality: Location:City:Destination*. Graffeg Limited.

Thees, H., Zacher, D., & Eckert, C. (2020). Work, life and leisure in an urban ecosystem – co-creating Munich as an Entrepreneurial Destination. *Journal of Hospitality and Tourism Management*, *44*, 171–183. doi:10.1016/j.jhtm.2020.06.010

Tonne, C., Adair, L., Adlakha, D., Anguelovski, I., Belesova, K., Berger, M., & Adli, M. (2021). Defining pathways to healthy sustainable urban development. *Environment International*, *146*, 106236.

Tretter, M., Pechlaner, H., & Märk, S. (2014). Spaces of inspiration and innovation and the role of creativity: The cases of Graz and Ingolstadt. *International Journal of Innovation and Regional Development*, *5*(4/5), 443–457.

UNESCO & World Bank. (2021). *Cities, culture, creativity: Leveraging culture and creativity for sustainable urban development and inclusive growth*. https://elibrary.worldbank.org/doi/abs/10.1596/35621

Vannieuwenhuyze, J. (2020). Dashboards for input-evaluation of policy programs: Lessons learned from AN Antwerp dashboard for garden streets. *ISPRS Annals of the Photogrammetry, Remote Sensing and Spatial Information Sciences*, *6*, 173–179. https://doi.org/10.5194/isprs-annals-VI-4-W2-2020-173-2020

Vogel, I. (2012). *ESPA guide to working with Theory of Change for research projects: Ecosystem services for alleviation of poverty*. www.espa.ac.uk/files/espa/ESPA-Theory-of-Change-Manual-FINAL.pdf

Volgger, M. (2019). Staging genius loci: Atmospheric interventions in tourism destinations. In M. Volgger & D. Pfister (Eds.), *Atmospheric turn in culture and tourism: Place, design and process impacts on customer behaviour, marketing and branding*. Emerald Publishing.

Weiss, C. H. (1995). Nothing as practical as a good theory: Exploring theory-based evaluation in complex community initiatives for children and families. In J. Connell, A. C. Kubisch, L. B. Schorr & C. H. Weiss (Eds.), *New approaches to evaluating community initiatives for children and families: Concepts, methods and contexts*. Aspen Institute.

Wiegerink, K., & Huizing, J. (2022). City hospitality. In D. Buhalis (Ed.), *Encyclopedia of Tourism Management and Marketing*. Edward Elgar Publishing.

Wu, Y., Long, H., Zhao, P., & Hui, E. C. M. (2022). Land use policy in urban-rural integrated development. *Land Use Policy, 115*, 106041. doi:10.1016/j.landusepol.2022.106041

Yin, R. K. (2009). *Case study research: Design and methods (Vol. 5)*. Sage.

Zacher, D., & Gavriljuk, E. (2021). Developing resilience understanding as a tool for regional and tourism development in Bavaria. In R. Wink (Ed.), *Economic resilience in regions and organisations*. Springer.

Zohar, H., Simeone, L., Morelli, N., Martelloni, L., & Marmo, D. (2022). Using Theory of Change to support participatory visual mapping in urban transformation projects. *The 23rd DMI: Academic design management conference proceedings: Design as a strategic asset*. https://vbn.aau.dk/en/publications/using-theory-of-change-to-support-participatory-visual-mapping-in

6 Scenario Planning as a Tool to Future Proof the Visitor Economy After the COVID-19 Pandemic

Opportunities for Sustainability and Digitalisation

Albert Postma, Jasper Heslinga and Stefan Hartman

1 Introduction

Worldwide, the tourism industry was highly affected by the COVID-19 outbreak and related measures by national governments, which created a crisis and led to uncertainty about future developments (Cheer, 2020; Van Leeuwen, Klerks, Bargeman, Heslinga & Bastiaansen, 2020). Because of these uncertainties, many tourism-related businesses, destination management organisations (DMOs) and other tourism-related organisations needed to reorient their perspectives on the future. In the short term, their focus has been, quite understandably, on survival and a quick recovery to business as usual. However, we argue that offering future perspectives on the mid- and longer term is crucial for sustainable recovery of the visitor economy in the long run. Scenarios provide a useful tool to foresee possible futures, potentially including more societally desirable options than the business-as-usual situation. Scenarios enable tourism businesses, organisations and DMOs to deal with uncertainty and to anticipate possible futures proactively in order to become more resilient and future proof. The aim of this chapter is to provide some context to scenario thinking and to present four scenarios of the post-COVID-19 visitor economy, including how they were developed, and how they could be used by the tourism industry to anticipate future uncertainties that arise from the pandemic. One of the scenarios embodies a future visitor economy with a strong focus on virtualisation of travel. The use of this scenario will be elaborated in more detail.

2 Scenario Thinking, Alternative Approaches

As an approach, scenario thinking is ideally suited for understanding a complex crisis such as the corona crisis and a complex sector such as the visitor economy. Scenario planning is rooted in complexity thinking (Hines & Bishop, 2015; Lindgren & Bandhold, 2009; Postma & Yeoman, 2021), which involves understanding emergent patterns and structures (Hartman, 2016, 2021). To develop scenarios, experts

DOI: 10.4324/9781003365815-10

share their expertise in order to understand the complex connections and emergent patterns within the system and the forces that drive the (post-pandemic) future of the visitor economy.

In media reports concerning the COVID-19 pandemic the concept of scenarios was used frequently, yet in different ways. Three interpretations of scenarios can be identified, each with a major difference in meaning and use: predictive scenarios, explorative scenarios and goal-based scenarios (Van Rijn & Van der Burgt, 2012; Yeoman, Postma & Hartman, 2022).

The first type of scenario is based on forecasting. Historical data of, for example, the number of COVID-19 infections in a country are statistically extrapolated into the future to make prognoses or projections for the quarters or the years to come. These scenarios assume a singular predictable surprise-free future in which trends are inevitable ("what will happen"). Uncertainties about the future are denied and interpreted as a statistical bandwidth around a central projection in a line chart. Often the highest demarcation of the bandwidth is referred to as the best-case scenario, and the lowest as the worst-case scenario (Postma, 2013). By "playing" with the assumptions in the calculations, alternative forecasts can be created. Usually, such so-called extrapolative or predictive scenarios have a range of a maximum of five years (Van Rijn & Van der Burgt, 2012), although the United Nations World Tourism Organisation (UNWTO) scenarios in which the future development of global tourism has been projected since 2010 had a range of 20 years (UNWTO, 2017). The notion of forecasting and predictive scenarios originates from the period after World War I, but was boosted after World War II in the USA in an attempt to make the Cold War more predictable.[1] The approach is rooted in the belief of a makeable society, a domesticable future and blueprint planning, and is based on a positivist scientific viewpoint (Gidley, 2017; Postma, 2013; Postma, Hartman & Yeoman, 2024). In the domain of tourism forecasting has been applied since the 1960s and 1970s in Europe and the United States, mainly to make predictions of the demand for leisure travel (Postma, 2013).

In response to the mechanic and predictable approach of the future, a counter-movement started to emerge in Europe and elsewhere. This approach was more critical, more multidisciplinary and more human-centred. The emerging world view of open possibilities, complexity and adaptivity, and the rise of a post-positivist and pluralist scientific viewpoint resulted in a belief of multiple futures (Gidley, 2017).[2] This was accompanied by the use of scenarios to explore the future instead of predicting it ("what could happen") (Van Rijn & Van der Burgt, 2012). With this new scenario concept, future uncertainty is not denied; rather, future uncertainties with the largest impact on the issue under investigation are taken as a starting point (Postma, 2013). In most cases complex reality is reduced to two key uncertainties.[3] These are used as two dimensions in a scenario cross to frame and explore four alternative futures (Schwartz, 1996). If two key uncertainties in the organisational or business environment are chosen for the axes, the scenario cross frames four "environmental scenarios". If two internal uncertainties or dilemmas are key, the result is four "strategic or internal scenarios". It is also possible to combine an internal and an external uncertainty, which results in four system scenarios.

The latter make a direct link between the business or organisation and its environment (Gausemeier, 1998). In all three cases the scenarios are crafted out in the form of key points or a narrative that is positioned within the future, preferably supported with illustrations to appeal to the user as much as possible (Postma, 2015). In tourism the interest in making predictions of a single future declined during the 1980s and 1990s to the advantage of making wider explorations of multiple possible futures with more qualitative input. This shift in interest was not only caused by the fact that the travel forecasts of the 1960s and 1970s often proved to be inaccurate or incorrect, but also by the slowdown of demographic, economic and mobility developments. These suppressed the interest in dealing with issues of future growth, while the growing complexity and dynamics of contemporary society raised the awareness that tourism policy and planning could not solely or mainly rely on quantitative forecasts. Ian Yeoman was one of the first to apply the new approach in a tourism context when he worked for Visit Scotland during the early 2000s (Postma, 2013).

Normative, goal-oriented or aspirational scenarios are a final category of scenarios with which the future can be addressed. This approach has grown from several sources from the 1970s. Aspirational scenarios represent alternative strategic directions for an organisation or business to achieve future objectives ("what should happen") (Bezold, 2009). If they focus on preservation, they usually have a planning term of three months to five years. If focused on change and transformation, they may have a time horizon of five to 50 years (Van Rijn & Van der Burgt, 2012).

During the COVID-19 pandemic two interpretations of scenario thinking have dominated the public discourse and media reports of national governments and national institutions around the globe: predictive scenarios and goal-based scenarios. For example, in the Netherlands, the national government frequently presented an update of the trends and predictions of the expected development of COVID-19 infections (predictive scenarios), and a roadmap (and its adjustments) to lead the country through the crisis (goal-based scenarios). The aim of these goal-based scenarios in the Netherlands (but also elsewhere) was to control the virus as much as possible, not overburdening healthcare and protecting vulnerable people in society (see www.nctv.nl). However, to prepare for the post-COVID-19 era, explorative scenarios were considered to offer a perspective and guidance to actors in the global visitor economy in general, and to the tourism industry in particular; a perspective many actors were calling for.

Explorative scenarios are consistent descriptions of alternative futures states and/or developments (Van der Heijden, 2009). They are based on an analysis of the complex forcefield that impacts upon the issue under investigation and identification of the forces that are perceived to drive its future. The driving forces of change that are considered the most powerful and the most uncertain are used as two dimensions that frame four alternative, extreme, yet plausible futures (Schwartz, 1996). Explorative scenarios are well-suited to enhance our understanding of a complex crisis such as the corona pandemic and a complex sector such as the visitor economy (Hines & Bishop, 2015; Lindgren & Bandhold, 2009; Postma &

Yeoman, 2021). The scenarios as presented in this article explore "what could happen" (explorative scenarios), not "what will happen" (predictive scenarios), nor "what should happen" (goal-oriented or aspirational scenarios).

3 Applying Scenario Thinking to the COVID-19 Pandemic

Directly after the COVID-pandemic started to hit the visitor economy, the Centre of Expertise in Leisure, Tourism and Hospitality (CELTH) took the initiative to develop a set of explorative scenarios for the global visitor economy. A team of researchers from three universities associated with CELTH employed media scanning to develop a list of uncertainties concerning the impact of the pandemic on the visitor economy. Media scanning is "a simple and popular method for continuous cover of environmental changes and for occasional overview and inspiration at the start of a scenario planning process" (Lindgren & Bandhold, 2009, p. 63), and simultaneously "is a powerful way to stimulate creativity and bring in new perspectives and ideas" (Lindgren & Bandhold, 2009, p. 128). For about four weeks, in April and May 2020, the researchers identified articles in national and international newspapers and magazines, physical and online, in which experts presented their viewpoints on the nature, extent and possible consequences of the pandemic. Special attention was given to concerns and considerations from scientific experts such as sociologists, economists, philosophers and psychologists. The researchers formed a "creative future group" (Lindgren & Bandhold, 2009, p. 153) and used their expertise to search for emerging patterns and to classify and sort the findings into five cause-and-effect clusters, each representing a complex process in its own right. The driving force of each of these processes was established. Thus, the analysis resulted in five forces that were considered to drive the post-COVID-19 future of the visitor economy: the attitude of national states, the attitude of the (semi) public sector, the attitude of large (multinational) businesses and corporations, the attitude of citizens in their role as consumer, and the length and depth of the crisis. Based on the perceived level of impact and the perceived level of unpredictability of each driver, the authors choose the latter two as being the most impactful and the most unpredictable at the same time. These two so-called key uncertainties were chosen to frame the scenarios to be developed. The uncertainty concerning these two driving forces and their impact was framed by two plausible extremes to be reached after five years.

The first key uncertainty, "the attitude and role of the citizen", raised several questions. Would a new consciousness have emerged after the pandemic? Would people have developed a different perception of the malleability of the world, and of the environment and climate? Would the way people live together have changed from the point of view of social hygiene? Would more involvement, unity and solidarity have emerged? Would local connectedness at street, neighbourhood, district, city and / or regional level have increased and would new regional identities have emerged? Would the citizen have become more creative, and would they have increased their ability to improvise? Would the function of social media versus traditional media have changed? Would citizens make different choices as a consumer?

I - perspective ⟵ Attitude and role of the citizen ⟶ *We* - perspective

Figure 6.1 Key uncertainty "attitude and role of the citizen" and its two plausible extremes
Source: Postma et al. (2020)

Would they put their focus on individual and especially material prosperity or on public welfare? What role would they allow the government in relation to privacy? How would the gap and the tension between generations have developed? Based on these questions the authors arrived at two extremes to frame the "limits of the plausible" (Figure 6.1).

- *Extreme 1:* I. People have not learned from the crisis and have fallen into old patterns. Values underlying views on nature and the environment have remained the same. Nature is considered malleable, and humankind considers itself the dominant species. Society is highly individualised, focused on "the self", material needs and individual prosperity. People have an unlimited drive to consume. People want to keep their private sphere under their own control. They are rebelling against the other resulting in high levels of polarisation (race, ethnicity, gender, social class, generations e.g. Gen Z vs. Gen Y & Z, youth vs. elderly). Social media functions as an outlet for discontent about the other. The public sector and authorities (government, science, police, teachers, etc.) do not get any respect.
- *Extreme 2:* We. The crisis has brought people to repentance. They have become fully aware of the inseparable relationship between man and nature and its effects on health. This has led to a collectivist society that is focused on "the group" and aimed at collective well-being and quality of life. Consumption has become attuned to this. The common interest is paramount. People consciously think about the implications of their own actions on others ("social hygiene"). In the street, neighbourhood, district, city and region people demonstrate commitment, togetherness and solidarity, regardless of race, ethnicity, social class, generation, gender or age. Along with nature, the environment and others, conventional media is also re-evaluated. Social media are social again, facilitating genuine connections between people. People are willing to hand over some control to governments, despite the use of technological tools (drones, facial recognition, apps). Public sector and authorities (government, science, police, teachers, etc.) enjoy full respect.

The other key uncertainty, "the length and depth of the crisis", generated its own specific questions. Would the virus surge over and over, resulting in successive lockdowns? When would a vaccine be widely available, and would social distancing be lifted? Would the public sector continue to provide (financial) support? How would government's debts develop and what consequences would that have? Would there be inflation or deflation, and to what extent? How would the level of prosperity and consumer confidence develop? What would the macroeconomic situation in

Short & shallow ⟵ Length and depth of the crisis ⟶ Long & deep

Figure 6.2 Key uncertainty "length and depth of the crisis" and its two plausible extremes

Source: Postma et al. (2020)

major tourism generating countries look like? The responses to such questions are embedded in two extremes that were considered plausible (Figure 6.2).

- *Extreme 1:* Short/shallow recession. A COVID-19 vaccine is widely available in 2020/2021, lockdowns and social distancing have come to an end and from 2021 onwards the economy will recover.
- *Extreme 2:* Long/deep recession. A COVID-19 vaccine will only become available to the world population in 2022 or later. This allows the COVID-19 virus to return annually in waves as a seasonal flu. Governments keep calling for new lockdowns. The global economy remains under pressure, and recovery will only be achieved after 2025.

4 Four Post-COVID-19 Scenarios for the Global Visitor Economy

The two key uncertainties and their extreme outcomes form a scenario cross that frames four alternative yet plausible post-COVID futures, which were labelled as: "business as usual", "survival of the fittest", "business as unusual" and "responsible tourism" (Figure 6.3). Each of these scenarios describes a future of the visitor economy with its specific features. The scenarios should not be regarded as predictions, but as explorations of extreme macro circumstances that represent the plausible borders of developments to come. It is likely that the post-COVID-19 era will feature a combination of characteristics of the four scenarios (Postma, Heslinga & Hartman, 2020). While the media scanning has continued since the publication of the scenarios, early warning signals of all four scenarios have been identified. Current tourism development shows rates of recovery that go beyond tourism levels from before the pandemic. Nevertheless, signals also point at features of the other scenarios.

Scenario 1 – "Business as usual". As soon as the crisis comes to an end, the tourists lapse into their old behaviour. The demand for travel has accumulated into a reservoir that suddenly "empties". Businesses sense their opportunities, fully respond to the reborn demand and flourish like never before. Because the recovery period has ended, the focus remains on further economic growth. Many companies are taken over by large international chains (conglomerations), but there are also opportunities for new niches. Both travellers and businesses feel unrestrained in their behaviour. The developments result in an overstrained visitor economy, even heavier ecological pressures and negative social impacts. Both the positive and the negative consequences of travelling continue unabated. The mutual distrust of, and fear between, countries within the EU and beyond has sharpened relations. This has

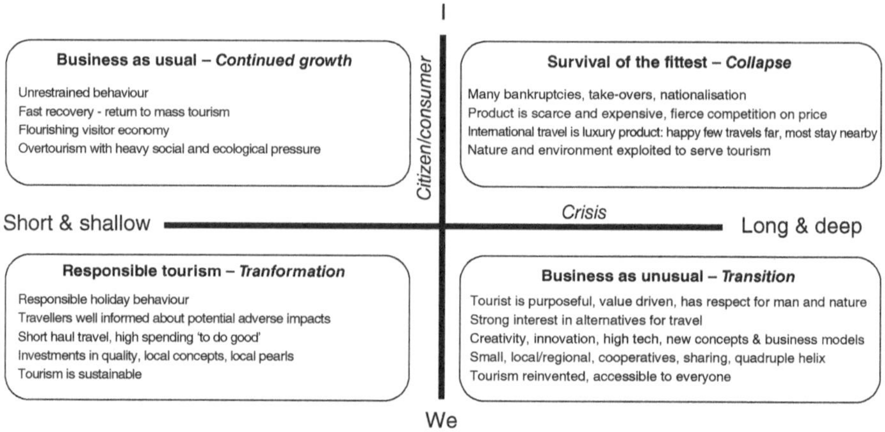

Figure 6.3 Four scenarios for the visitor economy post-COVID-19

Source: Postma et al. (2020)

led to the reintroduction of border controls in the EU and stricter border controls outside the EU.

Scenario 2 – "Survival of the fittest". Citizens continue to hold on to their "right" to vacation in faraway places, which means that the need for travel remains strong. However, the economic recession caused by the pandemic makes it financially impossible for most people to meet their needs. Because people are not able to visit faraway places, they are looking for alternatives close to their home / home country. Long-haul travel is only affordable for a select few. The sector, however, remains rigid in its approach. The battle for the reduced number of holidaymakers is expressed in fierce competition. Because of the crisis, many airlines (especially low-cost airlines), tourism-related businesses and catering businesses have gone bankrupt. This even goes for vital companies because they have constantly put their money into new investments. A few large investors and players dominate the scarce market. To prevent over-restructuring in the sector, to prevent fragmentation and still maintain a minimal supply for its own citizens, governments have nationalised important vital players such as airline companies and railway companies; also a number of hotel, bungalow and camping chains have been taken over by the national governments.

Scenario 3 – "Business as unusual". The long recession and fierce travel restrictions have forced holidaymakers to meet their holiday needs in a different way. Governments, companies, knowledge institutions and citizens have joined forces to help meet this need as much as possible and to reinvent themselves. These parties all contribute with knowledge, subsidies, expertise and manpower. "Everything becomes fluid under pressure" is a saying that also makes its mark here. Creativity flourishes and numerous innovations ensure a total revolution in tourism, both in terms of products and services and in terms of revenue, exploitation and management and business models. This represents a fundamental break from the past.

It is the era of high-tech tourism that is accessible to anyone who needs it. The visitor economy gives rise to local and regional value and production chains with legal entities such as the cooperative as a "renewed" exploitation model, providing purpose for society and a focus on circular production.

Scenario 4 – "Responsible tourism". Although relatively short, the recession has opened the eyes of the tourist. Citizens have increasingly realised that globalisation and the international travel associated with it has largely contributed to the spread of COVID-19 and to the recession. Holidaymakers have become more aware of the consequences of their travels and are taking more conscious and responsible choices based on transcending values. Consumers are purposefully choosing destinations close by, inhibiting the unrestrained growth of pre-crisis international tourism, and so there is a strong demand for purposeful products and services. To make safe and responsible choices, consumers rely on sound (scientific) information. Travellers have no problem with the fact that this information has been obtained through careful monitoring of their behaviour, knowing that, ultimately, this benefits themselves as a traveller, the community and the destination.

5 Implications: How to Make Use of the Scenarios

The scenarios can assist the tourism industry in a number of ways: they can be used as an analytical framework, for agenda setting and to raise awareness of the importance of resilience. These three uses are discussed below.

First, the set of possible and plausible scenarios presented in Figure 6.1 can be used as an analytical framework to map how the impact of COVID-19 and (collective) responses play out. Here, the axes of the framework should be interpreted as continua, not only as a 2x2 matrix with four options. This means a wide range of positions within the framework are possible. The framework can be applied to the business level, to subsectors or branches, or to entire destinations or countries. Users of the framework can pinpoint 1) the current position in the framework, 2) the desired future position, 3) the most realistic/likely future position, as well as 4) map the position of a business, a branch or a destination at various points in time as a means to identify possible pathways of development.

Second, the framework could be used for agenda setting. Each scenario is plausible and can be used to develop a vision on desirable futures to achieve and undesirable futures to avoid (Hartman, 2021). "Business as usual" might be interesting for a quick recovery of the tourism industry, but also runs the danger of recurring overtourism situations in the near future. Following the motto of "build back better" alternative futures that are more in line with sustainable development models of tourism can also be promoted. The framework helps to map various options, enable stakeholders to take a position, engage in agenda setting and develop anticipatory strategies for futures to achieve and futures to avoid (Heslinga, Groote & Vanclay, 2020).

Third, the scenario framework helps to raise awareness that unforeseen and/or autonomous shocks and stresses, such as COVID-19, may have severe impacts and push businesses, industries and destinations out of balance. More fundamentally, it means that the industry is "caught up in a continual process of adaptation to respond and anticipate ongoing pressures (e.g., rise of sharing economy, climate change, overtourism, COVID-19 'corona virus') that challenge [...] their structures, functions, identities and practices of agents" (Hartman, 2020, p. 2). The emerging challenge is to build resilience as a strategy to cope with continual pressures (Heslinga, 2022). It is important to be prepared for possible change by closely monitoring pressures; mapping their impacts, e.g. by means of scenario planning; and developing the capacity to adapt to changing circumstances. As such, the use of scenarios draws attention to the importance of taking uncertainty seriously, as well as developing the resilience to proactively manage possible disruption (discussed in-depth for instance in Hartman, 2018).

The main purpose of the explorative scenarios described above was to inspire tourism businesses and organisations to understand the complexity of the corona crisis and its potential implications for the visitor economy, to create a new perspective, and to take anticipatory measures to become more future proof. The original report (see Postma, Heslinga & Hartman, 2020) maps out the scenarios based on eight attributes: guiding principle, visitor, businesses, markets, key issues, tourism strategy, risks and key values (cf. Yeoman & McMahon-Beattie, 2014). Here, we must emphasise that the tourism industry is very diverse, meaning that the impact of the virus, the measures and the crisis may play out very differently across the industry.

In the context of this book, the third scenario, named "business as unusual", needs further consideration, because it challenges the tourism industry to reinvent itself. It is a plausible future that is framed by a long and deep recession caused by COVID-19 and a collective attitude of citizens. The scenario represents a tourism context that is entirely different from how it used to be, and a future in which a combination of sustainability and virtual reality play a major role. The scenario is characterised by the following interrelated features:

- *Guiding principle*: The progress of a fundamental transition. A transition marks the end of an existing and the beginning of a new period of relative stability and coherence (Bishop, 2005). Transition is a concept that is central to systems theory. It refers to social, institutional and technological change in the societal sub-system of the visitors' economy, with the focus on a shift from being unsustainable to sustainable, enabled by collaborative disruptive interventions by the stakeholders (Farla, Markard, Raven & Coenen, 2012; Hölscher, Wittmayer & Loorbach, 2018; Loorbach, Frantzeskaki & Avelino, 2017).
- *Visitor*: A new type of tourist and guest, demanding a purposeful holiday. They are "quality tourists", representing renewed values, for example concerning their relationship with nature. This is expressed in, for example, more attention to animal welfare and a drive to reduce the ecological footprint. They see the use of technologically advanced augmented, virtual and mixed reality as one of the solutions to achieve this. This type of tourist is still relatively few in number and spends relatively well.

- *Market*: The industry provided new forms of tourism, such as eSports, Virtual Reality (VR) tourism and Mixed Reality (MR) tourism. The hospitality offer is meaningful for the visitor, which means that it satisfies their fundamental needs. To reduce the ecological footprint, the sector makes intensive use of local and regional circular products. The new type of visitor does not travel far and therefore the share of domestic tourism and day tourism is large. The group of businesses and organisations representing the new market and accommodating to the new type of visitors is relatively small in size (early adopters), young and dynamic.
- *Key issues*: The recovery costs for business in leisure and tourism are substantial and are mainly recovered from the community. Businesses need not just to adapt, but to change fundamentally. Those who stay with their traditional offerings miss the boat and disappear from the market, either because they go bankrupt or because they move to another sector. Alternatively, governments may take over the operations to prevent business from disappearing.
- *Strategy*: Representatives from the industry need to collaborate within the quadruple helix (knowledge institutions, governments, businesses, community) in order to co-create and innovate products, services and revenue models. The industry is committed to tourism in line with authenticity and sense of place for high-quality and credible offerings. Thus, the drive for value has outweighed profit maximisation.
- *Risks*: To keep up with the dynamic consumer market, the industry needs a high degree of temporality and pop-up character of activities. It is a challenge to balance this with the sustainable needs of the visitors. Laggards who cannot keep up with (the pace of) development may fail. It is likely that many of these belong to generations before Generation Y (born before the 1980s).
- *Values*: Attitudes towards nature and others have changed, not only among the new type of visitor, but also among representatives of the industry. They realise that living respectfully with each other and with other life on Earth is important for the health of the entire planet. This scenario represents a transition to regenerative tourism, aimed at achieving positive changes in society rather than minimising negative effects of tourism; damage to nature is reduced, in some areas and in some periods nature is given more rest to recover.

6 Conclusion

In this chapter three approaches to scenario planning are compared. In the complex and dynamic society of today, explorative scenarios can be used to envision multiple extreme yet plausible futures that, together, delimit the likely playing field for tourism businesses and organisations. In this chapter four explorative scenarios for the post-pandemic visitor economy were presented. Media scanning and expert judgement were used as input to identify two driving forces of change that are most impactful and at the same time most uncertain. These so-called key uncertainties, the length and depth of the crisis and the attitude of the consumer, define four scenarios: business as usual (continued growth of the visitor economy), survival of the fittest (collapsed visitor economy), business as unusual (visitor economy transitioned) and responsible tourism (visitor economy transformed). To become more resilient to changes in the environment, the tourism industry can make use of explorative scenarios such as those presented in this chapter. The use of such scenarios can help businesses and organisations to deal with uncertainties, to gain more perspective, to be better prepared, and to initiate actions to seize emerging opportunities and become less sensitive to future threats. However, developing scenarios is not a one-time event. They need maintenance to stay ahead of the curve.

Notes

1 In 1945 the RAND Corporation was founded as a leading think tank in the USA.
2 In 1973 the World Futures Studies Federation was established.
3 It is also possible to explore the future with more than two key uncertainties. Based on the discrete "scores" that each uncertainty could have, a series of so-called morphological scenarios is created. To handle the large number of combinations dedicated computer software can be used (Godet & Durance, 2011).

References

Bezold, C. (2009). Aspirational futures. *Journal of Futures Studies*, *3*(4), 81–90.

Bishop, P. C. (2005). Framework forecasting: Managing uncertainty and influence in the future. Second Prague Workshop on Futures Studies Methodology, Charles University, Czech Republic.

Cheer, J. M. (2020). Human flourishing, tourism transformation and COVID-19: A conceptual touchstone. *Tourism Geographies*, *22*(3), 514–524. doi:10.1080/14616688.2020.1765016.

Farla, J., Markard, J., Raven, R. P. J. M., & Coenen, L. (2012). Sustainability transitions in the making: A closer look at actors, strategies and resources. *Technological Forecasting and Social Change*, *79*(6), 991–998. doi:10.1016/j.techfore.2012.02.001.

Gausemeier, J., Fink, A., & Schlake, O. (1998). Scenario management: An approach to develop future potentials. *Technological Forecasting and Social Change*, *59*(2), 111–130.

Gidley, J.M. (2017). *The future: A very short introduction*. Oxford University Press.

Godet, M., & Durance, P. (2011). *La prospective stratégique: Pour les entreprises et les territoires* (2nd Ed.). Dunod.

Hartman, S. (2016). Towards adaptive tourism areas? A complexity perspective to examine the conditions for adaptive capacity. *Journal of Sustainable Tourism*, *24*(2), 299–314. doi:10.1080/09669582.2015.1062017.

Hartman, S. (2018). Resilient tourism destinations? Governance implications of bringing theories of resilience and adaptive capacity to tourism practice. In E. Innerhofer, M. Fontanari & H. Pechlaner (Eds.), *Destination resilience: Challenges and opportunities for destination management and governance*. Routledge.

Hartman, S. (2020). Adaptive tourism areas in times of change. *Annals of Tourism Research*. doi:10.1016/j.annals.2020.102987.

Hartman, S. (2021). Destination governance in times of change: A complex adaptive systems perspective to improve tourism destination development. *Journal of Tourism Futures*. doi:10.1108/JTF-11-2020-0213.

Heslinga, J. H. (2022). Resilient destinations. In D. Buhalis (Ed.), *Encyclopedia of tourism management and marketing*. Edward Elgar Publishing. doi:10.4337/9781800377486

Heslinga, J. H., Groote, P. D., & Vanclay, F. (2020). Towards resilient regions: Policy recommendations for stimulating synergy between tourism and landscape. *Land*, *9*(2), 44. doi:10.3390/land9020044.

Hines, A., & Bishop, P. (2015). *Thinking about the future: Guidelines for strategic foresight* (2nd ed.). Hinesight.

Hölscher, K., Wittmayer, J. M., & Loorbach, D. (2018). Transition versus transformation: What's the difference? *Environmental Innovation and Societal Transitions*, *27*, 1–3. doi:10.1016/j.eist.2017.10.007.

Lindgren, M., & Bandhold, H. (2009). *Scenario planning the link between future and strategy* (2nd ed.). Palgrave Macmillan.

Loorbach, D., Frantzeskaki, N., & Avelino, F. (2017). Sustainability transitions research: Transforming science and practice for societal change. *Annual Review of Environment and Resources*, *42*(1), 599–626. doi:10.1146/annurev-environ-102014-021340.

Postma, A. (2013). Anticipating the future of European tourism. In A. Postma, I. Yeoman & J. Oskam (Eds.), *The future of European tourism*. European Tourism Futures Institute.

Postma, A. (2015). Investigating scenario planning – a European tourism perspective. *Journal of Tourism Futures*, *1*(1), 46–52. doi:10.1108/JTF-12–2014–0020.

Postma, A., Hartman, S., & Yeoman, I. (2024). *Scenario planning and tourism futures: Theory building, methodologies and case studies*. Emerald.

Postma, A., Heslinga, J., & Hartman, S. (2020). *Four future perspectives of the visitor economy after COVID-19*. Centre of Expertise in Leisure, Tourism and Hospitality.

Postma, A., & Yeoman, I. S. (2021). A systems perspective as a tool to understand disruption in travel and tourism. *Journal of Tourism Futures*, *7*(1), 67–77. doi:10.1108/JTF-04–2020–0052.

Schwartz, P. (1996). *The art of the long view*. Doubleday.

UNWTO. (2017). *UNWTO tourism highlights*. www.e-unwto.org/doi/book/10.18111/9789284419029

Van der Heijden, K. (2009). *Scenarios: The art of strategic conversation* (2nd ed.). Wiley.

Van Leeuwen, M., Klerks, Y., Bargeman, B., Heslinga, J. H., & Bastiaansen, M. (2020). Leisure will not be locked down – insights on leisure and COVID-19 from The Netherlands. *World Leisure Organization*, *62*(4), 339–343.

Van Rijn, M., & Van der Burgt, R. (2012). *Handboek scenarioplanning: toekomstscenario's als strategisch instrument voor het managen van onzekerheid* [Handbook scenario planning; future scenarios as a strategic instrument for managing uncertainty] (2nd rev. ed.). Kluwer.

Yeoman, I., & McMahon-Beattie, U. (2014). New Zealand tourism: Which direction would it take? *Tourism Recreation Research*, *39*(3), 415–435. doi:10.1080/02508281.2014.11087009.

Yeoman, I., Postma, A., & Hartman, S. (2022). Scenarios for New Zealand tourism: A COVID-19 response. *Journal of Tourism Futures*, *8*(2), 177–193. https://doi.org/10.1108/JTF-07-2021-0180.

7 Moratoria as Possible Governmental Regulations for Degrowth

A Critical Discourse Analysis of the Bed Capacity Limit in the Local Media of South Tyrol (Italy)

Szymon Kielar and Anna Scuttari

1 Introduction

Pre- and post-pandemic tourism phenomena in the twenty-first century are often connected with issues of overtourism (Andriotis, 2021). Before 2020, international tourism flows were increasing almost steadily until the COVID-19 outbreak (UNWTO, 2019, 2020) and a modification of the locals' life quality and resource availability was perceivable in many touristic hotspots, leading to episodes of *tourismophobia* (Milano, 2018). This "evidently excessive visitation" (Milano, Novelli, Cheer, 2019, p. 1857) was labelled overtourism, and it was defined as "a situation in which the impact of tourism, at certain times and in certain locations, exceeds physical, ecological, social, economic, psychological, and/or political capacity thresholds" (Peeters et al., 2018, p. 22). As a response, many approaches were proposed in the literature to address overtourism issues: from a paradigm shift in destination management to the use of taxation in political economy, from geo-tracking approaches in human geography to de-marketing strategies to re-orientate tourism flows (Milano, Novelli & Cheer, 2019).

Over the pandemic years, the disruption of several neoliberal mechanisms of global tourism opened up a novel debate about post-pandemic tourism and the new normal (Ioannides & Gyimóthy, 2020). Recovery versus degrowth scenarios were considered in this phase as alternative possibilities, both in tourism and in the wider economy (Andriotis, 2021; Habicher, Windegger, von der Gracht & Pechlaner, 2022). Further, the degrowth discourse – which was not new to academic scholars (see, e.g. Andriotis, 2014, 2018; or the 2019 Special Issue of the *Journal of Sustainable Tourism*, Volume 27(12) entitled "Tourism degrowth") – gained more attention, especially for the tourism practitioners (Sigala, 2020). The role of governments and destination management organisations (DMOs) in this phase was fundamental to lead and facilitate the necessary changes as a response to the emergency (Scuttari, Ferraretto, Stawinoga & Walder, 2021). In addition, long-term structural changes at destination level were also advocated in the public along with academic debate to avoid rebound effects after recovery (Sigala, 2020).

DOI: 10.4324/9781003365815-11

These strategies to reorientate the tourism economy require a long-term vision and might fail if there is no acceptance from local stakeholders (Scuttari, Pechlaner & Erschbamer, 2021).

Acknowledging the urgency to tackle overtourism phenomena and the lack of specific knowledge about stakeholder involvement in degrowth transitions during and after the COVID-19 era, this contribution aims to investigate the public discourses around a site-specific degrowth strategy, the bed capacity limit (or *moratorium*) developed in South Tyrol (Italy) up to 2019 and ultimately adopted in 2022.

After a brief theoretical framework about degrowth, moratoria and the role of media in representing stakeholders' perspectives, the exemplary case of South Tyrol is presented. Critical Discourse Analysis (CDA) and its application are then illustrated, followed by a presentation of the main results of the newspapers' analysis. Despite its exploratory nature, the case study illustrates a fascinating dynamic between regional tourism policy, lobbying mechanisms, and the mediating role of media, as the next sections will show.

2 Theoretical Background: Degrowth Strategies and Their Representation in Local Media

Degrowth – conceptually developed in France as a follow-up to the *Limits to Growth* report (Meadows, Meadows, Randers & Behrens, 1972) – is defined in general terms as "a process of political and social transformation that reduces a society's throughput while improving the quality of life" (Kallis et al., 2018, p. 292). It represents an attempt to deconstruct the "growth" paradigm as an unquestioned development option for postmodern economies, while at the same time critically evaluating and re-politicising the sustainability discourse (Asara, Otero, Demaria & Corbera, 2015; Kallis, Demaria & D'Alisa, 2014). It questions the real possibility to achieve a harmonic balance between economic expansion, environmental protection and social wellbeing and critically examines the contradictory features of the green growth paradigm (Fletcher et al., 2019).

In tourism, degrowth aims for the "voluntary downsizing of the tourism industry" (Andriotis, 2021, p. 6) and is leveraging inclusiveness, social justice, and equity to achieve long-term development. Fitzpatrick and colleagues argue that the main issues around degrowth policy proposals in tourism are *limits to growth* – i.e. moratoria, quotas and taxes – and *mindset shifts* – e.g. from "the right to visit" to "the right to live" (Fitzpatrick, Parrique & Cosme, 2022). Degrowth in tourism should therefore critically consider power and privilege relationships between visitors, local communities and other stakeholders, with the aim to enable a balance between them (Higgins-Desbiolles, Carnicelli, Krolikowski, Wijesinghe & Boluk, 2019). Overall, the degrowth paradigm in tourism "encourages qualitative development", avoiding quantitative expansion "to the detriment of natural capital" (Hall, 2010, p. 131). This is why global phenomena such as peer-to-peer accommodation are critically analysed in the degrowth discourse, as they have accelerated capital accumulation while provoking remarkable social and environmental impacts (Fletcher, Murray Mas, Blanco-Romero & Blázquez-Salom, 2019).

To pursue degrowth in tourism, Higgins-Desbiolles, Carnicelli, Krolikowski, Wijesinghe and Boluk (2019) adjust Latouche's recommendations and suggest the following priority actions:

- *Re-evaluate and shift values* away from commodification and exploitation and towards more social justice and community involvement.
- *Re-conceptualise entrenched capitalist concepts* through the cooperation between new business values and governmental regulation for sustainability.
- *Restructure production* to increase its linkages to local communities.
- *Redistribute* the right to travel and the freedom of mobility *at the global, regional and local scale.*
- *Re-localise the economy* not only ensuring the usage of local products, but also avoiding tourism-dependent economies.
- *Reduce, reuse and recycle resources* both in tourism production and consumption.

The application of these principles at the micro-level has translated into different measures to contain tourism flows, e.g. growth moratoria, management plans, eco-taxes and other special taxes, and several capacity limits (see Blázquez-Salom, 2006, as cited in Fletcher et al., 2019).

2.1 Moratoria as One Possible Governmental Regulation to Limit Growth

Among the possible governmental regulations to foster degrowth in tourism, one option refers to the so-called "growth moratoria". Hernández-Martín, Álvarez-Albelo and Padrón-Fumero (2015, p. 882) define a "moratorium" as temporary "[c]apacity controls on tourism accommodation [… i.e.,] as legal measures avoiding or limiting the introduction of new tourism beds". The moratorium is thereby a form of administratively imposed physical limitation that is supposed to implicitly guarantee that ecological, social, psychological and/or political capacity thresholds are not exceeded and, consequently, overtourism is prevented.

Moratoria, therefore, follow the aim to create more quality tourism and to increase revenue per overnight for locals. However, while containing tourism numbers, they might negatively affect the popularity of the destination. Further, prices might rise too much, and bed capacities might become exhausted (Pechlaner, Herntrei & Kofink, 2009). Additionally, if not understood and accepted by local tourism stakeholders, the limits set by the moratoria might be circumvented, making the flow restriction strategy completely or partially ineffective. In the past, several countries have already set limitations to the introduction of new bed capacities (Sharpley, 2007). Their effectiveness is assessed below.

The island of Cyprus, for instance, experienced rocketing numbers in their annual tourist arrivals, rising around 700% within 13 years (Sharpley, 2003; Witt, 1991). Therefore, a moratorium on new accommodations was established in 1989 to stop quantitative growth of bed capacities. However, this intervention proved to be ineffective, because most constructions had already been approved by local authorities before the legal framework was set and were not affected by the moratorium. However, one year later, the introduction of a spatial plan helped to regulate the uncontrolled quantitative tourism growth (Sharpley, 2003, 2007). Although there is not enough evidence about the specific long-term effects of the moratorium in Cyprus, a recent comparative study on European islands has shown that the island reduced its territorial density index (beds/surface in km^2) in the period 2007–2019, while showing lower growth rates of arrivals and overnights than the average values of other European islands (Ruggieri & Calò, 2022).

A more prominent example of a tourism moratorium is that of Lanzarote, a Canary Island with the status of a Biosphere Reserve by UNESCO (Extramedia Consultores, 1999; UNESCO, n.d.). This former agricultural land is now dominated by tourism, and it experienced rising numbers of arrivals and space consumption for tourism purposes (Extramedia Consultores, 1999; Inchausti-Sintes & Voltes-Dorta, 2020; Parreño Castellano, González Morales & Hernández Luis, 2018). After establishing a spatial plan, a moratorium was set in 1991 beginning with a maximum amount of 90,000 beds until 2002, and from the following year on, a maximum of 110,950 beds (Parreño Castellano, González Morales & Hernández Luis, 2018). In this case, the measure did not meet expectations completely, because corruption problems emerged and some municipalities bypassed the moratorium by summing up and redistributing their accommodation capacities (Parreño Castellano, González Morales & Hernández Luis, 2018). After the discovery of these kinds of legal loopholes, some adjustments were made and a series of moratoria were introduced from 2001 on, the last being in force until 2013 (Inchausti-Sintes & Voltes-Dorta, 2020). As is usually the case for moratoria, exceptions were established in the law-making process: for rural areas, five-star hotels and rejuvenated hotels (Bianchi, 2004; Inchausti-Sintes & Voltes-Dorta, 2020). While pointing out that the moratorium was not effective in limiting the use of land in the Canary Islands, Simancas-Cruz (2015, as cited in Inchausti-Sintes & Voltes-Dorta, 2020) shows that it has still restricted the growth of tourism supply, while improving its quality and generating a modest positive effect on gross domestic product, employment and social welfare.

Moratoria have been combined with other degrowth measures in some European destinations. For instance, Barcelona has imposed restrictions against new hotel buildings since 2015 (Camatti, Bertocchi, Carić, van der Borg, 2020). In addition, a tourism tax was introduced based on the category of hotel in which the guests stay (Bel, Joseph, Mazaira-Font, 2022). A similar – although not yet academically investigated – policy was implemented in Mallorca, as in 2018 an absolute bed capacity limit and a zoning system were imposed by law (Agencia Estatal Boletín Oficial del Estado (BOE), 2017) and a law for a circular economy in tourism was later implemented (BOE, 2022).

The examples mentioned show that moratoria can represent a first step towards a top-down management and regulation of tourism flows. However, the involvement of stakeholders in the creation of a new mentality and a common vision for the destination might be crucial to prevent a circumvention of the imposed rules. The next section will introduce the role of media in the communication and public debate about moratoria in relation to stakeholder involvement.

2.2 The Role of Media in Spreading Sustainability-Related Policies

The role of media in framing public debates and setting policy agendas has been investigated for several decades (Innes, 1996). In fact, public policy studies (see, e.g. McCombs & Shaw, 1972) analyse the so-called "agenda-setting" process and show that news has the power to work as "agenda setter", to frame key messages of complex policies and to ultimately influence public opinion about them. Further, media have the potential to activate pressure on governments' formal agendas to speed up, change or shape decision-making processes. Ultimately, they have the chance to portray an issue from the point of view of different stakeholders that may (or may not) have equal power to express their voice in the public discourse (Velasco González & Carrillo Barroso, 2021). This is why, in more general terms, media can be used to indirectly assess the social construction of a problem or issue (Fischer & Gottweis, 2012).

The role of media in shaping tourism-related policymaking has been underexplored so far (see, e.g. Hall, 2003; Schweinsberg, Darcy & Cheng, 2017). Indeed, a comprehensive literature review on media studies in tourism highlights that the main focus has been on travel motivation, destination image, tourism marketing, sustainable tourism, and social relationships in tourism, but not in policymaking (Qian, Wei & Law, 2018). Some academic contributions have dealt with overtourism and its perception (Velasco González & Carrillo Barroso, 2021; Phi, 2020; Araya López, 2021), highlighting that there is a simplistic portrayal of the phenomenon in the public discourse. To the authors' knowledge, only one paper – that of Valdivielso and Moranta (2019) – exists so far on the social construction of the degrowth discourse in tourism through media. However, their CDA developed around the case of the Balearic Islands is only partly based on media, as a wide spectrum of empirical sources is included. Results show two contrasting types of discourses: on the one hand, the interpretation of tourism degrowth as a "greenwashing" mechanism that implements only partially effective policies for decongestion, while still implicitly assuming growth; on the other hand, a politicised movement that embraces profound value shifts and concerns for social justice, based on a contestation or revision of tourism growth as such. The research design of this paper was based on this academic knowledge.

3 Method: Critical Discourse Analysis

3.1 South Tyrol: A Case Study for Moratoria

Located in the north of Italy, South Tyrol is known for its agricultural products and tourism offer, both leveraging local traditions (Eichinger, 1996). The accommodation and food service sector accounts for 11.4% of local added value (ISTAT, 2022) and tourism flows have been developing massively since the 1970s. Due to this steady growth, the local government introduced a moratorium on accommodations in 1979, which caused a limitation of bed availability. Between 1997 and 1998, after some liberalisations, qualitative growth of accommodation facilities was enabled (Pechlaner, Herntrei & Kofink, 2009) and the maximum number of around 230,000 beds was set (ASTAT, 2021). These past measures represent the background for the recent elaboration of the "Space-and-Landscape" law, introduced

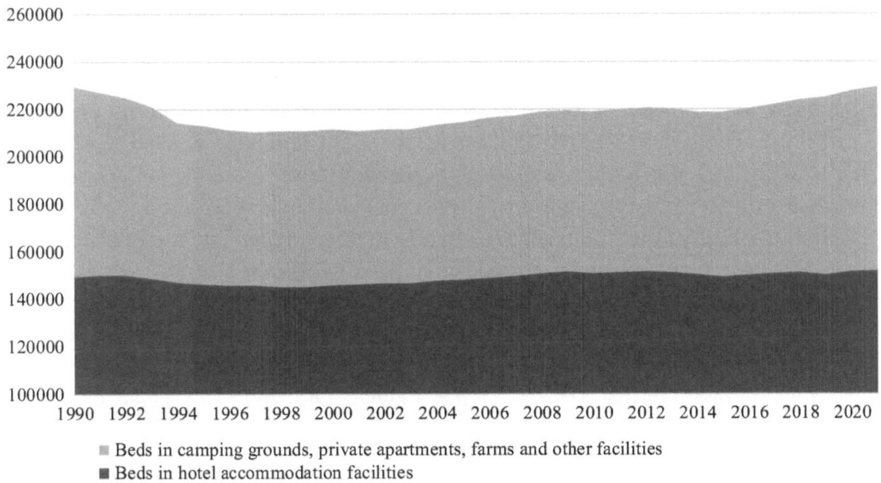

Figure 7.1 Bed capacity in South Tyrol by accommodation type (1990–2021)

Source: Provincial Institute for Statistics (ASTAT), own elaboration

in 2018 to protect locals and preserve the region from deterioration of its culture and environment (Autonomous Province of Bolzano South Tyrol, 2018). According to this law, municipalities are obliged to establish a concept for future tourism actions based on their needs and in compliance with the law (Pechlaner et al., 2022). To implement this law, a planning instrument called the "Regional-Tourism-Development-Concept" was set to frame a series of political measures that can support sustainable development and a mindset change in the tourism economy (Eurac Research, 2022; Pechlaner et al., 2022). Despite these regulatory measures, the numbers of annual overnight stays constantly rose to almost 34 million in 2019, and they showed high resilience after the unexpected drop during the COVID-19 pandemic years (ASTAT, 2022). As Figure 7.1 shows, bed capacities were quite stable from the early 90s until 2015, while between 2015 and 2021 there was an increase in capacity of about 5%, especially affecting camping grounds, private apartments, farms and other facilities. As a response to this trend, political and public debates on the moratorium took place up to 2019. The corresponding regulation came into force in 2022 (Governor's Decree of September 26th, no. 25, 2022). Within this planning instrument, a bed exchange system was set up to achieve a redistribution mechanism and reallocate bed capacities from closed accommodations to new accommodations (Art. 9). Further, special treatment was given to farm holiday providers, as they were excluded from the moratorium (Art. 11).

3.2 *Method: Critical Discourse Analysis*

Given the importance of local and stakeholder acceptance of moratoria, as well as the relevance of media in promoting or preventing their acceptance, CDA was chosen as a research method. Fairclough (2010) describes CDA as an important part of the critical evaluation of society, as language is understood as “a form of social practice” (Fairclough, 1989, p. 20). The assumption of CDA is that every written piece incorporates an ideology and exerts a certain kind of power and influence on the audience (Johnson & McLean, 2020), producing forms of consent and support of these ideologies. Thereby, analysing the ways language is used to produce consent about specific ideologies, CDA highlights how political, institutional or other forms of power might support (or prevent) social change in the media audience. This qualitative method has been already used to investigate the relationships between tourism practices and external social frameworks (see e.g. Mayer, Bichler, Pikkemaat & Peters, 2021; Qian, Wei & Law, 2018).

Six newspapers – some available online, some in print – were selected to represent the heterogeneous panorama of the local press: *Dolomiten – Tagblatt der Südtiroler*, *Südtirol Online – stol.it*, *Südtirol News*, *Die Neue Südtiroler Tageszeitung*, *ff – Das Südtiroler Wochenmagazin*, and *salto.bz*. Due to the dominance of the German-speaking population in South Tyrol (Autonome Provinz Bozen – Südtirol, n.d.), the press in Italian language was deliberately excluded from this first analysis. The articles were selected based on the German equivalents of the search terms “bed stop”, “bed limit” and “Regional-Tourism-Development-Concept”. It is important to mention that all articles are presenting the moratorium before it came into force in August 2022 (Governor’s Decree of September 26th, no. 25, 2022). Therefore, it might be assumed that the media discourse had the potential (or at least the implicit aim) to influence the course of the law implementation. All articles found (n = 154) from 22 March 2000 until 30 April 2022 were saved and screened. The articles generally refer to the public debate around the last moratorium, and only rarely about the previous ones. After the first selection, 129 paper articles were considered relevant for the topic of analysis, of which 13 are readers’ letters. The latter were intentionally included in the analysis to include the perspective of the locals. The coding procedure was performed using the software MAXQDA. Following the CDA steps illustrated by Fairclough (2010), codes were abstracted to form five discourses which will be presented in the following section.

4 Results: Discourses About the Moratorium

4.1 Language Features and Connotation of the Moratorium

The linguistic characteristics of the analysed material can be divided into three parts: the touristic situation in South Tyrol, references to the bed stop, and the general wording. Within the first, many metaphors are used. The importance of tourism in this region is described as the “Achilles’ heel” (Aschbacher, 2021) which will suffer “multiple organ failure” (Heiss, 2020) if numbers shrink. Emerging or existing problems connected to (over)tourism relate mostly to land use and traffic congestion (e.g. “continue to build bed fortresses into the landscape” (Benedikter, 2018), “Even more ground sealing and concrete is poison for the landscape” (Benedikter, 2021), “even now the virus’ mouth is watering again” (Costa, 2021), “The German’s favourite toy is […] the car” (Tötsch, 2022) or an “increasing traffic avalanche" (Dolomiten, 2022b)). Furthermore, some expressions bring attention to the changing situation and the need to reform tourism practices ("Homework to catch up" (Larcher, 2021) / "Homework to do" (Larcher, 2022a), "do not want to be a hawker's tray" (Mair, 2022)). Sometimes they are connected to cynical expressions of criticism or irony against politicians for not dealing sufficiently with those issues ("The regional government did not feel like formulating a necessary implementing provision" (Larcher, 2021), "In political language, this is then called 'sustainability' and 'authentic experiences'" (Aschbacher, 2021)).

The topic of the bed stop itself is used with mostly figurative comparisons and expressions which are predominantly negatively connoted ("death blow" (Larcher, 2022b; Mair, 2022), "outrageous unfair favouritism" (Gitzl, 2021)). Some of them are retrieved from military language, e.g. "Power struggle with the farmers' association" (Dolomiten, 2022c) / "Power struggle for more beds" (Dolomiten, 2022a) or "the tourist marching direction for South Tyrol" (Schwarz, 2021).

Overall, the language is negatively connoted. "Bed fortresses" (Dolomiten, 2022a) symbolise the bigger hotels in the region and "to not lump them together" (Pitro, 2022) means that South Tyrol should not be affected by the bed stop entirely. The criticism towards the new measures is openly mentioned, but some headlines also indicate different points of view, including support ("No to the bed stop" (Die Neue Südtiroler Tageszeitung [DNST], 2017), "In favour of total bed stop" (Pliger, 2020), "South Tyrol's bed stop and go" (Südtirol News, 2022)). In the following, the five discourses standing out from the CDA are presented.

4.2 Discourse 1: The Bed Stop as an Insufficient Measure for South Tyrol Remaining Attractive

This first discourse implies criticism predominantly from politicians of the Green Party and the People's Party of South Tyrol, and some tourism experts. Mostly subliminally mentioned, the bed stop is perceived as a non-sufficient measure to solve the existing problems such as day-trip tourism or overfilled streets. A sophisticated visitors/mobility management is perceived here as more relevant than the introduction of strict limits to tourism nights. If a moratorium is introduced, traditional accommodation structures and characteristics, which form a vital part of South Tyrol for inhabitants and visitors, are seen as endangered. If implemented, the moratorium should be adapted to municipal features, based on the different tourist intensities. In addition to the imposition of limits, the creation of acceptance of the population towards tourism is deemed as essential for the future and wealth of the region.

4.3 Discourse 2: The Bed Stop as an Unreasonable Measure

This second discourse deals with the rationales why the introduction of the bed stop is unreasonable. It is argued that farm holidays and private room renting are already regulated by other laws and do not need more restrictions. Moreover, especially during the pandemic, the number of tourists decreased, therefore a bed stop is maintained to be irrational. Setting quantitative criteria only is believed to be disproportionate if compared to the contextual (pandemic) situation. Further, maximum bed numbers are also seen as unreasonable, and it is posited that the innovative “bed exchange” mechanism will not bring the expected effects because bed capacities will be dynamically shifted among municipalities, but they will be ultimately needed in each region. The main critiques come from the hospitality association, mayors and the farmers’ association.

4.4 Discourse 3: The Bed Stop as an Unfair Measure

This discourse groups the reasons why it is maintained that the bed stop is an unfair measure from an equality perspective. Among them is the fact that peer-to-peer accommodations (especially Airbnb accommodations) were not affected by the bed stop, although they are responsible for an increasingly expensive and scarce living space for locals. The issue is even more relevant (and unfair) if considering that the peer-to-peer accommodation supply quintupled within a short time, while the traditional accommodation sector was not experiencing a similar growth in time. Furthermore, farm holiday providers are criticised that they will benefit from special exceptions and treatments, compared with traditional accommodation facilities. This criticism of farm holiday providers goes further, as some accommodations of this kind seem to be too luxurious, and some stakeholders argue that they should be subject to different fiscal regimes. According to this discourse, a powerful farmers’ lobby is presumed to alter legislation. Among the voices supporting this discourse are the hospitality associations and several mayors.

4.5 Discourse 4: The Bed Stop as a Restrictive Developmental Measure

In the fourth discourse, family-led accommodations, the youth and the hospitality association criticise the moratorium as being non-compatible with innovation, progress and economic sustainability. This discourse indirectly tackles intergenerational equity and green growth founding principles for sustainable development and claims that hotel businesses need to remain profitable to be attractive for future generations. Based on this narrative, economic success is a necessary condition to become ecologically friendly and therefore comply with sustainability goals. A moratorium is associated with the risk of a loss of tourist businesses and a decreasing popularity of South Tyrol as a destination. A bed stop might only be right in the case of an emergency, and since this is not an emergency situation, it is maintained that development restrictions are inappropriate for this specific case. The future is associated with progress and not with past conditions in terms of tourism development.

4.6 Counter-Discourse: The Bed Stop as an Indispensable Measure for South Tyrol

The only counter-discourse standing out in the analysis argues for the introduction of the bed stop as an indispensable measure for the South Tyrolean region. Significant groups forming this opinion are, among others, the Green and the People’s political parties, the DMOs, scientists, climatologists and the hospitality associations. The argument is based on the fact that South Tyrol has the highest tourist density in the Alps. Further, it is perceived that the maximum number of big hotels has already been reached. Therefore, restrictive actions will show positive effects for the future and are indispensable. The highest number of tourists was recorded in 2019 and the status quo needs to be changed towards a more sustainable situation (economically, ecologically, socially). Tourism will become resistant to crises and climate changes. Also, a bed stop will show effects against overmobility in the region and improve acceptance of tourism practices by locals.

4.7 Comparative Analysis

The five discourses identified via CDA and discussed here are interconnected. Table 7.1 shows their features and highlights some relevant quotations from the news, while Figure 7.2 illustrates their connections. Discourses 1 and 5 seem to highlight two sides of the same concept: although moratoria are necessary and

Table 7.1 Discourse overview, including citations and main features

Discourse	*Most significant citations*	*Features*
D.5 (*The bed stop as an indispensable measure for South Tyrol*)	• "Concerning larger businesses, we have probably reached the zenith" (Dolomiten, 3 March 2022) • "as a means of balancing between small and large businesses", "the highest density in the whole Alpine region" (DNST, 16 January 2020) • "Making tourism crisis-proof and climate-proof" (Dolomiten, 27 December 2021) • "One needs the acceptance of the population" (ff, 8 August 2019)	• In favour of bed stop • Max tourism amount reached • Crisis resistant • Climate resistant • Tackle overmobility • Need for acceptance
D.1 (*The bed stop as an insufficient measure for South Tyrol remaining attractive*)	• "One needs the acceptance of the population" (ff, 8 August 2019) • "With a bed stop, one is making it too easy: 'Why don't we talk about day tourism, tourist hotspots or overfilled Dolomites' passes. We have been talking about streams of visitors for 30 years'" (ff, 24 February 2022) • "One needs management [in hotspots] so that these places remain worth experiencing for locals as well as for visitors" (salto.bz, 17 January 2020) • "A bed stop alone isn't enough. Of course, this isn't feasible now in times of the COVID crisis but in the medium and long-term there is no way around it. 'Tourism must set limits to itself for economic efficiency', said the regional councillor […]. Apart from mere economic efficiency, there is also the protection of the landscape and the climate, the quality of life of inhabitants and the general welfare" (salto.bz, 5 March 2021)	• No solution of existing problems • Sophisticated visitors/mobility management needed • Traditions/ characteristics endangered • Need for acceptance

(*Continued*)

Table 7.1 (Continued)

Discourse	*Most significant citations*	*Features*
D.3 (*The bed stop as an unfair measure*)	• "the offer on Airbnb has […] quintupled and is to be limited […] also to ensure the availability of living space for locals" (Dolomiten, 28 April 2022) • "Farm holidays are exempt from the bed stop 'if necessary for the survival of the business'. What's meant by this is unclear" (Dolomiten, 28 April 2022) • "Luxury hotels disguised as farms to save taxes. 'Why should farmers be able to build […] and small pensions not?'" (salto.bz, 21 December 2021) • "Until the concept is approved in the upcoming days, the two lobbies are still putting a lot of pressure on the regional government" (salto.bz, 17 January 2020)	• Inclusion of Airbnb needed • Exception for farm holiday providers criticised • Farms seem like 5* hotels • Powerful farmers' lobby
D.4 (*The bed stop as a restrictive developmental measure*)	• "The cap of 150 beds per establishment is acceptable. 'What has to be guaranteed, however, is that small family-run businesses can develop to a profitable size'. Otherwise, he said, there is a risk of weakening or even losing these businesses in the upcoming years" (Dolomiten, 30 April 2022), "because the successors hardly see any perspectives anymore" (Dolomiten, 12 February 2022) • "We have young people who are well educated and motivated. It cannot be that those who have not yet managed to have a top business are denied any chance to have one. That is planned economy" (Dolomiten, 9 February 2022) • "A limitation is an emergency measure, but not the solution, because small businesses must also be able to develop" (salto.bz, 17 January 2020) • "The way things look now, we are falling back into the 1980s, when there was a complete urban stop" (Dolomiten, 30 December 2021)	• Importance of profitability and attractiveness • Youth without job perspectives • Tourism for South Tyrol is very important • Legitimacy for the measure in emergency case only • Progress, not decrease
D.2 (*The bed stop as an unreasonable measure*)	• "'unfitting timing'. In times where tourism is at a standstill […] the regional councillor comes along with his bed stop!" (ff, 17 June 2021) • "But scepticism remains: 'With the bed stop', the Greens say, 'it is a manoeuvre that isn't too far from a label fraud'. Among other things, they point to the new 'Space-and-Landscape' law, in which the limit of 229,088 beds was cancelled nationwide. And the Space-and-Landscape law ranks legally higher than the regional government's tourism concept" (ff, 17 June 2021) • "An only quantitative criterion, […] also seems disproportionate […]" (Dolomiten, 5 June 2020) • "The planned bed exchange for the redistribution of beds that become available makes no sense. 'As soon as someone gives up, the business is immediately bought up by others to get those beds'" (Dolomiten, 23 December 2021)	• Already existing rules • Not as many tourists as pre-COVID • Inappropriateness of exclusively quantitative criteria • Bed stop renewal as unnecessary • "Bed exchange" mechanism as a useless instrument

Source: Own elaboration.

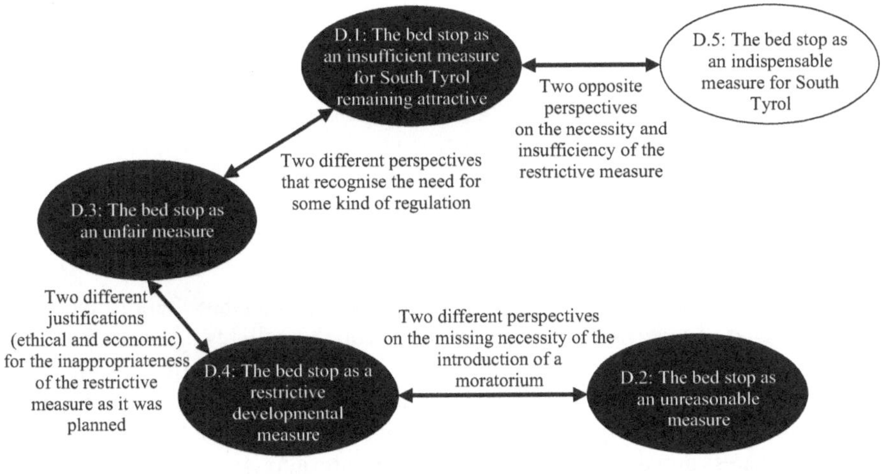

Figure 7.2 Interrelationships among discourses

Source: Own elaboration

useful, they are not sufficient alone to foster a change in the tourism ecosystem. Discourses 3 and 4 argue that the prospected growth limitation is not reasonable and – what is more – not ethically well distributed among the groups of interest. Discourse 2 mostly criticises the way the moratorium is implemented and focuses on the loopholes in the implementation procedure. Together with Discourse 4 they both complement each other by criticising the planned measure. In relation to the discussions on unfairness (D.3) and insufficiency (D.1), there is an attempt to challenge the implementation of the moratorium as originally planned. If the measure were to be modified in accordance with the demands, it is likely that these critics would alter their stance on the issue, given that the fundamental concept of regulating tourism has already been acknowledged.

In general, it needs to be said that some critiques that emerged in the local press were already highlighted in previous studies about moratoria (see, e.g. the potential failure of a bed exchange system analysed in Parreño Castellano, González Morales & Hernández Luis, 2018), while other narratives (e.g. the negative effects of moratoria on intragenerational equity and on sustainable growth) are novel to the degrowth debate. The analysis of the main actors giving a voice to the five discourses reveals the key role of some specific stakeholders. Representatives from the Green Party, scientists and the local DMO describe the moratorium as a necessary but not sufficient initiative to regulate tourism phenomena. Conversely, the tourism businesses, represented through the hotel associations, mostly stress the importance of economic sustainability over social equity and environmental protection and highlight the inability of the policy instrument to deal with special types of accommodations (e.g. peer-to-peer or farm holiday accommodations). Only a minority of them supports the moratorium as a necessary intervention. This indirectly recalls the juxtaposition between strong and weak conceptions of sustainability (see Neumayer, 2003). While the supporters of the former criticise the measure for being too simplistic and insufficiently bound up with a deeper change of values in society, proponents of the latter conception identify the intervention itself as a threat to current growth values and as a potential form of unfairness. Indirectly, the critiques also highlight the contradictory features of the moratorium, as it excludes those tourism accommodation providers with the most relevant rise in capacity over the last few decades.

5 Conclusion

This contribution has focused on the role of media as agenda-setters to pave the path for a debate on degrowth issues in South Tyrol. Overall, results show that both the tourism phenomenon and its limitation are negatively connoted in the local press. The media discussion around the moratorium in South Tyrol is also set from a predominantly negative point of view, highlighting its potential ineffectiveness, inappropriateness and unfairness. Further, the moratorium is represented as an isolated governmental intervention, ignoring the complexity of the tourism development strategy for South Tyrol (Eurac Research, 2022). From this first analysis, the media seem to report scepticism around the governmental intervention, without supporting the degrowth purposes deliberately.

Based on the framework of Higgins-Desbiolles et al. (2019), the analysis of the case of South Tyrol shows that there is an ongoing (but still unresolved) *reconceptualisation of capitalist concepts*, driven by the public hand and its regulations, but not yet corresponding to a value shift for local tourism businesses. In other terms, *limits to growth* have been imposed using a top-down approach (and adopting some exceptions), but *mindset and value shifts* have not completely happened so far (Fitzpatrick, Parrique & Cosme, 2022).

Such a situation has the potential to generate rebound effects or strategies to circumvent the moratorium (see e.g. Parreño Castellano, González Morales & Hernández Luis, 2018) and should be avoided by taking care of a collective change in attitude of the tourism ecosystem and more comprehensive restrictions, including peer-to-peer and other complementary accommodation facilities. Within this context, the question arises whether media might play an active role in consensus-building around planning issues, or whether social planning as such can only happen through the direct involvement of stakeholders in the planning process (Innes, 1996). Further research could investigate if, how, and when possible narratives associated with pro-sustainability regulation should be better communicated by governments to foster a constructive debate. Policy communication through media would then assume a crucial role to trigger a mindset shift of the tourism ecosystem. A second issue relevant to the degrowth debate in the media is the ethical responsibility of journalists (McCombs, 1997). Even in this context, there is no evidence of empirical research in tourism to the authors' knowledge.

5.1 Limitations

Despite the novel and enlightening outcomes of the present contribution, some limitations stand out concerning the empirical design. The qualitative approach is in line with a very novel topic to explore, but it is also associated with a low generalisability of results. Therefore, quantitative designs assessing causality between media debate and policy outcomes might be relevant in future research. Additionally, the use of a single case study is associated with low external validity of results. To overcome this weakness, a comprehensive overview of existing examples of moratoria and their features has been provided and similarities and differences between them have been highlighted. Finally, the empirical data is only based on newspapers written in the German language. Although this is the language used by the large majority of the local population, the inclusion of newspapers in Italian language and cross-language comparisons might be relevant in a second step of research.

References

Agencia Estatal Boletín Oficial del Estado (BOE). (2017). Boletín Oficial Del Estado Núm. 223 [Official State Gazette No. 223]. www.boe.es/boe/dias/2017/09/15/pdfs/BOE-A-2017-10539.pdf

Agencia Estatal Boletín Oficial del Estado (BOE). (2022). Boletín Oficial Del Estado Núm. 197 [Official State Gazette No. 197]. www.boe.es/eli/es-ib/l/2022/06/15/3/dof/spa/pdf

Andriotis, K. (2014). Tourism development and the degrowth paradigm. *Turisticko Poslovanje*, 13, 37–45. doi:10.5937/TurPos1413037A

Andriotis, K. (2018). *Degrowth in tourism: Conceptual, theoretical and philosophical issues*. CABI.

Andriotis, K. (2021). *Issues and cases of degrowth in tourism*. CABI. doi:10.1079/9781789245073.0000.

Araya López, A. (2021). A summer of phobias: Media discourses on "radical" acts of dissent against "mass tourism" in Barcelona. *Open Research Europe*, *1*, 66. doi:10.12688/openreseurope.13253.1

Asara, V., Otero, I., Demaria, F., & Corbera, E. (2015). Socially sustainable degrowth as a social-ecological transformation: Repoliticizing sustainability. *Sustainability Science*, *10*(3), 375–384. doi:10.1007/s11625-015-0321–9

Aschbacher, A. (2021). Ehrlich sein reicht nicht [Being honest is not enough]. *ff – Das Südtiroler Wochenmagazin*. www.ff-bz.com/politik-wirtschaft/politik/2021-24/ehrlich-sein-reicht-nicht.html

ASTAT. (2021). Südtirol in Zahlen [South Tyrol in numbers]. https://astat.provinz.bz.it/downloads/Siz_2021(16).pdf

ASTAT. (2022). Statistisches Jahrbuch für Südtirol [Statistical Yearbook for South Tyrol]. https://astat.provinz.bz.it/downloads/JB2021(7).pdf

Autonome Provinz Bozen – Südtirol. (n.d.). Autonomie für drei Sprachgruppen [Autonomy for three language groups]. https://autonomie.provinz.bz.it/de/autonomie-fur-drei-sprachgruppen

Bel, G., Joseph, S., & Mazaira-Font, F. A. (2022). The effects of moratoriums on hotel building: An anti-tourism measure, or rather protection for local incumbents? *Applied Economics Letters*. doi:10.1080/13504851.2021.1927958

Benedikter, T. (2018). Touristische Überbelastung: Was tun? [Tourist overload: What to do?]. *salto.bz*. www.salto.bz/de/article/05092018/touristische-ueberbelastung-was-tun

Benedikter, T. (2021). Bettenobergrenze ein Ablenkungsmanöver? [Bed limit as a distraction?]. *salto.bz*. www.salto.bz/de/article/05032021/bettenobergrenze-ein-ablenkungsmanoever

Bianchi, R. V. (2004). Tourism restructuring and the politics of sustainability: A critical view from the European periphery (The Canary Islands). *Journal of Sustainable Tourism, 12*(6), 495–529. doi:10.1080/09669580408667251

Blázquez-Salom, M. (2006). Calmar, contenir i decréixer. Polítiques provades (1983–2003) i possibles de planificació urbanística [Calm, contain and decline. Proven policies (1983–2003) and possible urban planning]. *Territoris, 6*, 163–181.

Camatti, N., Bertocchi, D., Carić, H., & van der Borg, J. (2020). A digital response system to mitigate overtourism. The case of Dubrovnik. *Journal of Travel & Tourism Marketing, 37*(8–9), 887–901. doi:10.1080/10548408.2020.1828230

Costa, M. (2021). Lieber Arno [Dear Arno]. *salto.bz*. www.salto.bz/de/article/26042021/lieber-arno

Die Neue Südtiroler Tageszeitung. (2017). "Nein zur Betten-Obergrenze" ["Not to the bed limit"]. *Die Neue Südtiroler Tageszeitung*. www.tageszeitung.it/2017/12/01/nein-zur-betten-obergrenze

Dolomiten. (2022a). Machtkampf um mehr Betten für die Kleinen [Power struggle for more beds for the little ones]. *Dolomiten – Tagblatt der Südtiroler*. www.wiso-net.de/document/DOL__a26655e2ff6373b7d3a46447ad8a268b81e0c4ed

Dolomiten. (2022b). Zur Entwicklung des Tourismus [The development of tourism]. *Dolomiten – Tagblatt der Südtiroler*. www.wiso-net.de/document/DOL__abc0aff19d7316d731c583666af02d6dce2964d0

Dolomiten. (2022c). Machtkampf mit dem Bauernbund [Power struggle with the farmers' association]. *Dolomiten – Tagblatt der Südtiroler*. www.wiso-net.de/document/DOL__806bdd24eee08b05e1b5df05a137e6212035de00

Eichinger, L. M. (1996). Südtirol [South Tyrol]. In R. Hinderling, L. M. Eichinger & R. Harnisch (Ed.), *Handbuch der mitteleuropäischen Sprachminderheiten*. Narr.

Eurac Research. (2022). *LTEK2030+: Ambition Habitat South Tyrol – On the way to a new tourism culture*. www.eurac.edu/en/institutes-centers/center-for-advanced-studies/projects/ltek2030plus

Extramedia Consultores. (1999). La Moratoria Turística de Lanzarote: Aspectos sociales [Lanzarote's moratorium on tourism: Social aspects]. https://datosdelanzarote.lztic.com/media/item/docs/20060124121404330MoratoriaLanzarote.-Aspectos%20sociales.pdf

Fairclough, N. (2010). *Critical discourse analysis: The critical study of language* (2nd ed.). Longman Applied Linguistics. Longman.

Fairclough, N. (1989). *Language and power*. Longman.

Fischer, F. & Gottweis, H. (2012). *The argumentative turn revisited: Public policy as communicative practice*. Duke University Press.

Fitzpatrick, N., Parrique, T., & Cosme, I. (2022). Exploring degrowth policy proposals: A systematic mapping with thematic synthesis. *Journal of Cleaner Production, 365*, 132764. doi:10.1016/j.jclepro.2022.132764

Fletcher, R., Murray Mas, I., Blanco-Romero, A., & Blázquez-Salom, M. (2019). Tourism and degrowth: An emerging agenda for research and praxis. *Journal of Sustainable Tourism, 27*(12), 1745–1763. doi:10.1080/09669582.2019.1679822

Gitzl, F. (2021). Bettenstopp und Seilbahnbau [Bed stop and cableway construction]. *Dolomiten – Tagblatt der Südtiroler*. www.wiso-net.de/document/DOL__d7c3654c147dc36ce444f41d558add5c35054abb

Governor's Decree of September 26th, no. 25 (2022). Criteria and modalities for the collection, setting of the ceiling and the allocation of guest beds. http://lexbrowser.provinz.bz.it/doc/de/227939/dekret_des_landeshauptmanns_vom_26_september_2022_nr_25.aspx

Habicher, D., Windegger, F., von der Gracht, H. A., & Pechlaner, H. (2022). Beyond the COVID-19 crisis: A research note on post-pandemic scenarios for South Tyrol 2030. *Technological Forecasting and Social Change*, *180*, 121749. doi:10.1016/j.techfore.2022.121749

Hall, C. M. (2003). Tourism issues, agenda setting and the media. *e-Review of Tourism Research*, *3*(1), 42–45.

Hall, C. M. (2010). Changing paradigms and global change: from sustainable to steady-state tourism. *Tourism Recreation Research*, *35*(2), 131–143. https://doi.org/10.1080/02508281.2010.11081629

Heiss, H. (2020). Vor der Wende [Before the change]. *ff – Das Südtiroler Wochenmagazin*. www.ff-bz.com/politik-wirtschaft/politik/2020-28/vorwende.html

Hernández-Martín, R., Álvarez-Albelo, C. D., & Padrón-Fumero, N. (2015). The economics and implications of moratoria on tourism accommodation development as a rejuvenation tool in mature tourism destinations. *Journal of Sustainable Tourism*, *23*(6), 881–899. doi:10.1080/09669582.2015.1027212

Higgins-Desbiolles, F., Carnicelli, S., Krolikowski, C., Wijesinghe, G., & Boluk, K. (2019). Degrowing tourism: Rethinking tourism. *Journal of Sustainable Tourism*, *27*(12), 1926–1944. doi:10.1080/09669582.2019.1601732

Inchausti-Sintes, F., & Voltes-Dorta, A. (2020). The economic impact of the tourism moratoria in the Canary Islands 2003–2017. *Journal of Sustainable Tourism*, *28*(3), 394–413. doi:10.1080/09669582.2019.1677677

Innes, J. E. (1996). Planning through consensus building: A new view of the comprehensive planning ideal. *Journal of the American Planning Association*, *62*(4), 460–472. doi:10.1080/01944369608975712

Ioannides, D., & Gyimóthy, S. (2020). The COVID-19 crisis as an opportunity for escaping the unsustainable global tourism path. *Tourism Geographies*, *22*(3), 624–632. doi:10.1080/14616688.2020.1763445

ISTAT. (2022). Sequenza dei conti [Sequence of accounts]. http://dati.istat.it/Index.aspx?DataSetCode=DCCN_SQCT

Johnson, M. N., & McLean, E. (2020). Discourse analysis. In A. Kobayashi (Ed.), *International Encyclopedia of Human Geography* (2nd ed.). Elsevier.

Kallis, G., Demaria, F., & D'Alisa, G. (2014). Introduction: Degrowth. In G. D'Alisa, F. Demaria, & G. Kallis (Eds.), *Degrowth: A vocabulary for a new era*. Routledge.

Kallis, G., Kostakis, V., Lange, S., Muraca, B., Paulson, S., & Schmelzer, M. (2018). Research on degrowth. *Annual Review of Environment and Resources*, *43*(1), 291–316. doi:10.1146/annurev-environ-102017-025941

Autonomous Province of Bolzano South Tyrol (2018). Landesgesetz vom 10. Juli 2018, Nr. 9 Raum und Landschaft (Land Act of 10 July 2018, No. 9 Space and Landscape). http://lexbrowser.provinz.bz.it/doc/de/212899%c2%a710%c2%a720%c2%a740/landesgesetz_vom_10_juli_2018_nr_9/i_titel_span_allgemeine_bestimmungen_span/i_kapitel_span_gegenstand_und_zielsetzung_span/art_2_zielsetzung_span_span.aspx

Larcher, M. (2021). Der Bettenzauber [The bed magic]. *ff – Das Südtiroler Wochenmagazin*. www.ff-bz.com/politik-wirtschaft/politik/2021-15/bettenzauber.html

Larcher, M. (2022a). Griff nach den Sternen [Reaching to the stars]. *ff – Das Südtiroler Wochenmagazin*. www.ff-bz.com/politik-wirtschaft/politik/2022-13/griff-nach-den-sternen.html

Larcher, M. (2022b). Im Bremsgang [On the breaks]. *ff – Das Südtiroler Wochenmagazin*. www.ff-bz.com/politik-wirtschaft/politik/2022-13/im-bremsgang.html

Mair, G. (2022). Betten nach Plan [Beds according to the plan]. *ff – Das Südtiroler Wochenmagazin*. www.ff-bz.com/politik-wirtschaft/politik/2022-08/betten-nach-plan.html

Mayer, M., Bichler, B. F., Pikkemaat, B., & Peters, M. (2021). Media discourses about a superspreader destination: How mismanagement of Covid-19 triggers debates about sustainability and geopolitics. *Annals of Tourism Research, 91*, 103278. doi:10.1016/j.annals.2021.103278

McCombs, M. (1997). Building consensus: The news media's agenda-setting roles. *Political Communication, 14*(4), 433–443. doi:10.1080/105846097199236

McCombs, M. E., & Shaw, D. L. (1972). The agenda-setting function of mass media. *The Public Opinion Quarterly, 36*(2), 176–187.

Meadows, D. H., Meadows, D. L., Randers, J., & Behrens III, W. W. (1972). *The limits to growth: A report for the Club of Rome's project on the predicament of mankind* (2nd ed.). Universe Books.

Milano, C. (2018). Overtourism, malestar social y turismofobia. Un debate controvertido [Overtourism, social unrest and tourismophobia. A controversial debate]. *PASOS: Revista de Turismo y Patrimonio Cultural, 18*(3), 551–564. doi:10.25145/j.pasos.2018.16.041

Milano, C., Novelli, M., & Cheer, J. M. (2019). Overtourism and degrowth: A social movements perspective. *Journal of Sustainable Tourism, 27*(12), 1857–1875. doi:10.1080/09669582.2019.1650054

Neumayer, E. (2003). *Weak versus strong sustainability: Exploring the limits of two opposing paradigms*. Edward Elgar.

Parreño Castellano, J. M., González Morales, A., & Hernández Luis, J. Á. (2018). La (des) regulación territorial del crecimiento del alojamiento turístico. El ejemplo de Lanzarote, Islas Canarias [Territorial (de)regulation of the growth of tourist accommodation. The example of Lanzarote, Canary Islands]. *Ciudad Y Territorio Estudios Territoriales, 50*(195), 50–70.

Pechlaner, H., Herntrei, M., & Kofink, L. (2009). Growth strategies in mature destinations: Linking spatial planning with product development. *Tourism: An International Interdisciplinary Journal, 57*(3), 285–307.

Pechlaner, H., Innerhofer, E., Gruber, M., Scuttari, A., Walder, M., Habicher, D., Gigante, S., Volgger, M., Corradini, P., Laner, P., & von der Gracht, H. (2022). Ambition Lebensraum Südtirol: Auf dem Weg zu einer neuen Tourismuskultur. Landestourismusentwicklungskonzept 2030+ [Ambition Habitat South Tyrol: Towards a new tourism culture. Regional-Tourism-Development-Concept RTDC 2030+]. https://assets-eu-01.kc-usercontent.com/c1c45d5a-c794-01a3-3c24-89f77bf8cab4/376e0647-6114-4566-95bf-71621b9e54f7/LTEK_de_FINAL.pdf

Peeters, P., Gössling, S., Klijs, J., Milano, C., Novelli, M., Dijkmans, C., Eijgelaar, E., Hartman, S., Heslinga, J., Isaac, R., Mitas, O., Moretti, S., Nawijn, J., Papp, B., & Postma, A. (2018). *Research for TRAN Committee – overtourism: Impact and possible policy responses*. European Parliament. www.europarl.europa.eu/RegData/etudes/STUD/2018/629184/IPOL_STU(2018)629184_EN.pdf

Phi, G.T. (2020). Framing overtourism: A critical news media analysis. *Current Issues in Tourism, 23*(17), 2093–2097.

Pitro, S. (2022). "Dieser Knoten ist zu lösen" ["This node is to be solved"]. *salto.bz*. www.salto.bz/de/article/30032022/dieser-knoten-ist-zu-loesen

Pliger, V. (2020). "Bin für totalen Bettenstopp" ["I'm in favor of a total bed stop"]. *ff – Das Südtiroler Wochenmagazin*. www.ff-bz.com/politik-wirtschaft/wirtschaft/2020-24/bin-fuer-totalen-bettenstopp.html

Qian, J., Wei, J., & Law, R. (2018). Review of Critical Discourse Analysis in tourism studies. *International Journal of Tourism Research, 20*(4), 526–537. doi:10.1002/jtr.2202

Ruggieri, G., & Calò, P. (2022). Tourism dynamics and sustainability: A comparative analysis between Mediterranean islands – evidence for post-COVID-19 strategies. *Sustainability, 14*(7), 4183. doi:10.3390/su14074183

Schwarz, H. (2021). Die Betten-Diskussion [The bed discussion]. *Die Neue Südtiroler Tageszeitung*. www.tageszeitung.it/2021/05/13/die-betten-diskussion

Schweinsberg, S., Darcy, S., & Cheng, M. (2017). The agenda setting power of news media in framing the future role of tourism in protected areas. *Tourism Management, 62*, 241–252. doi:10.1016/j.tourman.2017.04.011

Scuttari, A., Ferraretto, V., Stawinoga, A. E., & Walder, M. (2021). Tourist and viral mobilities intertwined: Clustering COVID-19-driven travel behaviour of rural tourists in South Tyrol, Italy. *Sustainability, 13*(20), 11190. doi:10.3390/su132011190

Scuttari, A., Pechlaner, H., & Erschbamer, G. (2021). Destination design: A heuristic case study approach to sustainability-oriented innovation. *Annals of Tourism Research, 86*, 103068. doi:10.1016/j.annals.2020.103068

Sharpley, R. (2003). Tourism, modernisation and development on the Island of Cyprus: Challenges and policy responses. *Journal of Sustainable Tourism, 11*(2–3), 246–265. doi:10.1080/09669580308667205

Sharpley, R. (2007). A tale of two islands: Sustainable resort development in Cyprus and Tenerife. In S. Agarwal & G. Shaw (Eds.), *Managing coastal tourism resorts: A global perspective*. Channel View Publications.

Sigala, M. (2020). Tourism and COVID-19: Impacts and implications for advancing and resetting industry and research. *Journal of Business Research, 117*, 312–321. doi:10.1016/j.jbusres.2020.06.015

Simancas-Cruz, M. (2015). La moratoria turística de Canarias. La reconversión de un destino maduro desde la Ordenación del Territorio [The tourism moratorium in the Canary Islands. The conversion of a mature destination from Spatial Planning]. Servicio de Publicaciones de la Universidad de La Laguna.

Südtirol News. (2022). Südtirols Betten Stop and Go [South Tyrol's bed stop and go]. *Südtirol News*. www.suedtirolnews.it/wirtschaft/suedtirols-betten-stop-and-go

Tötsch, A. (2022). "Wir müssen eine Grenze ziehen" ["We have to draw a line"]. *salto.bz*. www.salto.bz/de/article/04032022/wir-muessen-eine-grenze-ziehen

UNESCO. (n.d.). *Lanzarote biosphere reserve, Spain*. UNESCO. https://en.unesco.org/biosphere/eu-na/lanzarote

UNWTO. (2019). *International tourism highlights, 2019 edition*. World Tourism Organization (UNWTO). doi:10.18111/9789284421152

UNWTO. (2020). *UNWTO world tourism barometer, May 2020: Special focus on the impact of COVID-19*. World Tourism Organization (UNWTO). doi:10.18111/9789284421930

Valdivielso, J., & Moranta, J. (2019). The social construction of the tourism degrowth discourse in the Balearic Islands. *Journal of Sustainable Tourism, 27*(12), 1876–1892. doi:10.1080/09669582.2019.1660670

Velasco González, M., & Carrillo Barroso, E. (2021). The short life of a concept: Tourismphobia in the Spanish media. Narratives, actors and agendas. *Investigaciones Turísticas, 22*, doi:10.14198/INTURI2021.22.1

Witt, S. F. (1991). Tourism in Cyprus: Balancing the benefits and costs. *Tourism Management, 12*(1), 37–45. https://doi.org/10.1016/j.tourman.2017.04.011

8 Focusing at the Very Local Scale to Measure, Understand and Manage Overtourism

Raúl Hernández-Martín, Julia Schiemann, Hugo Padrón-Ávila and Yurena Rodríguez-Rodríguez

1 Introduction

Destination congestion is one of the most important challenges currently faced by tourism. Despite being a long-standing issue, the problem of overtourism (UNWTO, 2018) gained particular relevance in the years prior to the pandemic, though during this crisis it was replaced by a zero-tourism context. However, overtourism seems to have regained renewed strength now that the "new normality" has been achieved in areas like Europe. This chapter highlights how overtourism problems are highly concentrated in the territory. Overtourism is also influenced by the relationship between local residents and tourists, and destination management. Therefore, it needs to be addressed with better data at a very local scale. Even within a single tourism municipality, it is common to find different areas which have different tourism products, different perceptions of visitors' impacts and different effects of overtourism. Thus, destination management must take into account the varying causes and effects of this phenomenon that has such distinct features in different places. The relationship between the visitors and local residents is a key point to analyse, bearing in mind that this interaction takes place in very localised spots. The sustainable governance of regions specialised in tourism must consider the different situations that can be found throughout a territory, depending on tourist concentrations, interactions with the local population, the relevance of tourist supply (accommodation, restaurants or natural and built attractions), the spatial dispersion of tourists and the phenomenon of vacation homes.

In the next section, we focus on the relevance of the territory to analyse tourism congestion on a very local scale. The third section identifies local tourism destinations for statistical and decision-making purposes. Then, we provide some insights from overtourism indicators in some local destinations on the island of Tenerife and, finally, we provide some conclusions.

2 The Territorial Dimension of Overtourism

Overtourism is a new term for an old phenomenon (Capocchi et al., 2020) mainly related to the conflict between local residents and tourists, patterns of tourism development, the blurring concept of carrying capacity (Milano et al., 2019) and

DOI: 10.4324/9781003365815-12

the debate over a tourist area life cycle (Butler, 2019). All this literature on overtourism shares a focus on the local scale of analysis and the need for better information. However, because of a lack of data, empirical insights into overtourism are mainly related to cities and regions, while there is little attention being paid to districts, neighbourhoods or parts of municipalities which, in fact, suffer most from excessive tourism. In their bibliometric analysis of overtourism, Buitrago and Yñiguez (2021) found 38 scientific papers aimed at measuring overtourism. The territorial scope of these papers were cities (40%), coasts (20%), events (4%), rural (4%), while 32% were not focused in a particular tourist typology. However, in the case of Barcelona, which is one of the best-known cases of the phenomenon, *Observatori del Turisme a Barcelona* (2020) shows marked differences in perceptions of overtourism in the ten districts of the city. Considering these differences is key in the management of overtourism, as we will see in this chapter. Overtourism is very often related to a district, neighbourhood or local destination within a municipality.

Using administrative boundaries to measure the phenomenon is a source of biases. Overtourism is not an issue of national, regional or even municipal scale; the problem is particularly present in certain districts or *spots* in the main international cities, attractions, cultural or coastal destinations (Peeters et al., 2018). In fact, overtourism and zero tourism caused by the pandemic can be considered opposite situations sharing a common feature: both require detailed spatiotemporal information to be soundly managed. However, the main destination types such as cities, nature, rural, coastal and cultural face different problems included under the label of overtourism.

One of the main sources of tourism growth in the last decade has been the ability of cities to attract visitors. Cities around the world have become leading tourism destinations and an object of tourism management analyses and planning (Florido-Benítez, 2022). Nevertheless, there is a lack of tourism statistical information at the city level, which makes it difficult to analyse and manage tourism development. Not surprisingly, the main statistical report comparing tourism flows in main cities around the world is released by a global financial firm (Mastercard, 2020), given the lack of public bodies or official statistical offices involved in measuring tourism at the city scale. There have, however, been efforts to provide a framework for city tourism statistics (Wöber, 2000), but with few results until now. In the case of coastal and nature-based destinations, natural assets such as beaches or protected areas have been a source of congestion and conflict between local residents and tourists. While in the case of cultural, natural and rural destinations, the fragility of the immaterial heritage or lifestyles can also lead to conflicts between local residents and visitors. However, in any of these situations, the conflicts tend to be concentrated on the main sites that characterise the trip: the place of accommodation and/or the attractions visited by tourists.

From a more general perspective, the territory is a crucial dimension of tourism that has been rather neglected in tourism statistical information. Despite tourism destinations having been considered one of the most important concepts for tourism economics (Candela & Figini, 2012), there is scant information on the situation of local tourism destinations, as this information is mainly collected at the national and regional scales following international recommendations (United Nations et al., 2010a, 2010b).

A close spatial look at tourism statistics from the demand point of view, focusing on the movements of visitors, displays a clear perspective regarding concentration. Despite data on regional arrivals in a significant tourist continent like Europe not showing very high degrees of concentration, the concentration is much more intense when we use lower territorial scales of analysis. For example, at a municipal scale, tourism distribution is highly polarised. In the case of Spain, a major tourist destination, 16 municipalities out of 8,131 accounted for more than 50% of tourism overnights in 2019 (National Institute of Statistics, 2023). Moreover, if we look even closer at tourism flows, we can see an increased concentration of tourists in certain districts within each municipality, as Batista e Silva et al. (2018) showed for Europe using geolocated data.

This concentration of tourism means that local tourism destinations, infra-municipal areas where tourists and the tourism industry tend to concentrate, are crucial for tourism analysis and measurement. Therefore, there is a need for a widely accepted definition of a local tourism destination for statistical purposes. In addition, it is necessary to integrate such a definition into the existing framework of tourism statistics. The need for precision when identifying tourist destinations, as noted by Blázquez-Salom et al. (2021), is related to the fact that overtourism is site-specific, showing different problems and solutions and demanding different indicators. Following a literature review carried out by these authors, overtourism, even though not clearly defined, has several features. These features include high tourist/resident ratios, environmental pollution, housing, degradation, overloaded infrastructure, trivialisation, loss of identity, and a drop in the quality of life related to an excess of tourists. These issues occur not only in different regions or cities but also in different local destinations within a municipality. That is why the analysis of overtourism needs, even within a municipality, the identification of different local tourism areas with different problems.

3 The Identification of Local Tourism Destinations

In recent years, measuring tourism at the local level to achieve better decision-making has become increasingly important. Authors such as Dredge (1999), Lew and McKercher (2006), and Pearce (1999, 2001) consider local destinations as fundamental units of analysis in tourism, as it is where supply and demand intersect. One of the difficulties in managing overtourism is related to the blurred concept of a destination or local destination. Therefore, in this section, we provide some tools to delineate local tourism destinations. The identification of local tourism destinations and establishing their boundaries is a necessary first step to provide statistical information at the local level.

There are not many works on the delimitation of tourist destinations at a local scale, despite the United Nations World Tourism Organisation (UNWTO, 2004) recognising the importance of delimiting tourist destinations and providing guidelines to do so. This absence of literature on the delimitation of tourism destinations at the local level may be due, among other causes, to the fact that the development of tourism statistics has normally focused on national and regional scales. Despite the above, there is extensive literature on the delimitation of areas in other disciplines such as urban planning, sociology, public health, geography, etc. (Cutchin et al., 2011; Suttles, 1972). These works delimit functional areas (districts, neighbourhoods, etc.) with different research purposes and are based on a series of criteria.

Data, information and knowledge on tourism destinations are vital, given the proven relevance of having adequate governance and public–private coordination in them (Candela & Figini, 2012). This knowledge requires a powerful tourism information system at the local level, which contains both traditional statistical information and other data from emerging sources (credit card payment data, use of applications, etc.). These new sources have the advantage that, in general, they can be georeferenced, a very useful feature for their disaggregation at the local scale.

This research is part of the work carried out in collaboration with the Institute of Statistics of the Canary Islands (ISTAC) (Hernández-Martín et al., 2016) and the International Network on Regional Economics, Mobility and Tourism (INRouTe & UNWTO, 2012). As a result of this work, a methodology for the delimitation of local tourism destinations has been designed. This methodology has been applied in the Canary Islands, where 47 local tourism destinations have been identified.

The way in which local tourism destinations are delimited starts from the identification of a series of criteria to group characteristic tourism industries in such a way that the resulting areas have a certain degree of internal homogeneity (Rodríguez-Rodríguez & Hernández Martín, 2018). In the case of the Canary Islands, these criteria are: 1) concentration of establishments of tourism characteristic activities; 2) homogeneity of tourism supply characteristics; 3) stability of boundaries over time; 4) dynamism and flexibility; 5) feasibility and relevance; and 6) public and private support (Hernández-Martín et al., 2016).

These criteria must be applied in a sequential process consisting of three phases (Rodríguez-Rodríguez & Hernández Martín, 2018). The initial phase requires the application of criterion 1, concentration of establishments of tourism characteristic activities, to differentiate between tourism and non-tourism areas. In the second phase, criterion 2 is applied, homogeneity of the characteristics of the tourist supply, which enables the identification of the boundaries of local destinations. Finally, in a third phase, it is verified that the resulting areas meet the four remaining criteria; if not, the boundaries must be readjusted. The criteria described in this section have been applied by a consensus of experts (regional and local authorities, destination management organisations, business associations, etc.).

In this chapter, we will focus on the case of Tenerife. It is the most relevant island in terms of territory and tourism figures within the Canary Islands. In fact, the Canary Islands is the European region (NUTS 2) with the highest figures of overnight stays in tourism accommodation in the period before the pandemic. In Tenerife, 22 local tourism destinations have been identified within six of its 31 municipalities (ISTAC, 2015). For the whole Canary Islands, these areas represent

1.7% of the land yet concentrated 89.3% of overnight stays in hotels and apartments in 2021 (Hernández-Martín & León-González, 2022).

4 Overtourism Indicators for Local Destinations: Some Insights from Tenerife

The measurement of overtourism is not a straightforward process. Following Buitrago and Yñiguez (2021), overtourism has been measured by indicators, surveys, interviews and tools employing user-generated content in social networks. In the case of indicators, the most used have been tourism intensity, which refers to the ratio of visitors or beds to the number of residents, and tourism density, which compares visitors, overnight stays or tourism establishments to the total surface area of a certain place. Depending on data availability, this place is often a municipality or region.

The availability of statistics at the very local scale, infra-municipal, has been encouraged by both the availability of big data and the increasing need for destination managers to have site-specific information, particularly in places where tourism congestion problems need to be faced. One of the factors that has led to overtourism is the increasing relevance of second homes and vacation homes in certain destinations. In the case of Tenerife, three indicators related to overtourism can be highlighted. First, the relative relevance of tourism both in terms of density and intensity. Second, the relevance of vacation homes in terms of the local population (intensity). Third, visits to certain spots are being monitored to better understand tourist behaviour and better manage these points of interest (Wall, 1997; Padrón-Ávila & Hernández-Martín, 2017).

The first analysis is related to the proportion of tourists to local residents in tourism micro-destinations of Tenerife. Figure 8.1 displays the results and shows very different figures depending on the scale of analysis and the characteristics of the local destination. Tourist equivalent population published by ISTAC is the average number of tourists staying in traditional tourist accommodation (hotels and apartments) every day. This figure can be related to the number of local residents living in official local tourism destinations according to census data. In the case of the whole island, the percentage of tourist equivalent population to local residents is 10.1% following ISTAC official statistics (Hernández-Martín & León-González, 2022). However, this figure is much higher for certain tourism municipalities such as Adeje, where tourist equivalent population to local residents reaches 75%, or the municipality of Arona, where the figure is 34%. However, the highest figures and the most relevant differences appear when we provide the figures for the main local destinations of the island (see Figure 8.1). There are three local tourism destinations, shown in the upper-left part of the figure, where the number of equivalent tourist population is four to five times higher than the number of residents. This type of local destination is markedly tourist-oriented. Currently and in the future, these destinations may show some problems related to overtourism. It is worth noting that this situation is completely different from the local tourism destinations shown on the lower-right side of Figure 8.1. In these destinations, there is a mix of

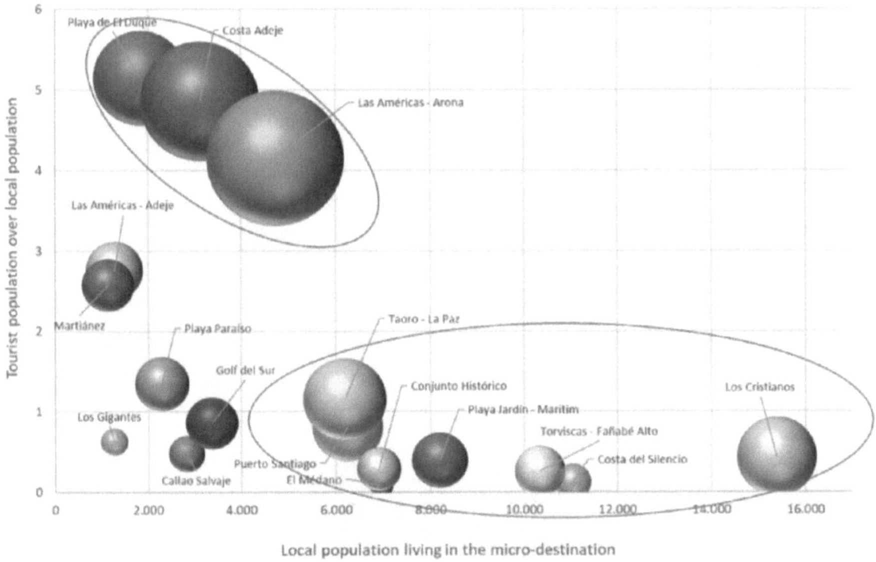

Figure 8.1 Tourist vs. local population in the main micro-destinations in Tenerife, 2019

The area of the bubble represents tourist equivalent population

Source: Own elaboration, based on Instituto Canario de Estadística (ISTAC, 2015).

local residents and tourists, and local resident numbers are normally higher than the tourists staying on a certain day. In this kind of destination, the interaction of tourists and local residents tends to be higher (from the point of view of tourists, but may not be from the point of view of residents), with potential positive effects of interaction but also some potential conflicts for the access to resources, such as beaches. Neither type of destination is better from the perspective of avoiding overtourism, but the challenges for management may be completely different.

Another important source of tourism growth and potential conflict in destinations is the growth of vacation homes promoted by well-known platforms. While the growth of traditional accommodation establishments has been slow in Tenerife, the soaring growth of vacation homes has been a concern for destination managers. Beyond the possible positive effects of this type of accommodation from the point of view of both hosts and guests, there is a potential conflict from its uncontrolled growth, particularly as regulation has not provided an even distribution of benefits and costs. In Figure 8.2, we can see insightful results for the second main tourism municipality, Arona, located in the south of Tenerife. The ratio between vacation homes bed places and resident population by census sections shows marked differences between different zones of the municipality. The area in dark red corresponds to the main tourist area, where there is a low local population, and where there are 3,712 places in vacation homes, which represents 25.8% of all those in the municipality. However, this census section only holds 4.8% of the population registered in the municipality. Moreover, 66.2% of the official population registered in this

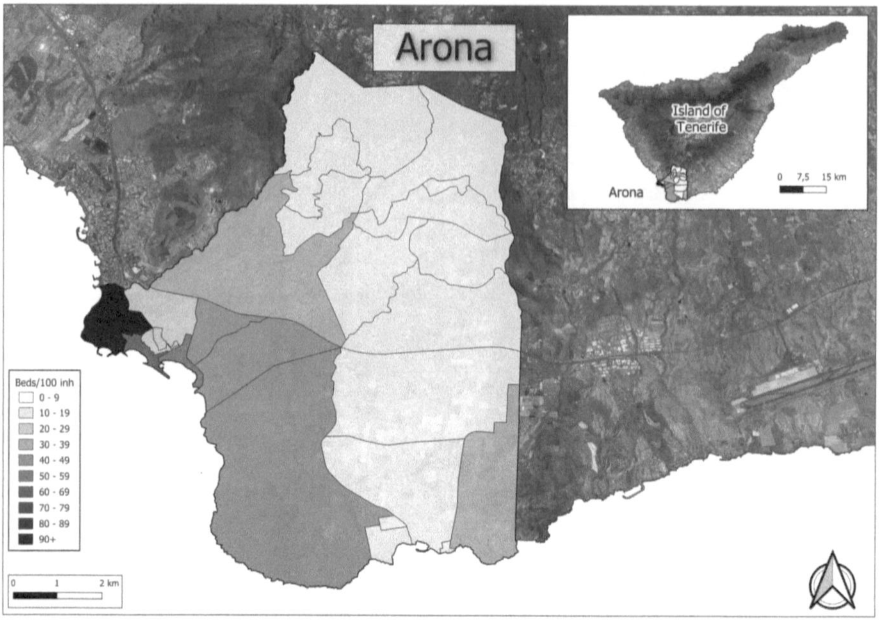

Figure 8.2 Ratio between holiday homes bed places and resident population by census sections in the municipality of Arona (Tenerife, Canary Islands), 2022

Source: National Institute of Statistics of Spain. Own elaboration

area have foreign nationality. The two census sections that are close to this, along the coastline, show around 40–50% of vacation homes over population registered in the census. Note that in the census districts located on the right side of the municipality and inland, the ratio of vacation home places to official residents falls to around 10%. The ratio variations in parts of Arona show how important it is to manage these areas separately. In a further step, this map can be compared with another timeframe to obtain the growth rate, as this also influences potential conflicts caused by uncontrolled growth.

Finally, the prevention and management of overtourism at a very local scale of analysis benefits from the availability of Big Data. Gathering trustworthy information regarding the excess of carrying capacities and other issues linked to overtourism is a complex matter. However, recent contributions have suggested the use of new sources of information to gather these data. Bertocchi et al. (2021) suggest the use of Big Data sources to analyse the places visited by tourists and their impact on destinations better. Regarding this type of analysis, Li et al. (2018) points to a wide range of sources that can be used to gather this information, such as web pages, mobile phone data, credit card transactions, city-sensors and user-generated content in social media. This data can show not only the number of tourists visiting a place but also the spatial or temporal distribution of the tourism supply and demand at a very local scale. An example of the use of this type of data can be

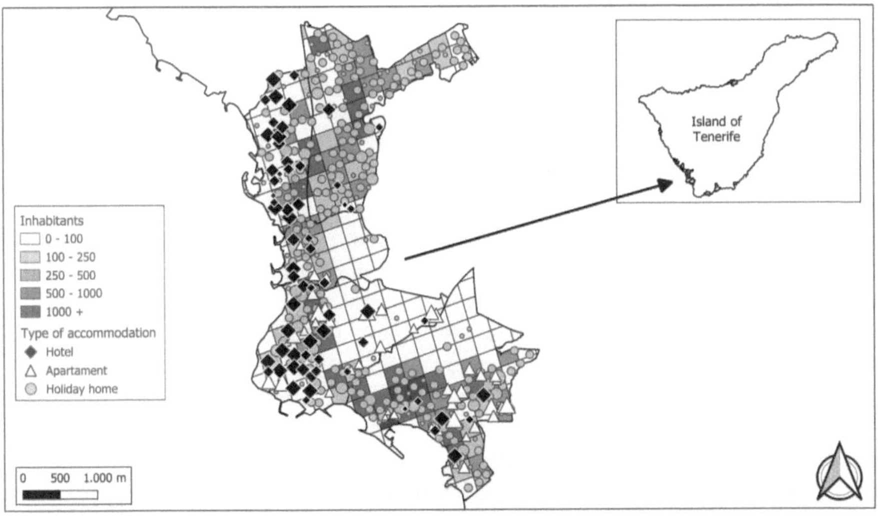

Figure 8.3 Spatial distribution of accommodation establishments and local population in local tourism destinations in South Tenerife, 2023

Source: Tourism Register of the Government of the Canary Islands, June 2023 and Institute of Statistics of the Canary Islands, Tourism Accommodation Survey and Population Census, 2021

Note: Includes Los Cristianos, Las Américas-Arona, Las Américas-Adeje, Costa Adeje, and Torviscas & Fañabé Alto

seen in Figure 8.3, which shows the spatial distribution of tourism accommodation establishments (hotels, apartments and rental homes) and the density of local population for 250 metre grid cells. This kind of data is being used in planning to prevent overtourism as areas with high density of tourism accommodation and local population can be identified.

Figure 8.3 highlights the potential of these new sources of information to detect and analyse overtourism. However, their use is quite recent, so they still present certain drawbacks that should be considered. Jang and Park (2020) indicate that their use can help account for the number of tourists visiting different spots in a destination. Furthermore, some studies have found that geotagging places on social media tends to produce a call effect on other tourists (Alonso-Almeida et al., 2019). Thereby, if managers want to solve issues related to overtourism, they should be cautious. Promoting practices such as sharing pictures may cause an increase in the number of visitors to certain fragile places, worsening overtourism issues.

5 Implications: Management and Governance of Overtourism in Local Destinations

The data provided in the previous section shows that the adaptation of different necessities concerning different types of destinations is highly relevant, as different intensities and densities in micro-destinations influence their problems.

This chapter aims to overcome the lack of information on the situation of local tourism destinations. Sustainable governance of places specialised in tourism requires better data at the local scale for management purposes.

Governance relates to a broader category than government and often takes place without state involvement by encouraging private actors to create institutions or by backing tourism arrangements. However, there is no consensus about the term itself, as governance can be defined from different perspectives and disciplines (Kersbergen & Waarden, 2004). It includes a bottom-up approach with the participation of the local population in decision-making. Residents are part of the destination, the tourism product and the experience of tourists. Locals perceiving overtourism in a destination may lead to antagonism, tourism phobia and a less favourable image of the destination, as in Barcelona, Venice or Dubrovnik (Pérez-Garrido et al., 2022). Governance mixes this bottom-up with a top-down approach. Through defining indicators specific to the local destinations, the status quo can be shown clearly, as well as targets and critical thresholds. Indicators at the local destination scale can also be used to promote more efficient regulations. Measurement is also critical to control, monitor and change the status quo. Opening data and sharing information among stakeholders can be used by decision-makers to implement regulations (see Coulmont et al., 2022; Klimecka-Tatar & Manuela, 2022).

The social pillar, related to local population needs and perceptions, can change from one place to another due to varying ratios of tourists to locals in the destination, as shown in Figure 8.1. Local tourism destinations may suffer from different types of overtourism (UNWTO, 2018). In Tenerife, some local destinations already have a relatively higher intensity of tourists in comparison with the local population. These destinations can be sorted into different groups depending on the visitors' preferences and interests. If someone is looking for a luxury experience with vibrant nightlife, they might prefer the upper-left side destinations. If someone is looking for a more relaxed, family-friendly atmosphere, they might prefer destinations on the lower-right side.

As we have seen, some local destinations in Tenerife have higher relative ratios of residents compared to tourists. These destinations are known for their more relaxed and laid-back atmosphere. However, in these areas, vacation homes can lead to excess tourist concentrations and lack of control of tourist arrivals. This, in turn, can create congestion and saturation in tourist infrastructures and public spaces. There is a need for effective governance to deal with the increasing tourist arrivals caused by vacation homes (Simancas-Cruz et al., 2017). Using imaging adequately enables the detection and analysis of overtourism and the implementation of measures for the sustainable development of the destination. Regarding possible overtourism caused by vacation homes, it is important for destination managers and policymakers to implement strategies that balance the needs and interests of both tourists and local communities.

A collaborative network of local stakeholders can bring benefits to destinations, but joint governance and measurement tools are also needed, such as sustainability indicators. This study focuses on Tenerife, but the methodology is not only applicable to the Canary Islands but also to other destinations, whether coastal, urban or nature-based.

6 Conclusions

This paper presents a new vision of overtourism focusing on the relevance of measurement at a very local scale of analysis. The management and governance of overtourism need better data for decision-making. Stakeholders at the local destination and public decision-makers can benefit from monitoring information and identifying problems related to overtourism, as there are many different types of local destinations with varying problems.

The insights from Tenerife, an important actor in the European tourism market, allow us to improve our knowledge on how to measure the relationship between tourists and local residents, the role of vacation homes and what tourist spots may suffer from overcrowding. This approach and these insights into overtourism are useful for better management and governance of overtourism issues for local destinations worldwide, whatever their particular characteristics.

References

Alonso-Almeida, M-M., Borrajo-Millán, F., & Yi, L. (2019). Are social media data pushing overtourism? The case of Barcelona and Chinese tourists. *Sustainability 11*(12), 3356.

Batista e Silva, F., Marín Herrera, M. A., Rosina, K., Ribeiro Barranco, R., Freire, S., & Schiavina, M. (2018). Analysing spatiotemporal patterns of tourism in Europe at high-resolution with conventional and big data sources. *Tourism Management*, *68*, 101–115.

Bertocchi, D., Camatti, N., & van der Borg, J. (2021). Tourism peaks on the Three Peaks: Using big data to monitor where, when and how many visitors impact the Dolomites UNESCO World Heritage Site. *Rivista Geografica Italiana*, *3*. https://doi.org/10.3280/rgioa3-2021oa12532

Blázquez-Salom, M., Cladera, M., & Sard, M. (2021). Identifying the sustainability indicators of overtourism and undertourism in Majorca. *Journal of Sustainable Tourism*, *31*(7), 1694–1718.

Buitrago, E. M., & Yñiguez, R. (2021). Measuring overtourism: A necessary tool for landscape planning. *Land*, *10*(9), Article 9.

Butler, R. W. (2019). Overtourism and the tourism area life cycle. In R. Dodds & R. Butler (Eds.), *Overtourism: Issues, realities and solutions*. De Gruyter.

Candela, G., & Figini, P. (2012). *The economics of tourism destinations*. Springer.

Capocchi, A., Vallone, C., Amaduzzi, A., & Pierotti, M. (2020). Is "overtourism" a new issue in tourism development or just a new term for an already known phenomenon? *Current Issues in Tourism*, *23*(18), 2235–2239.

Coulmont, M., Berthelot, S., & Martineau, F. (2022). Les rapports de responsabilité sociétale: Une étude des pratiques Canadiennes [Corporate responsibility reports: A study of Canadian practices]. *Canadian Journal of Administrative Sciences / Revue Canadienne des Sciences de l'Administration*, *39*(1), 112–126.

Cutchin, M. P., Eschbach, K., Mair, C. A., Ju, H., & Goodwin, J. S. (2011). The socio-spatial neighbourhood estimation method: An approach to operationalizing the neighborhood concept. *Health & Place*, *17*(5), 1113–1121.

Dredge, D. (1999). Destination place planning and design. *Annals of Tourism Research*, *26*(4), 772–791.

Florido-Benítez, L. (2022). The impact of tourism promotion in tourist destinations: A Bibliometric study. *International Journal of Tourism Cities*, *8*(4), 844–882.

Hernández-Martín, R., & León-González, C. (2022). *Canary Islands tourism sustainability: Progress report 2022.* Gobierno de Canarias, Santa Cruz de Tenerife.

Hernández-Martín, R., Simancas-Cruz, M. R., González-Yanes, J. A., Rodríguez-Rodríguez, Y., García-Cruz, J. I., & González-Mora, Y. M. (2016). Identifying micro-destinations and providing statistical information: A pilot study in the Canary Islands. *Current Issues in Tourism, 19*(8), 771–790.

Instituto Canario de Estadística (ISTAC). (2015). *Entidades y Núcleos Turísticos. Cuaderno Cartográfico 2015 [Entities and Tourist Areas. Cartographic Notebook. 2015].* Santa Cruz de Tenerife. https://datos.canarias.es/catalogos/estadisticas/dataset/3f4ad695-ddc8-427a-b864-a11de9d60946/resource/49e1926e-9530-41e6-9e93-a1574a815d84/download/cuaderno_cartografico_a4_web.pdf

International Network on Regional Economics, Mobility and Tourism (INRouTe) and World Tourism Organization (UNWTO). (2012). *A closer look at tourism: Sub-national measurement and analysis – towards a set of UNWTO guidelines.* UNWTO.

Jang, H., & Park, M. (2020). Social media, media and urban transformation in the context of overtourism. *International Journal of Tourism Cities, 6*(1), 233–260.

Kersbergen, K. V., & Waarden, F. V. (2004). Governance as a bridge between disciplines: Cross-disciplinary inspiration regarding shifts in governance and problems of governability, accountability and legitimacy. *European Journal of Political Research, 43*(2), 143–171.

Klimecka-Tatar, D., & Manuela, I. (2022). Digitization of processes in manufacturing SMEs-value stream mapping and OEE analysis. *Procedia Computer Science, 200*, 660–668.

Lew, A., & McKercher, B. (2006). Modeling tourist movements: A local destination analysis. *Annals of Tourism Research, 33*(2), 403–423.

Li, J., Xu, L., Tang, L., Wang, S., & Li, L. (2018). Big data in tourism research: A literature review. *Tourism Management, 68*, 301–323.

Mastercard. (2020). *Global destination cities index 2019.* Mastercard.

Milano, C., Novelli, M., & Cheer, J. M. (2019). Overtourism and tourismphobia: A journey through four decades of tourism development, planning and local concerns. *Tourism Planning & Development, 16*(4), 353–357.

National Institute of Statistics. (2023). *Hotel occupancy survey.* Spanish Statistical Office.

Observatori del Turisme a Barcelona. (2020). *Tourism activity report 2019.* Ayuntamiento de Barcelona. https://lc.cx/RdNx8U

Padrón-Ávila, H., & Hernández-Martín, R. (2017). Los puntos de interés turístico: Relevancia analítica, propuesta metodológica y caso de estudio [Tourist points of interest: Analytical relevance, methodological proposal and study case]. *PASOS: Revista de Turismo y Patrimonio Cultural, 15*(4), 979–1000.

Pearce, D. G. (1999). Tourism in Paris studies at the microscale. *Annals of Tourism Research, 26*(1), 77–97.

Pearce, D. G. (2001). An integrative framework for urban tourism research. *Annals of Tourism Research, 28*(4), 926–946.

Peeters, P., Gössling, S., Klijs, J., Milano, C., Novelli, M., Dijkmans, C., Eijgelaar, E., Hartman, S., Heslinga, J., Isaac, R., Mitas, O., Moretti, S., Nawijn, J., Papp, B., & Postma, A. (2018). *Research for TRAN Committee – Overtourism: impact and possible policy responses.* European Parliament, Policy Department for Structural and Cohesion Policies.

Pérez-Garrido, B., Sebrek, S. S., Semenova, V., Bal, D., & Michalkó, G. (2022). Addressing the phenomenon of overtourism in Budapest from multiple angles using unconventional methodologies and data. *Sustainability*, *14*(4), 2268.

Rodríguez-Rodríguez, Y., & Hernández Martín, R. (2018). Foundations and relevance of delimiting local tourism destinations. *Investigaciones Regionales – Journal of Regional Research*, *42*, 185–206.

Simancas-Cruz, M., Temes-Cordovez, R., & Peñabrrubia-Zaragoza, M. P. (2017). El alquiler vacacional en las áreas turísticas de litoral de Canarias [Holiday Rentals in the Coastal Tourist Areas of the Canary Islands]. *Papers de Turisme*, *60*, 1–24.

Suttles, G. D. (1972). *The social construction of communities*. University of Chicago Press.

United Nations et al. (2010a). *International recommendations for tourism statistics 2008*. United Nations.

United Nations et al. (2010b). *Tourism satellite account: Recommended methodological framework 2008*. United Nations.

UNWTO. (2004). *Indicators of sustainable development for tourism destinations, a guidebook*. UNWTO.

UNWTO. (2018). *"Overtourism"? – Understanding and managing urban tourism growth beyond perceptions*. UNWTO.

Wall, G. (1997). Tourism attractions: Points, lines, and areas. *Annals of Tourism Research*, *24*(1), 240–243.

Wöber, K. W. (2000). Standardizing city tourism statistics. *Annals of Tourism Research*, *27*(1), 51–68.

9 Sustainability Priorities Disclosures in Annual Reports and the Contribution of Museums to Sustainable Development and Overtourism

Elena Borin

1 Introduction

This paper discusses the sustainability priorities emerging in the non-financial reports of museums and their contribution to sustainable development and addressing overtourism. This research is related on the one hand to the debate on overtourism in cultural destinations, and on the other to the debate on the contribution of culture to sustainable development.

The debate on overtourism and its impact on cultural destinations has highlighted concerns about the degradation and overcrowding of cultural sites, changes in local lifestyles, and the economic benefits and distribution related to tourism (De Luca, Shirvani Dastgerdi, Francini & Liberatore, 2020; Frey & Briviba, 2021). Recently, the focus has also shifted to environmental sustainability issues, such as increased pollution and waste generation (Kvasnová & Marciš, 2022; Liberatore, Biagioni, Ciappei & Francini, 2022). Governance and planning play a crucial role in addressing overtourism, emphasising the need for effective destination management and community engagement (Maingi, 2019; Mihalic & Kuščer, 2022).

The debate on culture and sustainable development has recognised the role of cultural organisations in supporting social, economic and environmental development. Various interpretations exist regarding the relationship between culture and sustainable development, including culture as a separate pillar (Adams, 2010; Loach, Rowley & Griffiths, 2017; UCLG, 2010) or intertwined with other dimensions (CAE, 2019). Cultural organisations, including museums, are increasingly adopting sustainability reporting practices to demonstrate their contribution to sustainability and address overtourism challenges (Esposito & Fisichella, 2019). However, reporting frameworks specific to museums are lacking, leading to diverse approaches to sustainability reporting (Borin, 2023).

In this framework, it is relevant to understand how museums are interpreting their role in the two issues mentioned above and how they disclose information

DOI: 10.4324/9781003365815-13

related to the two topics. This paper aims to address this theme by answering two main research questions:

- What are the sustainability priorities emerging in the non-financial reports of museums?
- How do they provide insights into the targeted contribution of museums to increasing the sustainable development of a cultural destination, also concerning dealing with overtourism?

Using a qualitative case study analysis of a museum in Barcelona (selected since it is known for its overtourism challenges and efforts to address them) based both on secondary sources (mainly its annual reports, but also its website and other resources available online or shared by the organisation) and a semi-structured research interview with the manager in charge of sustainability reporting, the findings contribute to the understanding of sustainability reporting priorities in museums and the role of cultural organisations in sustainable development and overtourism.

The paper is organised into five sections. After this introduction (section 1), an in-depth analysis of the literature is presented. Section 3 describes the research design and methodology, while section 4 (divided into subsections) presents the results of the case study analysis. Finally, section 5 provides some concluding remarks, including the limitations of the research and potential future research developments.

2 Literature Review

The topic of overtourism and tourism management has been at the heart of academic, policy and professional debate over the last several decades. From the theoretical debate on cultural destinations, several points emerged that highlighted the peculiarities of tourism/overtourism management in cultural areas and that also resonate with the main debate on overtourism in general. Alongside concerns related to the management and measurement of the carrying capacity of tourism, one central aspect of the debate revolves around the impact of overtourism on cultural heritage (Bertocchi, Camatti, Giove & van der Borg, 2020; Scuttari, Isetti & Habicher, 2019). While some researchers argue that heavy tourism can lead to degradation, overcrowding, pollution and loss of authenticity in cultural sites (Innerhofer, Erschbamer & Pechlaner, 2019; Rickly, 2019; Seraphin, Ivanov, Dosquet & Bourliataux-Lajoinie, 2020), others emphasise the potential positive effects of tourism, such as increased preservation efforts and funding for cultural conservation, and propose potential strategies to mitigate its negative effects (Frey & Briviba, 2021; Høegh-Guldberg, Seeler & Eide, 2021; Murzyn-Kupisz & Hołuj, 2020; Postma & Schmuecker, 2017). A second main point is linked to the socio-cultural effects of tourism on cultural destinations (Seraphin, Ivanov, Dosquet & Bourliataux-Lajoinie, 2020; Zhuang, Yao & Li, 2019). Concerns include changes in local lifestyles, displacement of residents, gentrification, increased commercialisation and the commodification of local culture. Additionally, cultural appropriation and the loss of traditional practices have been discussed. In this framework, residents' perspectives and community engagement are major concerns (Musikanski, Rogers, Smith, Koldowski & Iriarte, 2019; Park, Choi & Lee, 2019; Szromek, Kruczek & Walas, 2019), and are considered an essential part of the debate. Academics highlight the importance of community engagement, participation, and empowerment to ensure that tourism development respects the interests and well-being of local communities (Chen & Rahman, 2018). Also, the debate on economic aspects of cultural tourism focused on the economic benefits to local economic growth (Atsiz & Akova, 2019; Zhao & Li, 2018); however, questions arose about the distribution of those benefits, the impact on local businesses, and the sustainability of tourism revenue in the long term (Rasoolimanesh, Taheri, Gannon, Vafaei-Zadeh & Hanifah, 2019). The debate often centred around finding a balance between economic growth and the protection of cultural heritage. Recently, mainly due to an increase in interest in climate change and environmental preservation, the focus has also shifted to environmental sustainability issues (Wall, 2019). Overtourism can have significant environmental impacts, including increased pollution, waste generation and strain on natural resources. The debate includes discussions on how to mitigate these effects and promote sustainable practices in cultural destinations.

Finally, the different concerns related to overtourism in cultural destinations have been connected to issues of governance, tourism management and planning (Mihalic & Kuščer, 2022; Pechlaner, Raich & Beritelli, 2010; Stevic & Breda, 2014). Specifically, the role of governance and planning in addressing overtourism is considered crucial: researchers discuss the need for effective destination management, regulation and strategic planning to ensure sustainable tourism development, visitor management and the involvement of local communities in decision-making processes. This would necessarily mean connecting governance structures to tourism management strategies, interpreted as potential solutions to address pressing problems (Goodwin, 2021). These may include implementing visitor quotas, managing visitor flows, diversifying tourism offerings, promoting alternative destinations and developing sustainable tourism practices.

In short, the academic debate on overtourism in cultural destinations is multifaceted and dynamic, and continues to evolve as new research and case studies emerge. Sustainability is also significantly related to the reflection on the role of cultural organisations in the sustainable development paradigm.

Therefore, to complete our understanding of overtourism in cultural destinations, we should consider the debate on the contribution of culture to sustainable development (Borin & Donato, 2022; CHCfE Consortium, 2015; Duxbury, Cullen & Pascual, 2012; Duxbury, Kangas & De Beukelaer, 2017; Nurse, 2006; UCLG, 2010; Vegheș, 2018; Yildrim et al., 2019). Indeed, it can give us relevant first insights into how sustainability issues caused by overtourism can be dealt with through cultural organisations. This debate is linked to a more holistic interpretation of sustainability which has been debated for more than three decades, also incorporating cultural dimensions and reflections. The relation between culture and sustainability is rooted in the concept of culturally sustainable development, which was first proposed by Throsby in 1995, redefining culture in economic and anthropological terms and emphasising its role in supporting or constraining social and economic development. Throsby's proposal sparked a debate on the contribution of cultural and creative fields to sustainable development over the following decades (Throsby, 1995). Initially, one aspect of the debate focused on the interplay between culture/cultural sustainability and the traditional dimensions of sustainability as expressed by Elkington's TBL –Triple Bottom Line model (Elkington, 1987) – resulting in an analysis of culture's transversality across different domains. In 2013, cultural networks launched the global campaign #culture2015goal, advocating for culture to be included in the United Nations' "Transforming Our World: the 2030 Agenda for Sustainable Development". While the explicit inclusion of culture in the agenda was not fully achieved, the importance of culture as an inherent and cross-cutting topic is noticeable within the 2030 Sustainable Development Agenda (UN, 2015). Other non-governmental organisations (NGOs) and policy organisations in the cultural and creative sectors have also emphasised culture's potential contribution to the Sustainable Development Goals (SDGs). Among them, Culture Action Europe (CAE) has been actively advocating for cultural organisations to increase awareness and promote transdisciplinary conversations and commitment. Their report "Implementing Culture within the Sustainable Development Goals", as well as other previous reports, identifies the contribution of cultural and creative organisations to each SDG (CAE, 2017 and 2019).

The subsequent campaign, called #culture2030goal, again led by a coalition of prominent cultural and creative networks and organisations (ICOMOS, CAE, International Music Council, International Federation of Library Associations and Institutions, International Federation of Coalitions for Cultural Diversity, and UCLG Culture Committee), aimed to mainstream culture across the global development agenda. Their objectives included the adoption of culture as a distinct goal in the post-2030 development agenda and the establishment of a worldwide agenda for culture, recognising it as the fourth pillar of sustainable development.

Another interpretation of the role of culture in the sustainability paradigm has also proposed that culture be considered the fourth pillar of sustainable development, alongside the social, economic and environmental pillars (Hawkes, 2001; Loach, Rowley & Griffiths, 2017; Nurse, 2006; UCLG, 2010). Cultural sustainability, defined as inter- and intra-generational access to cultural resources, is considered as important as the other three pillars (Throsby, 1995; UCLG, 2010; World Commission on Culture and Development, 1995). This holistic vision of sustainable development recognises culture's significance for economic, social and environmental issues (Adams, 2010). However, the idea of culture as the fourth pillar of sustainable development has faced criticism (Isar, 2017). For instance, Soini and Dessein argued that culture should not be seen as a separate pillar but rather as intertwined with the other dimensions of sustainable development (Soini & Dessein, 2016). In this regard, they propose three interpretations of the relationship between culture and sustainable development: culture in sustainable development, culture for sustainable development and culture as sustainable development.

These interpretations highlight, respectively, the recognition of culture as a fourth pillar, the instrumental role of culture in connecting the three pillars, and culture as the foundation for achieving sustainable development goals (Soini & Dessein, 2016), but their arguments remained rather vague (Borin, 2023).

In short, over the last 30 years, the reflection on culture and sustainability has opened a dialogue on the role of culture in supporting or constraining social, economic and environmental development. It has prompted discussions on the integration of culture into the global development agenda, and various interpretations have emerged regarding the relationship between culture and sustainable development. All these debates focus on tressing the role of culture mainly on social and cultural sustainability dimensions, although economic and environmental concerns have been partially considered.

Recently, starting from the above-mentioned theoretical and policy premises, several streams of research have developed more specialised investigations focusing on the significance of various fields within the cultural and creative sector (cultural heritage, creative industries, etc…), leading to research on sustainability in specific subsectors (Aageson, 2008; Dameri & Demartini, 2020). Among them, a significant group of researchers have investigated the museums and built cultural heritage field (CHCfE Consortium, 2015; Grazuleviciute-Vileniske, 2006; Mergos & Patsavos, 2007; Roders & Van Oers, 2011; Vegheş, 2018), also bringing to the fore considerations related to the need for cultural organisations and museums to be accountable and transparent about their sustainability impact (Goswami & Lodhia, 2014; Greco, Sciulli & D'Onza 2015; Pop & Borza, 2016; Pop, Borza, Buiga, Ighian & Toader, 2019; Wickham & Lehman, 2015). Relevant sustainability reporting practices have been developed mainly by museum associations and cultural NGOs, with museums in Anglo-Saxon countries leading the way while museums in other countries and contexts are lagging (Esposito & Fisichella, 2019). In particular, there have been specific initiatives and frameworks developed in the United Kingdom, Australia, Spain and Italy to promote sustainability in museums and encourage measuring and reporting (Museum Association, 2008 and 2009; National Museum Directors' Conference, 2009). Research has acknowledged that sustainability reporting in museums is complex due to the diverse nature of these organisations and the lack of an agreed-upon reporting framework (Adams, 2010; Esposito & Fisichella, 2019). Museums combine features of non-profit organisations with needs related to the public sector and educational sectors, making it challenging to define standards. So far, some cultural organisations, international NGOs, and agencies have issued specific guidelines or standards for sustainability reporting in museums (ICOM and OECD, 2019; Julie's Bicycle, 2017); however, no recognised association specialised in sustainability standards (such as GRI – Global Reporting Initiative or IR – Integrated Reporting) has paid attention to the peculiarities of the museum and cultural sector (Borin, 2023). For instance, the GRI guidelines and standards indicated in G4 Sector Disclosures for NGOs and the Sector Supplement for Public Agencies can serve as a reference for identifying specific disclosures relevant to the museum field (such as those related to aspects of sustainability reporting, including governance, economic sustainability, environmental disclosures, social disclosures, and public awareness and advocacy) but do not include the cultural fields in its list of prioritised sectors (Borin, 2023; GSSB, 2016).

Therefore, cultural organisations have swung between diverse solutions: they have either partially adopted existing sector-specific supplements related to other sectors (such as those for NGOs and non-profit organisations or public agencies); or have decided to use standards developed for the cultural organisations but not recognised by reporting authorities; or have developed their own standards, referring to a chosen interpretation of sustainability.

Notwithstanding the differences, sustainability reporting in the cultural and museum sector is becoming an increasingly used and required tool for legitimacy, accessing funding and in general testifying for museums' contribution to sustainability efforts: disclosures in annual, social or sustainability reports are necessary to acknowledge the ongoing efforts in museums and their interpretation of their contribution to sustainable development (Wickham & Lehman, 2015; Borin & Donato, 2022; Santos et al., 2019).

Understanding what non-financial disclosures are provided by museums could help us understand what sustainability aspects they are working on and how they are contributing to addressing diverse local challenges, including overtourism.

This chapter aims to address this topic, using an in-depth analysis of the sustainability reports of one of the most visited museums in Barcelona, a city that has been faced with overtourism for a long time, but whose efforts to solve the problem have also been internationally recognised.

3 Research Design and Methodology

To investigate the above-mentioned topics, a qualitative analysis was implemented on a case study of a museum in Barcelona (Spain). The qualitative approach is regarded as the best methodology for comprehending events that are still in progress and have a high level of complexity (Bluhm, Harman, Lee & Mitchell, 2011; Gummesson, 2006). The museum was selected among the museums of Barcelona (Spain): the city was chosen as the geographical area of research because it has been indicated in the literature as one of Europe's most striking examples of overtourism (alongside other cities such as Venice, Amsterdam and Dubrovnik) but also as a leader to tackle the overtourism problem (Álvarez-Sousa, 2021; Goodwin, 2017 and 2021; Milano, Novelli & Cheer, 2019). Furthermore, Spain is one of the richest countries in terms of cultural heritage, with a high number of UNESCO World Heritage Sites (WHSs) – Spain hosts 49 WHSs, ranking fourth in the world list – and it is among the top ten countries in terms of the number of museums in the world (UNESCO, 2022; Statista, 2022). According to the Barcelona Tourist Office, there are approximately 80 museums in the city of Barcelona.[1] These museums cover a wide range of topics, including art, history, science and technology, and attract millions of visitors every year. Some of the most popular museums in the city (ranking as the most visited ones) include the National Art Museum of Catalonia (MNAC), the Barcelona Museum of Contemporary Art (MACBA), the Picasso

Museum and the Joan Miró Foundation. Among these institutions, the MNAC is one of the most cited examples of sustainable museums, also thanks to its attention to designing and communicating its sustainable management practices and providing appropriate non-financial disclosures related to sustainability. It has therefore been selected as the case study for our analysis.

To guarantee a rigorous data analysis process, the case study was examined according to the guidelines provided by Yin (2018), which were combined with the protocols for content analysis delineated by Finn, White and Walton (2000), Hodson (1999) and Neuman (2003). The analysis consisted of two phases: the first phase was based on preliminary desk research on the information available on the website of the museum related to sustainability and tourism and was complemented by a research interview with the manager responsible for the sustainability accounting and reporting processes to get an overview of the case study and clarify missing aspects in the documentary analysis. The second phase consisted of a content analysis of the Annual Reports of the museum issued in the years 2018 to 2020. The report of the year 2018 was selected as the starting document since from that year the museum decided to also include the social sustainability report in its annual report; the report related to the year 2020 was the last one available at the time of the empirical analysis. To speed up the analysis, the reports were examined using software for content analysis (NVIVO) that helped identify semantic clusters and areas of interest in sustainability. It must be noted that some scholars distinguish between quantitative and qualitative approaches to content analysis. The contrast between the two approaches, according to Krippendorff (2004), is muddled because they both entail qualitative reading, even though quantitative approaches tend to use computers to speed up the process. Although some descriptive statistics are provided, the findings of our investigation are related mainly to the qualitative approach to content analysis, due to the objective of explaining a phenomenon in detail rather than delivering statistical information.

The case study report will be presented in the next section of this chapter, providing first an overview of the museum and then summarising the analysis of the annual reports. After this part, a brief discussion of the results will be provided.

4.0 Results of the Empirical Investigation

As stated above, the city of Barcelona is often indicated as an example of overtourism but also as a best practice for tackling the overtourism problem (Álvarez-Sousa, 2021; Goodwin, 2017 and 2021; Milano, Novelli & Cheer, 2019). Between 2012 and 2019, the number of international tourists visiting the city nearly doubled from 7.5 million to 14 million. This rapid tourism growth has resulted in a range of challenges, including overcrowding in popular tourist areas, increased traffic congestion, environmental degradation and pressure on local resources and infrastructure. The relationship between locals and tourists has become particularly tense: referring to Doxey's Irritation Index[2] (Doxey, 1975), the city appears to have entered the antagonism phase (Abril Sellarés, Azpelicueta Criado & Sánchez Fernández, 2015) with the organisation of several protests against tourists to express locals' hostility toward them (Farrés, 2015; Canto-Alamilla, 2016; Crespi-Vallbona & Mascarilla-Miró, 2018). Tourism is claimed to be one of the main causes of the rise in real estate prices since houses are being turned into hotels and other accommodation facilities, resulting in a scarcity of available properties[3] (Abril Sellarés, Azpelicueta Criado & Sánchez Fernández, 2015). In response to these challenges, the city has implemented various measures to manage overtourism and mitigate its negative impacts. Among the strategic objectives of Barcelona is the supervision of the social aspects of sustainable tourism; as a result, the local government worked on ensuring the city's liveability, a good cohabitation of tourists and locals, and paid particular attention to residents' well-being.[4]

Museums and cultural heritage organisations can be significant players in this scenario, given their interactions with both tourists and locals and the impact they might have on improving the cultural and social aspects and place branding, creation of identity and sense of place, as well as well-being and education (CHCfE Consortium, 2015). The following subsections of this paper explore the annual reports of the MNAC in Barcelona, to investigate if the priority of the institution is tackling these sustainability objectives and how they are measuring up with the challenges posed by overtourism.

4.1 MNAC – Museu Nacional D'art De Catalunya (National Art Museum of Catalunya)

The *Museu Nacional d'Art de Catalunya* opened as the *Museu d'Art de Catalunya* in 1934 in the Palau Nacional de Montjuc; in 1995 it took the current denomination of *Museu Nacional d'Art de Catalunya*. The museum houses the world's best collection of Romanesque mural paintings, Gothic art, major European Renaissance and Baroque as well as the most prominent Catalan Modernist artists and contemporary artworks donated by the Thyssen Bornemisza family, and a photography collection. It has an average of 900,000 visitors per year. From a legal point of view, it is an independent entity taking the form of a consortium formed by the *Generalitat de Catalunya*, the City Council of Barcelona, and the General State Administration. It is governed by a Board of Trustees, which is made up of representatives from the consortium members and the museum's management (the Directors and Secretary General), as well as individuals and private entities who contribute to the realisation of its statutory aims and objectives. The organisational model of the museum is composed of nine departments, managed by the museum's Director. The museum currently employs around 132 people; each year it launches a volunteer programme (about 30–40 people).

Over the years, MNAC has obtained several certifications testifying to its efforts for sustainability: specifically, the label issued by the Spanish agency AENOR,[5] the environmental certification ISO14001 (since 2011), and since 2012, the certification EMAS – The Eco-Management and Audit Scheme.[6] During the interview, it emerged that these labels necessitated a separate report, which was ultimately incorporated into the museum's Annual Report (*Memoria*). Finally, the organisation conforms to the principles of the World Charter for Sustainable Tourism +20, implementing the Responsible Tourism Policy and meeting the conditions of the BIOSPHERE accreditation, which the city of Barcelona and its institutions have received.

In 2019, MNAC issued its first Environmental Report (*Declaration Ambiental*), alongside its Social Responsibility Report. The interviewee explained that they initially tried to refer to GRI G4 guidelines and developed indicators referring to them, but soon gave up because of the distance of these standards (designed for private companies) from the unique characteristics, activities and impacts of a museum.

On average, sustainability reports are prepared over a period of 4–6 months for data collection and report preparation, involving both the human resources of the MNAC and an external agency.

Specifically, the data for the reports are initially compiled by a team from the museum with a background in business administration, project management or fundraising (one or two people, depending on the needs of the museum that year) with tasks related to the collection of information from the museum's departments and process supervision. Their reports are then officially elaborated by ECOGESA, a specialised consultancy located in Barcelona. Alongside this group, the museum established a Social Responsibility Committee which included experts from within and outside the museum, voluntarily; it has mainly a steering responsibility, but it is also in charge of supporting the identification of material issues. This Committee is managed by a Social Responsibility and Projects Coordinator. Since 2013, the reports have also involved collaboration with stakeholders.

The engagement with some stakeholder groups was organised through focus groups or questionnaires, which were also used for the materiality analysis. Structured dialogue was constant with the sponsors and Friends of the Museum association, and the organisation's internal staff; material issues were regularly updated depending on the strategy and cooperation with the museum itself. Other material difficulties were recognised informally via the day-to-day work of the various museum sections with their interest groups, particularly at the community level and with disadvantaged populations. During the interview, it emerged that some groups were particularly difficult to monitor. "It is complicated to establish a regular dialogue with some stakeholder groups, in particular visitors and tourists [...] We launch a questionnaire with them almost every three years, but at the moment this is the only monitoring implemented by our museum". Alongside these activities, material issues for these groups were identified based on the experience and perception of the museum staff who directly worked in contact with them.

The analysis of the Annual Reports (*Memòria*) shows that the museum progressively incorporated the different reports (Environmental Report and Social Responsibility Report), thus increasing the accessibility of the information related to the different dimensions of sustainability. As previously indicated, since 2018 the Annual Report included the Social Responsibility Report and since 2020 it also encompassed the Report on Environmental Sustainability (*Declaració Ambiental*).[7]

By comparing the three reports of the analysed period, some common patterns can be identified in the themes and the disclosed information. The Annual Reports of the years 2018 and 2019 are structured according to a similar model, though the 2019 report discloses more details than the 2018 one. Following a brief overview themed on general quantitative information (visitors, number of exhibitions, research initiatives, etc.), a significant portion of the reports is dedicated to collections and exhibitions, and programmes for the public, also presenting information about research and outreach activities, and educational programmes with a special focus on some targeted groups (often consisting of locals or disadvantaged groups). Volunteers (especially the Friends of Museum – *Amics del Museu*), sponsorship, and collaboration with local entities are other significant themes in both reports. From the semantic analysis, it emerged that in the 2018 and 2019 reports the main semantic clusters are those of development, outreach, research and education containing sub-themes related to exhibitions, educational and research activities (2% coverage in 2018 and 2.1% in 2019), collections preservation (1.3% in 2018 and 1.6% in 2019), and visitors and publics (1.2% in 2018 and 1.5% in 2019). Sustainability is a significant theme, although it must be noted that it is connected to the perspective of social sustainability (0.5% in 2018 and 0.6% in 2019). Other relevant semantic groups are those related to digital/online (0.4% and 0.6%) and related to volunteers (0.3% in both years) and sponsors (0.3% in both years).

The 2020 Annual Report is the longest of the three reports since it also discloses data previously presented in the Environmental Report. In particular, the section related to the museum overview is enriched with information about the museum building, environmental impact and use of materials alongside data on its governance system, internal organisational structure and sponsors. In the section Activities of the Museum (*Activitats del Museu*), there are detailed disclosures on the collections, research activities, exhibitions, programmes and public activities, with a special emphasis on families and training for museum professionals. The section on social responsibility (*Museu i la Responsabilitat Social*) has also been enlarged, with a full stakeholder and materiality analysis, as well as a portion dedicated to the communication methods established to manage their concerns and requests. The other sections mirror the structure of the previous reports, including sections dedicated to volunteers and collaboration with local institutions. The semantic analysis shows a strong focus on exhibitions, educational and research topics (2.2%), and collections (1.2%) but it has a higher coverage of the topics of public/publics (1.8%) and more consistently introduces social issues (1.4%), thus aligning the semantic coverage with the general shift of focus highlighted in the analysis of the structure of the report. Environmental topics also emerge (0.4%) along with the online/digital issue (0.4%), while the topics of volunteers and sponsors remain the same (0.3%).

4.2 Discussion: The Contribution of Museums to Sustainable Tourism

From the analysis of the empirical results, it emerges that the museum does not directly address the topic of sustainable tourism but rather sustainable development in general, focusing in particular on the two main aspects of cultural sustainability and social sustainability.

Regarding the topic of cultural sustainability, it appears to be the focus of the three analysed reports since it occupies the majority of their sections. As explained in the literature review, it refers to the capacity to preserve cultural heritage, but also to the activities to enhance cultural heritage for current and future generations.

Regarding social sustainability, it emerges that the museum interprets it as related to cultural sustainability. On the one hand, the reports include data on traditional social sustainability dimensions, such as gender balance and education and training for staff, but on the other, there is a deep focus on the use of cultural heritage for outreach, education and social engagement activities, especially with local disadvantaged groups. Therefore, cultural heritage is interpreted as a means for increasing the liveability of the local areas and as a tool for social inclusion and development.

The topic of environmental sustainability, though indicated as important during the interview with the manager and specifically addressed in the Environmental Report in the years 2018 and 2019, appears marginal in the 2020 reports, where the focus is still on topics of social and cultural sustainability. Economic sustainability is only partly addressed.

Therefore, museums are working on indirect aspects of overtourism, guaranteeing the preservation of their cultural heritage and aspects of cultural sustainability as they are linked to functions of social enhancement, sustainable social and cultural development, and integration of disadvantaged categories through arts and culture.

5 Concluding Remarks

This research aimed to understand the sustainability priorities emerging in the non-financial reports of museums, and how they provide insights into the contribution of museums to increasing the sustainable development of a cultural destination and addressing overtourism.

In the literature review, it emerged that overtourism poses significant challenges to cultural destinations, spanning from threats to cultural heritage preservation to loss of authenticity, local identity and liveability, as well as promoting an unsustainable model of local development. The role of cultural and creative organisations in reversing and mitigating the effects of unsustainable tourism and development has long been debated, interpreting culture as functional to other dimensions of sustainable development or as a fourth pillar of sustainability.

Using a case study analysis mainly based on the annual reports of a museum in the overtourism-plagued city of Barcelona, it was found that the museum prioritises mainly cultural and social sustainability issues, such as inclusion and participation, cultural enhancement and preservation, cultural identity creation, education and research. These first insights might indicate that museums interpret their role in the sustainable development of cultural destinations mainly in relation to the preservation of cultural heritage and in the enhancement, inclusion and engagement of the different communities of inhabitants and local stakeholders. By prioritising initiatives and actions (that are reflected in the related reporting disclosures) concerning these topics, museums show their potential to mitigate the negative effects of overtourism by making cultural destinations more liveable, acting on the development, inclusion, education and general well-being of local communities living alongside tourists.

These results are relevant first insights into the contribution of the museum sector to tackling overtourism, although they could be considered only initial results given the limited sample of the research.

Future research development could be to replicate the analysis on a broader sample of museums and in different overtourism locations in order to reach a deeper understanding of this topic.

Notes

1 More information is available on the official website of the Barcelona Tourist Office: www.barcelonaturisme.com/wv3/en/
2 George Doxey (1975) investigated and reviewed the various attitudes of residents regarding tourists and developed the Irritation Index, which is divided into four stages that describe the shift in mood that residents have against tourism. The index's initial phase, euphoria (i.e. confidence toward the newly established tourist trend and the economic possibilities it might entangle), is typically followed by a second phase characterised by apathy: in this phase, tourism is taken for granted, and exploitation of the sector's profits reaches a peak (Doxey, 1975). A third stage is defined as a period of irritation: the local tourism industry becomes overcrowded, and gentrification is spreading; public authorities typically take measures during this period, attempting to facilitate a harmonious coexistence between residents and tourists. The ultimate stage is antagonism, in which rallies are set up to express disapproval and occasionally resentment toward tourists, who are explicitly accused of causing or exacerbating the destination's issues.
3 More information is available on the website of the Barcelona Dirección de Turismo, Gerencia de Empresa y Turismo: https://professional.barcelonaturisme.com/es
4 More information is available on the website of the Barcelona Dirección de Turismo, Gerencia de Empresa y Turismo: https://professional.barcelonaturisme.com/es
5 IQNet SR10 certification guarantees the implementation of social responsibility management systems. MNAC was the first cultural public institution and museum to which this certification was awarded and in 2019 the certification was renewed for a further three-year period.
6 EMAS is a voluntary environmental management tool, designed by the EC – European Commission in 1993 to certify an organisation's environmental performance. It is primarily related to issues of sustainable management and environmental impact.
7 Source: museunacional.cat/es/memoria-de-actividades-e-informacion-estadistica.

References

Abril Sellarés, M., Azpelicueta Criado, C., & Sánchez Fernández, M. D. (2015). ¿Vale todo en turismo? Residentes frente a turistas. Estudio comparativo entre el barrio de La Barceloneta, Barcelona y la localidad de Magaluf, Calvià [Is everything allowed in tourism? Residents versus tourists. A comparative study of the districts of La Barceloneta, Barcelona, and the city of Magaluf, Calvià]. *Impulso al desarrollo económico a través del Turismo: VIII jornadas de investigación en turismo*, 493–510.

Adams, E. (2010). Towards sustainability indicators for museums in Australia. *Collections Council of Australia Ltd.* www.collectionscouncil.com.au/Default.aspx?tabid=802

Aageson, T.H. (2008). Cultural entrepreneurs: Producing cultural value and wealth. In H. K. Anheier & Y. Raj Isar (Eds.), *The cultural economy*. Sage.

Álvarez-Sousa, A. (2021). La percepción de los problemas del overtourism en Barcelona. [The perception of problems of overtourism in Barcelona] *RECERCA: Revista de pensament i anàlisi*, *26*(1), 59–92.

Atsiz, O., & Akova, O. (2019). Sociocultural impacts of tourism development on heritage sites. In D. Gursoy & R. Nunkoo (Eds.), *The Routledge handbook of tourism impacts*. Routledge.

Bertocchi, D., Camatti, N., Giove, S., & van der Borg, J. (2020). Venice and overtourism: Simulating sustainable development scenarios through a tourism carrying capacity model. *Sustainability*, *12*(2), 512.

Bluhm, D. J., Harman, W., Lee, T. W., & Mitchell, T. R. (2011). Qualitative research in management: A decade of progress. *Journal of Management Studies*, *48*(8), 1866–1891.

Borin, E. (2023). *Sustainability reporting in museums*. Hoepli.

Borin, E., & Donato, F. (2022). Cultural ecosystem approaches as key for new development paths: A reflection on management and governance implications. In E. Borin, M. Cerquetti, M. Crispí & J. Urbano (Eds.), *Cultural leadership in transition tourism: Developing innovative and sustainable models*. Springer.

CAE. (2017). *The value and values of culture*. https://cultureactioneurope.org/files/2018/02/CAE_The-Value-and-Values-of-Culture_Full.pdf

CAE. (2019). *Implementing culture within the Sustainable Development Goals*. https://cultureactioneurope.org/files/2019/09/Implementing-Culture-in-Sustainable-Development-Goals-SDGs.pdf

Canto-Alamilla, C. (2016). Análisis de los impactos socioculturales desde la perspectiva del residente que el turismo genera en el barrio de La Barceloneta, España [Analysis of sociocultural impacts generated by tourism out of residents' perspectives in La Barceloneta, Spain]. *Rotur: Revista de ocio y turismo*, *11*, 1–11.

CHCfE Consortium. (2015). *Cultural heritage counts for Europe*. www.encatc.org/culturalheritagecountsforeurope

Chen, H., & Rahman, I. (2018). Cultural tourism: An analysis of engagement, cultural contact, memorable tourism experience and destination loyalty. *Tourism Management Perspectives*, *26*, 153–163.

Crespi-Vallbona, M., & Mascarilla-Miró, O. (2018). La transformación y gentrificación turística del espacio urbano. El caso de la Barceloneta (Barcelona) [Transformation and tourist gentrification of urban spaces. The case of La Berceloneta (Barcelona)]. *Revista EURE – Revista de Estudios Urbano Regionales*, *44*, 133.

Dameri, R. P., & Demartini, P. (2020). Knowledge transfer and translation in cultural ecosystems. *Management Decision*, *58*(9), 1885–1907.

De Luca, G., Shirvani Dastgerdi, A., Francini, C., & Liberatore, G. (2020). Sustainable cultural heritage planning and management of overtourism in art cities: Lessons from atlas world heritage. *Sustainability*, *12*(9), 3929.

Doxey, G. V. (1975). A causation theory of visitor–resident irritants: Methodology and research inferences. In *Travel and Tourism Research Association's sixth annual conference proceedings* (pp. 195–198). The Travel Research Association.

Duxbury, N., Cullen, C., & Pascual, J. (2012). Cities, culture and sustainable development. In H. Anheier & Y.A. Isar (Eds.), *Cultural policy and governance in a new metropolitan age*. Sage.

Duxbury, N., Kangas, A., & De Beukelaer, C. (2017). Cultural policies for sustainable development: Four strategic paths. *International Journal of Cultural Policy*, *23*(2), 214–230.

Elkington, J. (1987). Triple bottom line: Sustainability's accountants. In M.V. Russo (Ed.), *Environmental management: Readings and cases* (2nd ed.). Sage.

Esposito, A., & Fisichella, C. (2019). Sustainability in museums practices evidence from Italian perspective. In *Excellence in Services, 22nd International Conference Proceedings*, Thessaloniki (Greece), 29 and 30 August 2019 (pp. 236–249). Perrotis College Thessaloniki.

Farrés, J. C. (2015). Barcelona noise monitoring network. *Proceedings of Euronoise 2015*, 31 May – 3 June 2015 (pp. 2315–2321). Barcelona Noise Monitoring Network.

Finn, M., White, E. M., & Walton, M. (2000). *Tourism and leisure research methods: Data Collection, analysis and interpretation*. Pearson Education.

Frey, B. S., & Briviba, A. (2021). Revived originals – A proposal to deal with cultural overtourism. *Tourism Economics*, *27*(6), 1221–1236.

Global Sustainability Standards Board (GSSB). (2016). *Consolidated set of GRI sustainability reporting standards*. GSSB.

Goodwin, H. (2017). The challenge of overtourism. *Responsible Tourism Partnership*, *4*, 1–19.

Goodwin, H. (2021). City destinations, overtourism and governance. *International Journal of Tourism Cities*, *7*(4), 916–921.

Goswami, K., & Lodhia, S. (2014). Sustainability disclosure patterns of South Australian local councils: A case study. *Public Money & Management*, *34*(4), 273–280.

Grazuleviciute-Vileniske, I. (2006). Cultural heritage in the context of sustainable development. *Environmental Research, Engineering and Management*, *37*(3), 74–79.

Greco, G., Sciulli, N., & D'Onza, G. (2015). The influence of stakeholder engagement on sustainability reporting: Evidence from Italian local councils. *Public Management Review*, *17*(4), 465–488.

GRI. (2021). *GRI sustainability reporting standards 2021*. Global Reporting Initiative.

Gummesson, E. (2006). Qualitative research in management: Addressing complexity, context and persona. *Management Decision*, *44*(2), 167–179.

Hawkes, J. (2001). *The fourth pillar of sustainability: Culture's essential role in public planning*. Common Ground.

Hodson, R. (1999). *Analyzing documentary accounts: Quantitative applications in the social sciences*. Sage.

Høegh-Guldberg, O., Seeler, S., & Eide, D. (2021). Sustainable visitor management to mitigate overtourism: What, who and how. In A. Sharma & A. Hassan (Eds.), *Overtourism as Destination Risk: Impacts and Solutions*. Emerald Publishing.

ICOM and OECD. (2019). *Culture and Local Development: Maximising the Impact. A Guide for Local Governments, Communities and Museums*. https://icom.museum/wp-content/uploads/2019/08/ICOM-OECD-GUIDE_EN_FINAL.pdf

Innerhofer, E., Erschbamer, G., & Pechlaner, H. (2019). Overtourism: The challenge of managing the limits. In H. Pechlaner, E. Innerhofer, & G. Erschbamer (Eds.), *Overtourism: Tourism management and solutions*. Routledge.

Isar, Y. R. (2017). Culture, sustainable development and cultural policy: A contrarian view. *International Journal of Cultural Policy*, *23*(2), 148–158.

Julie's Bicycle. (2017). *Museums' Environmental Framework*. https://juliesbicycle.com/wp-content/uploads/2022/01/Museums_Environmental_Framework_2017.pdf

Krippendorff, K. (2004). *Content analysis: An introduction to its methodology*. Sage.

Kvasnová, D., & Marciš, M. (2022). Eliminating overtourism in UNESCO destinations: A case study from Slovakia. In E. Borin, M. Cerquetti, M. Crispì & J. Urbano (Eds.), *Cultural leadership in transition tourism: Developing innovative and sustainable models*. Springer International Publishing.

Liberatore, G., Biagioni, P., Ciappei, C., & Francini, C. (2022). Dealing with uncertainty, from overtourism to overcapacity: A decision support model for art cities: The case of UNESCO WHCC of Florence. *Current Issues in Tourism*, *26*(7), 1067–1081.

Loach, K., Rowley, J., & Griffiths, J. (2017). Cultural sustainability as a strategy for the survival of museums and libraries. *International Journal of Cultural Policy*, *23*(2), 186–198.

Maingi, S. W. (2019). Sustainable tourism certification, local governance, and management in dealing with overtourism in East Africa. *Worldwide Hospitality and Tourism Themes*, *11*(5), 532–551.

Mergos, G., & Patsavos, N. (Eds.) (2007). *Cultural heritage and sustainable development: Economic benefits, social opportunities and policy challenges*. Άυλη Πολιτιστική Κληρονομιά της Ελλάδας.

Mihalic, T., & Kuščer, K. (2022). Can overtourism be managed? Destination management factors affecting residents' irritation and quality of life. *Tourism Review*, *77*(1), 16–34.

Milano, C., Novelli, M., & Cheer, J. M. (2019). Overtourism and degrowth: A social movements perspective. *Journal of Sustainable Tourism*, *27*(12), 1857–1875.

Murzyn-Kupisz, M., & Hołuj, D. (2020). Museums and coping with overtourism. *Sustainability*, *12*(5), 2054.

Museum Association. (2008). *Sustainability and museums: Your chance to make a difference*. Museum Association. www.museumsassociation.org/download?id=16398

Museum Association. (2009). *Sustainability and museums: Report on consultation*. Museum Association. www.museumsassociation.org/download?id=17944

Musikanski, L., Rogers, P., Smith, S., Koldowski, J., & Iriarte, L. (2019). Planet happiness: A proposition to address overtourism and guide responsible tourism, happiness, well-being, and sustainability in world heritage sites and beyond. *International Journal of Community Well-Being*, *2*(3–4), 359–371.

National Museum Directors' Conference. (2009). *NMDC guiding principles for reducing museums' carbon footprint*. www.nationalmuseums.org.uk/media/documents/what_we_do_documents/guiding_principle s_reducing_carbon_footprint.pdf

Neuman, W. L. (2003). *Social research methods* (5th ed.). Prentice Hall.

Nurse, K. (2006). Culture as the fourth pillar of sustainable development. *Small states: Economic review and basic statistics*, *11*, 28–40.

Park, E., Choi, B. K., & Lee, T. J. (2019). The role and dimensions of authenticity in heritage tourism. *Tourism Management*, *74*, 99–109.

Pechlaner, H., Raich, F., & Beritelli, P. (2010). Destination governance. *Tourism Review*, *65*(4), 4–85.

Pop, I., & Borza, A. (2016). Factors influencing museum sustainability and indicators for museum sustainability measurement. *Sustainability*, *101*(8), 1–22.

Pop, I., Borza, A., Buiga, A., Ighian, D., & Toader, R. (2019). Achieving cultural sustainability in museums: A step toward sustainable development. *Sustainability, 11*(4), 970.

Postma, A., & Schmuecker, D. (2017). Understanding and overcoming negative impacts of tourism in city destinations: Conceptual model and strategic framework. *Journal of Tourism Futures, 3*(2), 144–156.

Rasoolimanesh, S. M., Taheri, B., Gannon, M., Vafaei-Zadeh, A., & Hanifah, H. (2019). Does living in the vicinity of heritage tourism sites influence residents' perceptions and attitudes? *Journal of Sustainable Tourism, 27*(9), 1295–1317.

Rickly, J. M. (2019). Overtourism and authenticity. In R. Dodds & R. Butler (Eds.), *Overtourism: Issues, realities and solutions*. De Gruyter.

Roders, A.P. & Van Oers, R. (2011). Editorial: Bridging cultural heritage and sustainable development. *Journal of Cultural Heritage Management and Sustainable Development, 1*(1), 5–14.

Santos, B. T., Tolentino, J., Aquino, D. N., Malibiran, R., Lim-Ramos, C. D., Cheng, C., & Ngo, C. A. (2019). A museum information system for sustaining and analyzing national cultural expressions. *Proceedings of the 21st International Conference on Information Integration and Web-based Applications & Services* (pp. 444–447).

Scuttari, A., Isetti, G., & Habicher, D. (2019). Visitor management in world heritage sites: Does overtourism-driven traffic management affect tourist targets, behaviour and satisfaction? The case of the Dolomites UNESCO World Heritage Site, Italy. In H. Pechlaner, E. Innerhofer, & G. Erschbamer (Eds.), *Overtourism: Tourism management and solutions*. Routledge.

Seraphin, H., Ivanov, S., Dosquet, F., & Bourliataux-Lajoinie, S. (2020). Archetypes of locals in destinations victim of overtourism. *Journal of Hospitality and Tourism Management, 43*, 283–288.

Soini, K., & Dessein, J. (2016). Culture-sustainability relation: Towards a conceptual framework. *Sustainability, 8*(2), 167.

Statista. (2022). Museums in Spain – statistics & facts. www.statista.com/topics/11128/museums-in-spain/#topicOverview

Stevic, I., & Breda, Z. (2014). Tourism destination governance: The case of UNESCO world heritage site of Oporto City. *Revista Turismo & Desenvolvimento, 3*(21/22), 189–199.

Szromek, A. R., Kruczek, Z., & Walas, B. (2019). The attitude of tourist destination residents towards the effects of overtourism – Kraków case study. *Sustainability, 12*(1), 228.

Throsby, D. (1995). Culture, economics and sustainability. *Journal of Cultural Economics, 19*(3), 199–206.

UNESCO. (2022). World heritage convention: World heritage list. https://whc.unesco.org/en/list/

United Cities and Local Governments (UCLG). (2010). Culture: Fourth pillar of sustainable development. *United Cities and Local Governments (UCLG) Policy Statement*, 17. UCLG.

United Nations. (2015). *Transforming our world: The 2030 agenda for sustainable development*. https://sustainabledevelopment.un.org/post2015/transformingourworld

Vegheș, C. (2018). Cultural heritage, sustainable development and inclusive growth: Global lessons for the local communities under a marketing approach. *European Journal of Sustainable Development, 7*(4), 349.

Wall, G. (2019). Perspectives on the environment and overtourism. *Overtourism: Issues, realities and solutions*, *1*, 27–43.

Wickham, M., & Lehman, K. (2015). Communicating sustainability priorities in the museum sector. *Journal of Sustainable Tourism*, *23*(7), 1011–1028.

World Commission on Culture and Development. (1995). *Our creative diversity: Report of the World Commission on Culture and Development*. Programme and meeting document. UNESCO.

Yildrim, E., Baltà Portolés, J., Pascual, J., Perrino, M., Llobet, M., Wyber, S., & Guerra, C. (2019). *Culture in the implementation of the 2030 agenda*. Project Report Culture-2030Goal campaign. http://openarchive.icomos.org/2167/

Yin, R. K. (2018). *Case study research and applications* (6th ed.). Sage.

Zhao, J., & Li, S. M. (2018). The impact of tourism development on the environment in China. *Acta Scientifica Malaysia*, *2*(1), 1–4.

Zhuang, X., Yao, Y., & Li, J. (2019). Sociocultural impacts of tourism on residents of world cultural heritage sites in China. *Sustainability*, *11*(3), 840.

10 Addressing the Limits to Growth in Tourism

A Degrowth Perspective

Felix Windegger

1 Introduction

One of the key messages of the influential *Limits to Growth* report, commissioned and published by the Club of Rome some 50 years ago, is that endless material growth on a finite planet is impossible and will eventually lead to systemic collapse (Meadows et al., 1972). Inspired by this report and its focus on system thinking, more recently, a group of researchers have identified nine "planetary boundaries" marking the (environmentally) safe operating space for humanity (Rockström et al., 2009). Transgressing even one of these thresholds is linked to a substantial risk of non-linear, abrupt environmental shifts with far-reaching and possibly devastating consequences for the stability of the entire Earth system (Steffen et al., 2015). As of today, human activity – in particular in the Global North since the industrial revolution – has already contributed to six of these critical thresholds being exceeded (related to climate change, the loss of biosphere integrity, land-system change, altered biogeochemical cycles, novel entities and freshwater change)[1] (Persson et al., 2022; Wang-Erlandsson et al., 2022). This diagnosis is alarming and hints at the urgency to act and re-organise human development within planetary boundaries.

Degrowth is one possible answer to this challenge. Having emerged in the early 2000s in France, it offers a frame that connects diverse thoughts, ideas and positions related to a radical critique of the capitalist "growth paradigm" (Dale, 2012; see also Schmelzer, 2016), while also offering a new vision of a social-ecological transformation, based on a democratic and redistributive downscaling of the biophysical size of the global economy (Schneider et al., 2010; D'Alisa et al., 2014; Asara et al., 2015).[2] The diagnosis degrowth supporters start from is that the ideology of economic growth is not only economically unsustainable and ecologically catastrophic (Kallis et al., 2009), but also no longer improves social welfare and happiness (Asara et al., 2015; see also Jackson, 2009). In light of this, degrowth advocates argue for the abolition of economic growth as a social objective, a reduction of natural resource consumption and throughput of energy and raw materials, but also, and even more importantly, qualitative changes in the structures of the social metabolism that allow for a degrowth transformation to be socially sustainable and equitable (Schneider et al., 2010; Demaria et al., 2013).

DOI: 10.4324/9781003365815-14

In recent years, degrowth-related ideas have increasingly spilled over to tourism debates and research (see Hall, 2009; Andriotis, 2018; Fletcher et al., 2019; Higgins-Desbiolles et al., 2019). This is likely linked to an increasing awareness of the fact that conventional tourism development is Janus-faced insofar as, beyond its beneficial impacts, it is also associated with a variety of problematic issues. Among them are unsustainable levels of resource consumption and pollution, social problems and unrest linked to high tourism intensity and gentrification, a lack of economic diversification, precarious working conditions, the violation of workers' rights, as well as commodification and damage of cultural heritage (Higgins-Desbiolles & Whyte, 2015; Mowforth & Munt, 2016; Lenzen et al., 2018; Fletcher et al., 2021). As a result of these effects and the intensification of touristification processes over the past decades, debates on "overtourism" (Milano et al., 2019a), "carrying capacities" (Butler, 2020) and "tourismphobia" (Milano et al., 2019b) have surged in the public discourse and in tourism research. All of these terms and the underlying arguments are linked to a discontent with current tourism development paths and hint at a perceived point of saturation having been reached, be it in environmental or social terms.

Against this backdrop, the article at hand intends to explore the idea of limits to growth in tourism, which has so far been somewhat neglected in tourism discourses. It does so by, first, in section 2, examining the intricate relationship between tourism and growth, which is often taken for granted but rarely critically scrutinised. Then, in section 3, the sustainable tourism paradigm as the predominant solution strategy to sustainability-related challenges will be assessed vis-à-vis the idea of ecological limits. Finally, in section 4, degrowth will be introduced as an alternative to dominant development paradigms built around continuous economic expansion. Based on an extensive review of the burgeoning but still relatively sparse literature on degrowth and tourism, two existing conceptual models will be outlined in order to provide orientation on how a degrowth-inspired tourism development within planetary boundaries could look.

2 Tourism as a Capitalist Growth Engine

Prior to the COVID-19 pandemic, despite occasional shocks, global tourism had experienced a continued phase of exponential expansion. According to the World Tourism Organization (UNWTO), in less than two decades, annual tourist arrivals more than doubled from 687 million in 2000 to 1,460 million in 2019, averaging a staggering increase of 4.8% per year (UNWTO, 2020). During the same time, total tourism receipts tripled from US$481 billion to US$1,481 billion (UNWTO, 2020). This enormous growth was brought to an abrupt halt by the COVID-19 pandemic and the travelling restrictions put in place, resulting in the deepest crisis the tourism industry has ever faced. Indeed, in the years 2020 and 2021, the number of international arrivals was more than 70% below pre-pandemic levels, while tens of millions of tourism-related jobs were lost (UNWTO, 2021, 2022). Despite this major recession, already in 2022, many destinations were back on the same track as before the crisis. Globally, the World Travel and Tourism Council (WTTC) has projected that in 2023 the sector as a whole will have returned to pre-pandemic levels. More than that: over the next ten years (2022–2032), WTTC expects global tourism to increase at an annual growth rate of 5.8%, thus, once more, "outstripping global GDP" (WTTC, 2022).

The above considerations are revealing in two ways. On the one hand, they underline the spectacular growth the global tourism industry has experienced, which is projected to persist in the next decade. On the other hand, they hint at the way in which growth is typically referred to and conceived of in tourism discourses, namely as something inherently beneficial and desirable. Indeed, growth is perceived by most tourism actors on both a national and international scale as the ideal prescription against all sorts of problems and crises, largely ignoring the mounting evidence for its detrimental environmental and social consequences (Sharpley, 2020).

Based on a detailed discourse analysis, Torkington et al. (2020: 1046) show how various rhetorical devices have been used in tourism policy documents and by international tourism organisations to naturalise and reinforce a "growth is good" discourse. Their analysis unveils that even the term "sustainable" is nowadays often used to refer to the notion of continued growth, reducing "sustainable" to a mere synonym of "sustained" (Torkington et al., 2020: 1054). Similarly, Fletcher (2011) notes that almost all tourism studies published in recent years start with emphasising the spectacular growth of the industry over the last decades. This language and the underlying premises are now so commonplace that they are widely taken for granted and rarely ever questioned. This led Higgins-Desbiolles et al. (2019: 1927) to contend that global tourism, in its current form, is founded on a "pro-growth ideology", which, according to them, can be traced back to the "growth fetishism" (Hamilton, 2003) of neoliberal capitalism.

This claim builds on the growing body of literature suggesting that the quest for unrestrained economic growth is a "structural feature of capitalism in all its varieties" (Kallis, 2011: 875; see also Harvey, 2007; Magdoff & Foster, 2011). Indeed, in most modern societies, economic growth is widely considered an unconditional imperative and taken-for-granted societal goal (Jackson, 2009; Schmelzer, 2016). As degrowth scholars like Kallis (2011) have argued, many basic (e.g. financial, property, political and redistributive) institutions in capitalist societies are dependent on perpetuated growth, necessitating ever new ways to expand the economy in order to stabilise the economic and social system as a whole. Whenever growth stops – as was the case during the financial crisis of 2007/2008 and the COVID-19 pandemic – the entire edifice starts to tremble.

The fact that in the wake of both crises the stimulation of tourism growth has been relied on as a crucial instrument in broader economic recovery indicates the essential role the tourism industry plays in today's global political economy (Fletcher, 2011).[3] The vast bulk of tourism activity (including more alternative forms of travelling such as ecotourism) is closely associated with neoliberal capitalist modes of production, consumption and exchange and tied to mechanisms of commodification, privatisation, marketisation and deregulation (Fletcher, 2011). More than that, according to various scholars (Robinson, 2008; Fletcher, 2011;

Higgins-Desbiolles, 2018), in virtue of its significance as one of the largest and fastest-growing industries, tourism development is a primary means to sustain global capitalism in light of the fundamental contradictions threatening the latter's long-term survival – or, in Fletcher et al.'s (2019: 1794) words, a "capitalist fix".

From this political economic perspective, one of the reasons for the success of international tourism lies precisely in its crucial function to sustain global capitalism, which is mostly ignored in contemporary discussions on sustainability and tourism (Fletcher, 2011). This analysis reframes the understanding of the very problem of overtourism and, more generally, limits to tourism growth. Far beyond being a mere reaction to "anti-tourist" sentiments among discontent residents (see Milano et al., 2019b), critiques of overtourism, viewed through this lens, are to be conceived of as a "product of structural dynamics within the global capitalist system as a whole" (Fletcher et al., 2019: 1750; see also Mosedale, 2016).

Acknowledging the reciprocal relation between tourism development and the capitalist growth regime is essential to understand the complex interdependencies binding tourism in its current form to the idea and practice of continuous expansion, and, as a consequence, of the challenges when it comes to addressing biophysical limits in tourism. An important conclusion to be drawn regarding the overarching topic of this book, namely overtourism, is that a thorough and critical engagement of this phenomenon should not stop at discussions about carrying capacities and the appropriate number of guests visiting a specific place. Rather, it implies the necessity to rethink the political economy of tourism as a whole (Fletcher et al., 2019). This provides an important intersection with degrowth thought, whose supporters advocate for a re-politicisation of sustainability and the economy more broadly. Before going into the specifics of how an alternative, degrowth-inspired tourism development might look, in the following section the conventional response to sustainability challenges in tourism will be briefly examined.

3 How Sustainable Is Sustainable Tourism?

As a response to the growing awareness of environmental and social impacts of the travel and hospitality industry, already in the late 1980s and early 1990s the concept of sustainable tourism emerged (Du Pisani, 2006). In this context, a notable milestone was the Agenda 21 for the Travel and Tourism Industry (WTTC, UNWTO & Earth Council, 1995), a follow-up to the Rio Earth Summit of 1992, which intended to provide practical steps for governments and businesses to make the future of tourism more sustainable. Since then, the concept was further developed by numerous researchers and practitioners and is today widely accepted by most governments, corporations and civil societies alike as a strategy to promote socio-economic benefits, while simultaneously contributing to minimise the negative impacts of tourism on the environment and society (UNEP & UNWTO, 2005).

From the above definition it becomes clear that sustainable tourism intends to provide a compromise between the three typical dimensions of sustainability (i.e. ecology, society and economy). In doing so, it is very close to the broader notion of sustainable development, which had developed somewhat in parallel (Mowforth & Munt, 2016). However, for the latter, it is not always clear what sustainability means exactly in the context of sustainable tourism (Butler, 1999). In the case of sustainable development, this openness has been key to the success of the concept and its rise to become one of the predominant global development paradigms. Yet, at the same time, many critics have emphasised the lack of meaning related to the concept and the entailed ambiguity and malleability, which make it susceptible to co-option by governments and business actors to promote their own agendas, even if they conflict with environmental sustainability (Rolston, 2012; Victor, 2008). Similarly, following Sharpley (2020: 1934), global tourism entities like the UNWTO and WTTC were relatively quick to adopt the concept of sustainable development, not for the sake of environmental or social goals, but "arguably […] in order to 'greenwash' their explicit growth agendas".

Indeed, in both discussions, trade-offs and inherent conflicts between competing economic and environmental objectives have largely been ignored – as for instance in the Sustainable Development Goals – with priority being given almost exclusively to economic growth at the expense of environmental protection (Bramwell, 2006; Torkington et al., 2020). This is sometimes justified by pointing out that the problem, rather than being growth *per se*, is the way in which growth is managed (WTTC and McKinsey & Company, 2017; UNWTO, 2018). However, this reframing of the problem as a pure management issue adamantly ignores the mounting empirical evidence suggesting that an absolute decoupling of economic growth from resource use and emissions on the scale needed to address the impending environmental breakdown is highly implausible (Hickel & Kallis, 2019; Jackson & Victor, 2019; Parrique et al., 2019; Haberl et al., 2020; Wiedenhofer et al., 2020). Against this evidence, the persistent focus of policymakers and businesses in high-income countries on pursuing "green" or "sustainable" growth is to be rejected as based on the (conceptually and empirically) flawed assumption that sufficient decoupling is possible in an ever-expanding economy. More concretely, these findings suggest that the dominant solution strategies focused on technological innovation and increasing efficiency need to be urgently reoriented towards the pursuit of sufficiency, that is "the direct downscaling of economic production in many sectors and parallel reduction of consumption that together will enable the good life within the planet's ecological limits" (Parrique et al., 2019: 3).

As shown, instead of challenging hegemonic unsustainable social practices, structures and imaginaries, the sustainable development paradigm as well as the related concept of sustainable tourism have further consolidated a vision of progress based on perpetuated economic growth (Naess & Høyer, 2009; Cavagnaro & Curiel, 2012), prioritising business interests over social considerations and ignoring the importance of biophysical limits (Bramwell, 2006; Chakraborty, 2021). Consequently, an increasing number of scholars and practitioners have questioned their usefulness in the context of the pursuit of sustainable futures (Adelman, 2017). If the scientific evidence against decoupling is taken seriously and combined with the political economic analysis provided in section 2 emphasising the systemic nature of the growth imperative in tourism and beyond, more radical (in the etymological sense of the word, i.e. going to the "roots" of the problem) and determined alternatives to address the interrelated social-ecological crises are urgently needed.

4 Rethinking Tourism from a Degrowth Perspective

Having emerged in the 2000s in France, degrowth has gained increasing traction in the past two decades, evolving both as a lively field of interdisciplinary research and an international social movement with close ties to other social-ecological movements, such as postdevelopment, *Buen Vivir*, transition towns, eco-socialism and the commons movement (D'Alisa et al., 2014; Treu et al., 2020). What unites the "multiplicity" (Barca et al., 2019; see also Paulson, 2017) of actors in the "degrowth spectrum" (Eversberg & Schmelzer, 2018) is the identification of the capitalist growth imperative as the root cause of contemporary social and environmental challenges and injustice (Kallis et al., 2009; Asara et al., 2015; see also Jackson, 2009). Contrary to the sustainable development paradigm and its recent reincarnation, "green growth" (Dale et al., 2015), degrowth proponents reject the compatibility of ecological sustainability and perpetual economic growth. This rejection is based both on theoretical insights of ecological economics (Georgescu-Roegen, 1971; Daly, 1996) and the mounting empirical evidence already referenced in section 3 (see e.g. Parrique et al., 2019). Instead, they endorse the idea of a democratically deliberated reduction of absolute throughput of energy and raw materials to ensure well-being within planetary boundaries, most importantly in high-income countries of the Global North (Martinez-Alier, 2009; Kallis, 2011).

This shrinking of the economy would presuppose a "decolonization of the imaginary" (Latouche, 2009: 53), by which degrowth advocates like Kallis (2011: 877) mean an "active process of liberation" of social imaginaries from the "economism" dominant in capitalist societies and related ideologies and practices. Furthermore, to allow for such a reduction of material and energy throughput to be socially desirable, just and equitable, beyond a rupture with existing capitalist institutions, "socially sustainable degrowth"[4] (Asara et al., 2015) would require structural changes in all dimensions of society oriented towards principles such as social justice, care, sufficiency, environmental sustainability, conviviality, solidarity and collective self-determination (Barlow et al., 2022; D'Alisa et al., 2014). Systemic shifts of this kind would imply a web of transformations at various levels, from the local to the global scale, involving various actors (e.g. activists, practitioners, researchers, policy makers, trade-unionists, lay citizens) and strategies, ranging from oppositional activism to reforming existing institutions and building alternative ones (Demaria et al., 2013). Ultimately, a degrowth transformation would thus, in the eyes of its proponents, lead to an "altogether new, qualitatively different world" (Kallis & March, 2015: 362; see also Latouche, 2009; Schneider et al., 2010).

In recent years, there has been an increasing interest in relating degrowth ideas to tourism (Hall, 2009; Andriotis, 2018; Fletcher et al., 2019; Higgins-Desbiolles et al., 2019), not least due to surging debates on overtourism, as well as the perceived insufficiencies of conventional solution strategies to sustainability-related problems in tourism and beyond. The topics discussed in the tourism literature on degrowth include, among other things, the growth paradigm in (sustainable) tourism (Hall et al., 2015; Saarinen, 2018; Blázquez-Salom et al., 2019), the key role of tourism for global capitalism (Fletcher et al., 2019; Higgins-Desbiolles et al., 2019), alternative forms of travelling such as responsible, slow and community-based tourism (Andriotis, 2018; Chassagne & Everingham, 2019), decommodification practices and entrepreneurial lifestyles (Andersson Cederholm & Sjöholm, 2021), approaches to shift the focus from the rights of travellers to those of local residents (Higgins-Desbiolles et al., 2019) as well as sketches of new models of tourism development like "steady-state tourism" (Hall, 2009) or "degrowth-induced tourism development" (Andriotis, 2018). However, so far, these voices come mainly from the margins of the tourism scholarship and represent a minority within overall tourism discourses, which are still dominated by a conception of "'tourism as industry', engine of growth and development" (Higgins-Desbiolles & Everingham, 2022).

In the following, two conceptual approaches will be outlined which have been proposed in the growing but still sparse literature on degrowth and tourism: Higgins-Desbiolles et al.'s (2019) "local community-centred tourism" and Andriotis' (2018) "degrowth-induced tourism development". Both identify several elements constitutive of a degrowth-inspired tourism development from the perspective of the respective authors. Thereby, they provide an idea of how a degrowth transformation in the context of tourism might look in practice. While the two models do share some commonalities, they were selected to illustrate different approaches to degrowth in tourism, with Higgins-Desbiolles et al.'s (2019) model being strongly grounded in considerations of social justice, while Andriotis' (2018) model is primarily concerned with reducing detrimental sociocultural and environmental impacts of unbound tourism activity.

4.1 Local Community-Centred Tourism

A useful framework to include degrowth thinking in tourism is that of Higgins-Desbiolles et al. (2019), depicted in Figure 10.1. Arguing from a social justice point of view, the authors believe that in order for tourism to become truly sustainable, just and equitable, tourism as such needs to be reimagined and redefined. Most importantly, a degrowth-compatible tourism as envisaged by Higgins-Desbiolles et al. (2019) would focus primarily on the needs and interests of host communities, leading to a new understanding of tourism as "the process of local communities inviting, receiving and hosting visitors in their local community, for limited time durations, with the intention of receiving benefits from such actions" (2019: 1936). From this redefinition clearly follows that the rights of local communities would be placed above the rights of tourists to travel and the rights of tourism businesses to make profits.

As emerges from Figure 10.1, redefining tourism along these lines would imply that citizens are empowered to take a lead role in important tourism development

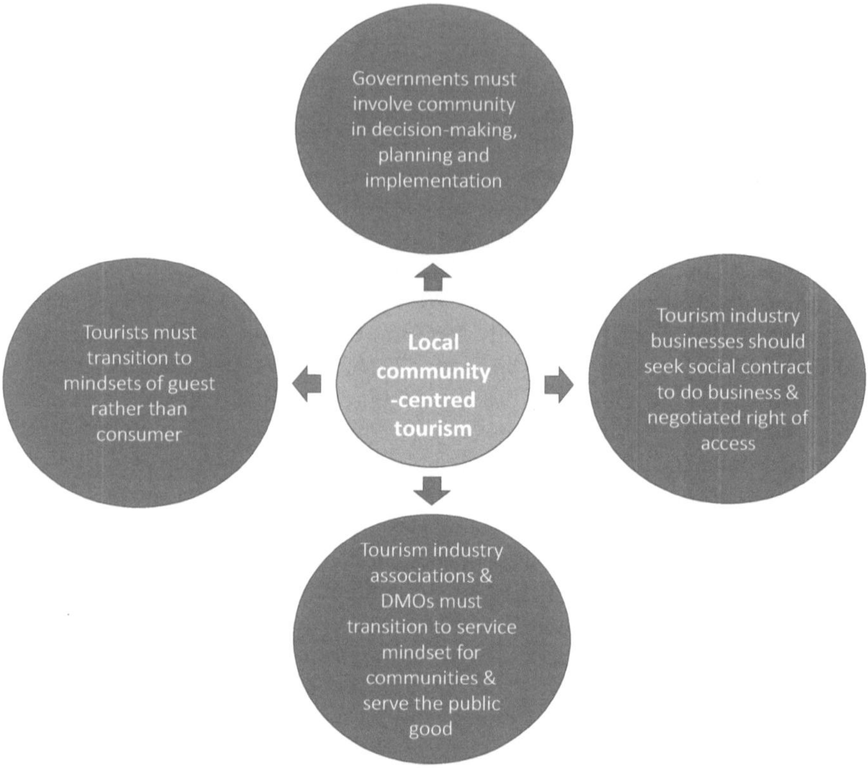

Figure 10.1 Community-centred tourism framework as mechanism for degrowing tourism
Source: Own elaboration, based on Higgins-Desbiolles et al. (2019: 1937)

planning and decision making, to make sure that in their communities only the types of tourism which they view as beneficial, and which they thus welcome, are pursued.

However, the role of other actors would also need to change substantially under such a reformed tourism. Tourists themselves, for instance, would need to change their attitudes, acknowledging that the places they visit are home to local communities, and thus strive to become responsible guests rather than privileged consumers. Such a shift in mindset (and, as a consequence, also of behaviours) could be fostered through education, codes of conduct and guidelines, through providing positive examples of meaningful experiences entailed by this type of travelling, but also through more negative tools such as penalties, closure or limited access to certain areas, and fines for infringements (Boluk et al., 2021).

Tourism governance, in the account of Higgins-Desbiolles et al. (2019), would have to be directed more towards the concerns and requirements of the local community, instead of acting primarily as facilitators of the growth and business agendas of the tourism industry. This latter role has partly been a consequence of neoliberalisation processes that have contributed to the development of government agencies from public services to mere marketing bodies. This trend would now need to be reversed, fostering a governance of tourism focused on community well-being.

Finally, the ways in which tourism businesses themselves operate would also need to be rethought. This concerns, most importantly, international and powerful business players, such as low-cost airlines, cruise companies but also firms like Airbnb[5] who have disrupted local economies and contributed significantly to the problems connected to overtourism (Boluk et al., 2021). Guaranteeing that companies not only extract benefits from but also contribute substantively to the well-being of the resident community takes on added significance from this point of view. In any case, locally owned and locally responsible tourism operators and businesses should be prioritised and fostered under such a new form of local community-centred tourism.

While Higgins-Desbiolles et al.'s (2019) vision is useful in showing the direction towards an alternative, degrowth-compatible tourism, it remains unclear how such a re-orientation of tourism might come about, especially when considering power relations and material as well as ideational path dependencies. Both aspects make a shift away from business-driven, profit-oriented tourism development very challenging.

Nonetheless, myriad examples and initiatives exist throughout the world hinting at the possibility to challenge and overturn this current paradigm. Among them are concrete examples of institutionalised attempts by governments to re-orient tourism towards the rights of residents in places like Guna Yala (Panama) (Snow, 2001) and Kangaroo Island (Australia) (Miller & Twining-Ward, 2005), new narratives and trends in tourism development such as *Buen Vivir*, localism and slow tourism (Fisher, 2018; Chassagne & Everingham, 2019) as well as policies and political programmes abandoning the measurement of prosperity based on the gross domestic product, for instance in Bhutan with its focus on gross national happiness or in New Zealand with its well-being agenda and the associated *tiaki* promise in tourism (Boluk et al., 2021). In addition, many not-for-profit social enterprises and tourism cooperatives exemplify viable alternatives to capitalist business models, perhaps best illustrated by the case of Hotel Bauen in Argentina, a recuperated business run collectively by its workers, serving both as a hotel and a free meeting place for workers groups (Higgins-Desbiolles, 2012). Finally, recent developments in cities affected by overtourism like Barcelona, Venice and Amsterdam also demonstrate viable opportunities for a more engaged local population proactively involved in shaping tourism to their own needs (Milano, Cheer & Novelli, 2019a).

4.2 Degrowth-Induced Tourism Development

Another way to conceptually sketch the basic features of a degrowth-inspired tourism development was provided by Andriotis (2018), whose monograph *Degrowth in Tourism: Conceptual, Theoretical and Philosophical Issues* provides the most elaborate engagement with the topic so far. In this book, Andriotis compares the environmental, sociocultural and economic impacts of degrowth-inspired tourism to that of conventional mass tourism. Through the analysis of concrete case studies, he demonstrates the manyfold benefits of a degrowth approach in tourism, among which are improved community welfare and a reduction of detrimental socioeconomic and environmental effects.

Figure 10.2 depicts the key actors, actions and processes which, according to Andriotis, form the backbone of a degrowth shift in tourism. Similar to

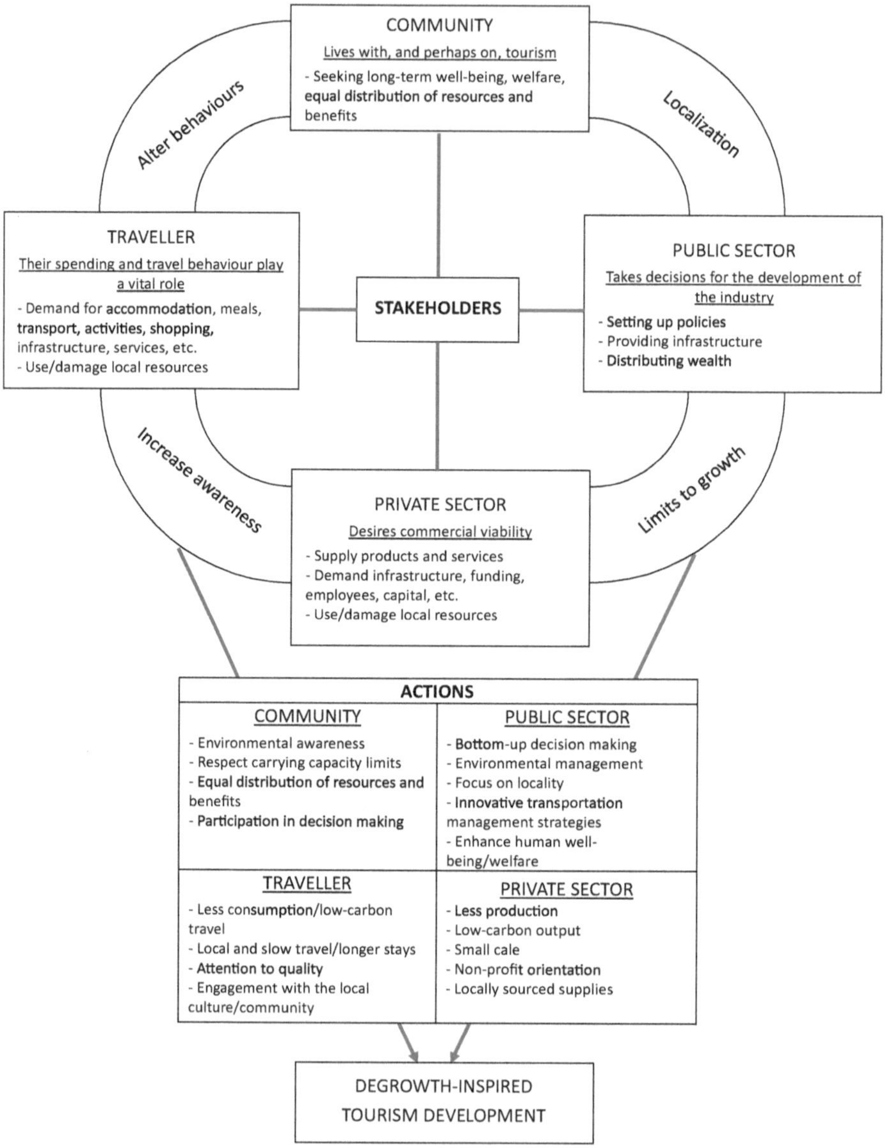

Figure 10.2 Model of degrowth-induced tourism development

Source: Own elaboration, based on Andriotis (2018: 191)

Higgins-Desbiolles et al.'s (2019) framework, four types of stakeholders are crucial for this to materialise: (a) the local community, who should be in control of local resources and involved in decisions on future development paths; (b) travellers, whose travel and spending behaviour should be oriented towards low environmental impact and a high quality of their travelling experience[6]; (c) tourism businesses, which should be predominantly small, locally owned enterprises that tend to use local materials, sell to local markets and adhere to strict environmental regulations; (d) local (as well as national) authorities, who should establish and protect the preconditions for an environmentally and socially friendly tourism development (e.g. create low-carbon mobility infrastructure, introduce governance methods based on participatory decision-making) and coordinate the activities of all stakeholders.

According to the model, all the stakeholders need to undertake actions aimed at structurally changing unsustainable modes of production and consumption as well as the ways in which people think, live, act and travel. In doing so, four fundamental principles are of paramount importance: (a) alter behaviour, which includes a shift in attitudes and lifestyles among all actors involved in favour of environmental and social concerns; (b) act and think locally, which implies an antidote to economic globalisation, favouring regional cycles and local distinctiveness; (c) increase awareness, which points to the importance of informing and educating all stakeholders regarding the negative consequences of conventional modes of production and consumption, and more environmentally and socially friendly alternatives; (d) limits to growth, acknowledging the physical limits to material growth and – if these limits have been surpassed – downscaling the industry accordingly.

Summing up, for Andriotis (2018) traditional growth-based models of tourism development (including sustainable tourism) have proven insufficient to bring about the radical changes necessary to align tourism with planetary and social boundaries. Therefore, efforts of all stakeholders need to be directed towards the "rightsizing" of tourism in destinations through the actions identified in Figure 10.2 and, eventually, reaching and maintaining a steady-state economy with only slightly fluctuating levels of consumption and production (2018: 200). The obvious difficulty in bringing about this type of radical change was already touched upon in relation to Higgins-Desbiolles et al.'s (2019) model. An additional issue worth exploring concerns the scope of degrowth-induced tourism development and its relation to broader societal structures.

According to Andriotis (2018) and many others, for many destinations, organised, high-impact mass tourism cannot be entirely avoided, despite the growing interest in alternative forms of travelling. From this follows that for serious infringements on social and ecological boundaries to be prevented, mass tourism needs to become socially equitable and environmentally sustainable. However, whether "sustainable" or "enlightened" mass tourism is truly possible or an inherent contradiction remains a subject of debate (Peeters, 2012; Weaver, 2015). What is clear, however, is that from a degrowth point of view, the quest for a future-proof tourism should not merely aim at degrowth-inspired travelling becoming an alternative practice in an otherwise growth-oriented world. More than that, it ought to be placed within the broader pursuit of a degrowth society, which implies rethinking the entire political economy of tourism and the system it is embedded in. After all, as shown in this chapter, there is good evidence that challenging the capitalist growth imperative pervading all dimensions of contemporary societies is necessary for a re-orientation of tourism development within social and ecological limits to be possible.

5 Conclusion

Against the backdrop of recent debates on overtourism, the main aims of this article were to, first, critically examine the intricate relationship between tourism development and growth and, second, introduce degrowth as an alternative framework to rethink tourism within social and ecological limits. With regards to the first aim, it has been shown that the global tourism industry is currently embedded in a wider political economy based on a structural growth imperative, while also functioning as a stabilising factor to this very system. Acknowledging this two-fold relationship helps to explain why the disentanglement of tourism and growth is so challenging. This becomes evident not least in the notion of sustainable tourism and its practical applications, which still rely firmly on the assumption that sustaining further growth of the industry is necessary for achieving sustainability. What is still largely neglected in these debates is the question of how far unrestricted growth can itself be an obstacle to sustainability. In this context, the existing body of knowledge about biophysical limits to material growth and already transgressed planetary boundaries as well as the mounting evidence against the absolute decoupling of economic growth from environmental impact provide a strong case for rejecting the current growth fixation in tourism and beyond. Thus, the articulation of more radical, alternative visions of sustainability becomes an urgent necessity. Degrowth constitutes a notable approach in this regard, which is underscored by its first mention in the recent report of the Intergovernmental Panel on Climate Change (IPCC, 2022). Building on the burgeoning literature on degrowth and tourism, the previous section of this chapter shed light on some of the core degrowth ideas and outlined two conceptual models exemplifying how a degrowth-oriented tourism development could look in practice. Even though these and other accounts linking degrowth and tourism are not yet fully developed, and some important questions remain to be addressed, they do provide a useful and promising starting point for future debates, research and action regarding limits to growth in tourism.

Notes

1 The remaining three planetary boundaries are linked to stratospheric ozone depletion, atmospheric aerosol loading and ocean acidification.

2 The term *décroissance* itself was first coined by André Gorz, an important predecessor to the degrowth movement, in 1972 in the context of a follow-up discussion to the *Limits to Growth* report, asking about the compatibility of the capitalist system with a possible degrowth of material production (see Demaria et al., 2013; Asara et al., 2015).

3 Within tourism economics, one of the most prominent ways to conceptualise the relationship between tourism and economic growth has been the tourism-led economic

growth (TLEG) hypothesis, affirming that as an export industry, tourism creates export revenue and, in doing so, contributes to economic growth (Brida et al., 2016).

4 The terms "sustainable" or "socially sustainable" degrowth are sometimes used by degrowth advocates to highlight the normative content or objective of degrowth as being environmentally and socially sustainable and just (see e.g. Asara et al., 2015).

5 The example of Airbnb is insightful insofar as it shows how innovative business ideas initially associated with the potential to favour local value creation over the interests of powerful hotel groups can themselves become drivers of socially detrimental outcomes (e.g. related to housing affordability).

6 Andriotis (2018) adds an interesting dimension to the discussion on degrowth and tourism when he identifies freedom (in its natural, material, economic, temporal, political and sociopsychological forms) as the main prerequisite for degrowth-inspired travel. For a detailed discussion of the idea of freedom in degrowth see Windegger and Spash (2022).

References

Adelman, S. (2017). The sustainable development goals: Anthropocentrism and neoliberalism. In D. French & L. Kotze (Eds.), *Sustainable Development Goals: Law, theory and implementation*. Edward Elgar.

Andersson Cederholm, E., & Sjöholm, C. (2021). The tourism business operator as a moral gatekeeper – the relational work of recreational hunting in Sweden. *Journal of Sustainable Tourism*. doi:10.1080/09669582.2021.1922425

Andriotis, K. (2018). *Degrowth in tourism: Conceptual, theoretical and philosophical issues*. CABI.

Asara, V., Otero, I., Demaria, F., & Corbera, E. (2015). Socially sustainable degrowth as a social-ecological transformation: Repoliticizing sustainability. *Sustainability Science*, *10*, 375–384. https://doi.org/10.1007/s11625-015-0321-9

Barca, S., Chertkovskaya, E., & Paulsson, A. (2019). The end of political economy as we knew it? From growth realism to nomadic utopianism. In E. Chertkovskaya, A. Paulsson & S. Barca (Eds.), *Towards a political economy of degrowth*. Rowman & Littlefield International.

Barlow, N., Regen, L., Cadiou, N., Chertkovskaya, E., Hollweg, M., Plank, C., Schulken, M., & Wolf, V. (Eds.) (2022). *Degrowth & strategy: How to bring about social-ecological transformation*. Mayfly Books.

Blázquez-Salom, M., Blanco-Romero, A., Vera-Rebollo, F., & Ivars-Baidal, J. (2019). Territorial tourism planning in Spain: From boosterism to tourism degrowth? *Journal of Sustainable Tourism*, *27*(12), 1764–1785. doi:10.1080/09669582.2019.1675073

Boluk, K. A., Krolikowski, C., Higgins-Desbiolles, F., Carnicelli, S., & Wijesinghe, G. (2021). Rethinking tourism: Degrowth and equity rights in developing community-centric tourism. In C. M. Hall, L. Landmark & J. J. Zhang (Eds.), *Degrowth and tourism: New perspectives on tourism entrepreneurship, destinations and policy*. Routledge.

Bramwell, B. (2006). Actors, power, and discourses of growth limits. *Annals of Tourism Research*, *33*(4), 957–978. doi:10.1016/j.annals.2006.04.001

Brida, J. G., Cortes-Jimenez, I., & Pulina, M. (2016). Has the tourism-led growth hypothesis been validated? A literature review. *Current Issues in Tourism*, *19*(5), 394–430. doi: 10.1080/13683500.2013.868414

Butler, R. W. (1999). Sustainable tourism: A state-of-the-art review. *Tourism Geographies*, *1*(1), 7–25. http://doi:10.1080/14616689908721291

Butler, R. W. (2020). Tourism carrying capacity research: A perspective article. *Tourism Review, 75*(1). 207–211. https://doi.org/10.1108/TR-05-2019-0194

Cavagnaro, E., & Curiel, G. (2012). *The three levels of sustainability*. Routledge.

Chakraborty, A. (2021). Can tourism contribute to environmentally sustainable development? Arguments from an ecological limits perspective. *Environment, Development and Sustainability, 23*, 8130–8146. https://doi.org/10.1007/s10668-020-00987-5

Chassagne, N., & Everingham, P. (2019). Buen Vivir: Degrowing extractivism and growing wellbeing through tourism. *Journal of Sustainable Tourism, 27*(12), 1909–1925. doi: 10.1080/09669582.2019.1660668

D'Alisa, G., Demaria, F., & Kallis, G. (2014). *Degrowth: A vocabulary for a new era*. Routledge.

Dale, G. (2012). The growth paradigm: A critique. *International Socialism, 2*, 134.

Dale, G., Mathai, M. V., & Puppim de Oliveira, J. A. (2015). *Green growth: Ideology, political economy and the alternatives*. Zed Books.

Daly, H. E. (1996). *Beyond growth: The economics of sustainable development*. Beacon Press.

Demaria, F., Schneider, F., Sekulova, F., & Martinez-Alier, J. (2013). What is degrowth? From an activist slogan to a social movement. *Environmental Values, 22*, 191–215. https://doi.org/10.3197/096327113X13581561725194

Du Pisani, J. A. (2006). Sustainable development – historical roots of the concept. *Environmental Sciences, 3*(2), 83–96. doi:10.1080/15693430600688831

Eversberg, D., & Schmelzer, M. (2018). The degrowth spectrum: Convergence and divergence within a diverse and conflictual alliance. *Environmental Values, 27*, 245–267.

Fisher, J. (2018). Nicaragua's Buen Vivir: A strategy for tourism development? *Journal of Sustainable Tourism, 1*. doi:10.1080/09669582.2018.1457035

Fletcher, R. (2011). Sustaining tourism, sustaining capitalism? The tourism industry's role in global capitalist expansion. *Tourism Geographies, 13*(3), 443–461. doi:10.1080/14616688.2011.570372

Fletcher, R., Mas, I. M., Blanco-Romero, A., & Blázquez-Salom, M. (2019). Tourism and degrowth: An emerging agenda for research and praxis. *Journal of Sustainable Tourism, 27*(12), 1745–1763. doi:10.1080/09669582.2019.1679822

Fletcher, R., Blanco-Romero, A., Blázquez-Salom, M., Cañada, E., Murray, I., & Sekulova, M. F. (2021). Pathways to post-capitalist tourism. *Tourism Geographies, 25*(2–3), 707–728. doi:10.1080/14616688.2021.1965202

Georgescu-Roegen, N. (1971). *The entropy law and the economic process*. Harvard University Press.

Haberl, H., Wiedenhofer, D., Virág, D., Kalt, G., Plank, B., Brockway, P., Fishman, T., Hausknost, D., Krausmann, F. P., Leon-Gruchalski, B., Mayer, A., Pichler, M., Schaffartzik, A., Sousa, T., Streeck, J., & Creutzig, F. (2020). A systematic review of the evidence on decoupling of GDP, resource use and GHG emissions, Part II: Synthesizing the insights. *Environmental Research Letters, 15*, 065003. doi:10.1088/1748–9326/ab842a

Hall, C. M. (2009). Degrowing tourism: Decroissance, sustainable consumption and steady-state tourism. *Anatolia, 20*(1), 46–61. doi:10.1080/13032917.2009.10518894

Hall, C. M., Gossling, S., & Scott, D. (2015). *The Routledge handbook of tourism and sustainability*. Routledge.

Hamilton, C. (2003). *Growth fetish*. Pluto Press.

Harvey, D. (2007). *The limits to capital*. Verso.

Hickel, J., & Kallis, G. (2019). Is green growth possible? *New Political Economy*. doi: 10.1080/13563467.2019.1598964

Higgins-Desbiolles, F. (2012). The Hotel Bauen's challenge to cannibalizing capitalism. *Annals of Tourism Research, 39*(2), 620–640. doi:10.1016/j.annals.2011.08.001
Higgins-Desbiolles, F., & Whyte, K. P. (2015). Tourism and human rights. In C. M. Hall, S. Gossling & D. Scott (Eds.), *The Routledge handbook of tourism and sustainability*. Routledge. doi:10.4324/9780203072332
Higgins-Desbiolles, F. (2018). Sustainable tourism: Sustaining tourism or something more? *Tourism Management Perspectives, 25*, 157–160.
Higgins-Desbiolles, F., Carnicelli, S., Krolikowski, C., Wijesinghe, G., & Boluk, K. (2019). Degrowing tourism: Rethinking tourism. *Journal of Sustainable Tourism, 27*(12), 1926–1944.
Higgins-Desbiolles, F., & Everingham, P. (2022). Degrowth in tourism: Advocacy for thriving not diminishment. *Tourism Recreation Research*. doi:10.1080/02508281.2022.2079841
Intergovernmental Panel on Climate Change (IPCC). (2022). *Climate change 2022: Impacts, adaptation and vulnerability*. https://report.ipcc.ch/ar6/wg2/IPCC_AR6_WGII_FullReport.pdf
Jackson, T. (2009). *Prosperity without growth: Economics for a finite planet*. Earthscan.
Jackson, T., & Victor, P. A. (2019). Unraveling the claims for (and against) green growth. *Science, 366*, 950–951.
Kallis, G. (2011). In defence of degrowth. *Ecological Economics, 70*, 873–880.
Kallis, G., & March, H. (2015). Imaginaries of hope: The utopianism of degrowth. *Annals of the Association of American Geographers, 105*(2), 360–368.
Kallis G., Martinez-Alier J. & Norgaard R. B. (2009). Paper assets, real debts: An ecological-economic exploration of the global economic crisis. *Critical Perspectives on International Business, 5*(1/2), 14–25. doi:10.1108/17422040910938659
Latouche, S. (2009). *Farewell to growth*. Polity Press.
Lenzen, M., Sun, Y.Y., Faturay, F., Ting Y.P., & Malik, A. (2018). The carbon footprint of global tourism. *Nature Climate Change, 8*, 522–528. https://doi.org/10.1038/s41558-018-0141-x
Magdoff, F., & Foster, J. B. (2011). *What every environmentalist needs to know about capitalism*. NYU Press.
Martinez-Alier, J. (2009). Socially sustainable economic de-growth. *Development and Change, 40*, 1099–1119.
Meadows, D. H., Meadows, D. L., Randers, J., & Behrens III, W. W. (1972). *The limits to growth*. Universe Books.
Milano, C., Cheer, J. M., & Novelli, M. (2019a). *Overtourism: Excesses, discontents and measures in travel and tourism*. CABI.
Milano, C., Cheer, J. M., & Novelli, M. (2019b). Overtourism and tourismphobia: A journey through four decades of tourism development, planning and local concerns. *Tourism Planning & Development, 16*(4), 353–357. doi:10.1080/21568316.2019.1599604.
Miller, G., & Twining-Ward, L. (2005). Tourism optimization management model. In G. Miller & L. Twining-Ward (Eds.), *Monitoring for a sustainable tourism transition*. CABI.
Mosedale, J. (Ed.). (2016). *Neoliberalism and the political economy of tourism*. Routledge.
Mowforth, M., & Munt, I. (2016). *Tourism and sustainability: Development, globalisation and new tourism in the third world* (4th ed.). Routledge.
Naess, P., & Høyer, K. G. (2009). The emperor's green clothes: Growth, decoupling, and capitalism. *Capitalism Nature Socialism, 20*(3), 74–95. doi:10.1080/10455750903215753

Parrique, T., Barth, J., Briens, F., Kerschner, C., Kraus-Polk, A., Kuokkanen, A., & Spangenberg, J. H. (2019). *Decoupling debunked: Evidence and arguments against green growth as a sole strategy for sustainability*. European Environmental Bureau.

Paulson, S. (2017). Degrowth: Culture, power and change. *Journal of Political Ecology*, *24*(1), 425–448. https://doi.org/10.2458/v24i1.20882

Peeters, P. (2012). A clear path towards sustainable mass tourism? Rejoinder to the paper "Organic, incremental and induced paths to sustainable mass tourism convergence" by David B. Weaver. *Tourism Management*, *33*(5), 1038–1041.

Persson, P., Carney Almroth, B. M., Collins, C. D., Cornell, S., de Wit, C. A., Diamond, M. L., Fantke, P., Hassellöv, M., MacLeod, M., Ryberg, M. W., Jørgensen, P. S., Villarrubia-Gómez, P., Wang, Z., & Zwicky Hauschild, M. (2022). Outside the safe operating space of the planetary boundary for novel entities. *Environmental Science & Technology*, *56*(3), 1510–1521. doi:10.1021/acs.est.1c04158

Robinson, P. (2008). *The case for community-led tourism development: Engaging & supporting entrepreneurial communities*. Leisure Studies Association Conference.

Rockström, J., Steffen, W., Noone, K., Persson, A., Stuart Chapin, F., Lambin, E. F., Lenton, T. M., Scheffer, M., Folke, C., Schellnhuber, H. J., Nykvist, B., de Wit, C. A., Hughes, T., van der Leeuw, S., Rodhe, H., Sörlin, S., Snyder, P. K., Costanza, R., Svedin, U., Falkenmark, M., Karlberg, L., Corell, R. W., Fabry, V. J., Hansen, J., Walker, B., Lieverman, D., Richardson, K., Crutzen, P., & Foley, J. A. (2009). A safe operating space for humanity. *Nature*, *461*, 472–475. doi:10.1038/461472a

Rolston, H., III (2012). *A new environmental ethics: The next millennium for life on earth*. Routledge.

Saarinen, J. (2018). *Sustainable growth in tourism? Rethinking and resetting sustainable tourism for development*. Routledge. doi:10.4324/9780429320590

Schmelzer, M. (2016). *The hegemony of growth: The OECD and the making of the economic growth paradigm*. Cambridge University Press. doi:10.1017/CBO9781316452035

Schneider, F., Kallis, G., & Martinez-Alier, J. (2010). Crisis or opportunity? Economic degrowth for social equity and ecological sustainability. Introduction to this special issue. *Journal of Cleaner Production*, *18*(6), 511–518.

Sharpley, R. (2020). Tourism, sustainable development and the theoretical divide: 20 years on. *Journal of Sustainable Tourism*, *28*(11), 1932–1946. doi:10.1080/09669582.2020.1779732

Snow, S. G. (2001). The Kuna General Congress and the Statute on Tourism. *Cultural Survival Quarterly*, 24(4). www.culturalsurvival.org/publications/cultural-survival-quarterly/kuna-general-congress-and-statute-tourism

Steffen, W., Richardson, K., Rockström, J., Cornell, S. E., Fetzer, I., Bennett, E. M., Biggs, E., Carpenter, S. R., de Vries, W., de Wit, C. A., Folke, C., Gerten, D., Heinke, J., Mace, G. M., Persson, L. M., Ramanathan, V., Reyers, B., & Sörlin, S. (2015). Planetary boundaries: Guiding human development on a changing planet. *Science*, *347*, 1259855. doi:10.1126/science.1259855

Torkington, K., Stanford, D., & Guiver, J. (2020). Discourse(s) of growth and sustainability in national tourism policy documents. *Journal of Sustainable Tourism*, *28*(7), 1041–1062. doi:10.1080/09669582.2020.1720695

Treu, N., Schmelzer, M., & Burkhard, C. (2020). *Degrowth in movement(s): Exploring pathways for transformation*. John Hunt Publishing.

UNEP & UNWTO. (2005). Making tourism more sustainable – a guide for policy makers. www.unep.org/resources/report/making-tourism-more-sustainable-guide-policy-makers

UNWTO. (2018). *"Overtourism"? Understanding and managing urban tourism growth beyond perceptions*. UNWTO.
UNWTO. (2020). *International tourism highlights, 2020 edition*. UNWTO. doi:10.18111/9789284422456
UNWTO. (2021). *UNWTO world tourism barometer* (English version), *19*(1), 1–42.
UNWTO. (2022). *UNWTO world tourism barometer* (English version), *20*(1), 1–40.
Victor, P. (2008). *Managing without growth: Slower by design, not disaster*. Edward Elgar.
Wang-Erlandsson, L., Tobian, A., van der Ent, R. J. et al. (2022). A planetary boundary for green water. *Nature Reviews Earth Environment, 3*, 380–392. doi:10.1038/s43017-022-00287-8
Weaver, D. (2015). Enlightened mass tourism as a "third generation" aspiration for the twenty-first century. In M. Hughes, D. Weaver & C. Pforr (Eds.), *The practice of sustainable tourism: Resolving the paradox*. Routledge.
Wiedenhofer, D., Virág, D., Kalt, G., Plank, B., Streeck, J., Pichler, M., Mayer, A., Krausmann, F., Brockway, P., Schaffartzik, A., Fishman, T., Hausknost, D., Leon-Gruchalski, B., Sousa, T., Creutzig, F., & Haberl, H. (2020). A systematic review of the evidence on decoupling of GDP, resource use and GHG emissions, Part I: Bibliometric and conceptual mapping. *Environmental Research Letters, 15*(6), 063002. doi:10.1088/1748–9326/ab8429
Windegger, F., & Spash, C. L. (2022). Reconceptualising freedom in the 21st century: Neoliberalism vs. degrowth. *New Political Economy*. doi:10.1080/13563467.2022.2149719
WTTC. (2022). *WTTC global summit Philippines*. Global Summit Report.
WTTC, UNWTO & Earth Council. (1995). *Agenda 21 for the travel and tourism industry: Towards environmentally sustainable development*. WTTC, UNWTO & Earth Council.
WTTC and McKinsey & Company. (2017). *Coping with success: Managing overcrowding in tourism destinations*. WTTC.

11 Acting Responsibly for Sustainable Tourism Development

The Evolution Towards Responsible Tourism

Sarah Schönherr and Mike Peters

1 Sustainable Development in Tourism Research

1.1 Sustainable Tourism Development Research Concepts

In tourism, sustainable development emerged in response to social and political debates about environmental concerns (Inskeep, 1998; WCED, 1987), following the development of sustainable development science in recent years (Munasinghe, 2007). However, tourism research in general has focused particularly on the economic and political structures of tourism systems that guide the development of the industry (Bramwell, Lane, McCabe, Mosedale & Scarles, 2008), neglecting the environmental impacts. Sustainable tourism development is a manifestation of tourism ecology by bringing the natural environment into the tourism discussion (Mihalic, 2020). Additionally, sustainable tourism development is represented by sociological debates on mass tourism (Cohen, 1984; Krippendorf, 1982) and is therefore not just an environmental issue (as depicted by Hall (2019)). It also requires socio-economic consideration, which is addressed in the framework of the three pillars of sustainability: a) economic, b) environmental and c) social sustainability (Becken, 2019; Hall, 2010; Higgins-Desbiolles, 2018; Mensah, 2019).

The debates about carrying capacity in the 1960s defined the beginning of research on sustainable tourism development (Butler, 1996; Doxey, 1975), and in the subsequent decades, tourism researchers focused on addressing the negative impacts of tourism development (Saarinen, 2006). Sustainable tourism development emerged as a paradigm that aimed to address the challenges of tourism research and practice (Hall, 2019; Liu, 2003). Tourism is often confronted with criticism of its development, as evidenced by the development of concepts around sustainable construction in recent decades (Hughes, 1995).

Debates about the concept of sustainable tourism dominated research on sustainable development in tourism at the beginning of the 1990s (Saarinen, 2006) and today tourism development is increasingly confronted with crises (such as the COVID-19 pandemic, climate change and economic crisis). Various challenges such as natural disasters, terrorist attacks or wars (Kuščer, Eichelberger & Peters, 2022; Weaver, Moyle, Casali & McLennan, 2022) introduce unpredictability,

DOI: 10.4324/9781003365815-15

which impacts tourism research and practice (de Sausmarez, 2007). Research in the field of sustainable tourism development focuses thus on the impact of crises, which are seen as an opportunity to introduce sustainability (Higgins-Desbiolles, 2021; Kuščer et al., 2022; Peters, Eichelberger & Pikkemaat, 2021).

Today, various concepts and approaches aim at sustainable tourism development, with eco-friendly (Ahmad, Kim, Anwer & Zhuang, 2020; Hwang & Moon, 2022), environmentally sustainable (Dolnicar, 2020; Zhang, Zhang, Song & Lew, 2019) and ecotourism (Lee & Jan, 2018a, 2018b), green tourism (Cheng, Chiang, Yuan & Huang, 2018; Wang, Wu, Wu & Pearce, 2018), pro-poor tourism (Knight, 2018), regenerative tourism (Bellato, Frantzeskaki & Nygaard, 2022; Cave & Dredge, 2020), low-carbon tourism (Becken, 2017; Lee & Jan, 2019), as well as sustainable (Bramwell, Higham, Lane & Miller, 2017; Hall, 2019; Hardy, Beeton & Pearson, 2002) and responsible tourism (Bramwell et al., 2017; Fang, 2020; Goodwin, 2011).

Most studies in the field of sustainable tourism development refer to the concept of sustainable tourism (Hall, 2019; Hall, Gossling & Scott, 2015; Liu, 2003), as it is seen as a subset of sustainable development and thus an integrated pathway to sustainable development (Bramwell et al., 2017). The three pillars of sustainability – social, environmental and economic – are addressed by both responsible and sustainable tourism approaches, although tourism has been criticised for prioritising economic sustainability (Becken, 2019; Hall, 2010; Higgins-Desbiolles, 2018; Mensah, 2019). However, sustainable tourism also faces criticism for considering cultural, political or spiritual dimensions, which are addressed in regenerative tourism, with the inclusion of holistic systems thinking through living systems (Cheer, 2020).

Most of the other concepts (green tourism, low carbon tourism, environmentally friendly tourism) particularly emphasise the pillar of environmental sustainability (Ahmad et al., 2020; Becken, 2017; Cheng, Li, Zhang & Cao, 2021; Dolnicar, 2020; Hwang & Moon, 2022; Lee & Jan, 2019; Wang et al., 2018; Zhang et al., 2019). While the concept of ecotourism addresses social and environmental concerns (Lee & Jan, 2018a, 2018b), pro-poor tourism emphasises the creation of benefits for the poor via addressing the economic sustainability pillar (Harrison, 2008; Knight, 2018).

Distinctions and differentiations between these concepts discussed are regarded as challenging (Hall, 2019; Liu, 2003), requiring integration and consideration of the divergent approaches rather than a focus on the margins to inform sustainable tourism development (Harrison, 2008).

1.2 Blurring of Sustainable and Responsible Tourism Discussion

The relevance of sustainable tourism has been present in research for quite some time (Bramwell et al., 2017), while the relevance of the concept of responsible tourism has emerged rather recently (Saarinen, 2021), although the concept was defined in tourism research already in the 1980s by Krippendorf (1982).

Although the concepts of sustainable and responsible tourism have evolved from ecological thinking, they both take cultural, social and natural resources into account (Dávid, 2011). Sustainable tourism aims to strengthen the environmental, economic and social pillars of sustainability (Higgins-Desbiolles, 2018; Kuščer & Mihalič, 2019; Mensah, 2019; Mihalic, 2016). Responsible tourism focuses on the protection of the natural environment and natural resources, includes respect for religions, cultures and traditions in the destination, and emphasises the creation of economic and social benefits for residents (Bramwell et al., 2008; Fang, 2020; Mathew & Sreejesh, 2017; Mihalic, 2016).

Both concepts are considered to have similar elements (particularly related to the three sustainability pillars), but they are also based on different ideological and social contexts (Saarinen, 2021). The differences between the two concepts are reflected by different perspectives in research. While in some cases the two concepts are considered equivalent (using both terms for the same focus), responsible tourism, for example, is also interpreted as a mere complement to sustainable tourism, an alternative manifestation of the sustainable tourism concept (Pope, Wessels, Douglas, Hughes & Morrison-Saunders, 2019). Other perspectives include responsible tourism as an outcome of sustainable tourism strategies (Mathew & Sreejesh, 2017), with responsible tourism serving as an implementation of sustainable tourism (Medina, 2005; Mihalic, Mohamadi, Abbasi & Dávid, 2021). In this sense, the concept of sustainable tourism is used as a theory, while the concept of responsible tourism refers to the responsible behaviours and actions of tourism stakeholders (Mihalic, 2016). In addition to this practical difference between responsible and sustainable tourism at the operational level (Saarinen, 2021), the difficulties of distinguishing the two concepts from each other constitute a further issue (Sharpley, 2013). Research has tried to combine both concepts into "responsustable" tourism (Mihalic, 2016), building on their contextual differences (Mihalic, 2016, 2020; Saarinen, 2021), with Mihalic et al. (2021), however, showing that no paradigm shift towards a combined perspective has developed to date.

Building on these theoretical developments of both concepts there is a need to distinguish between sustainable and responsible tourism. In this context, the structuration theory (Giddens, 1986) as advocated by Saarinen (2021) can be invoked. Following the assumption that responsible tourism represents the responsible behaviour and actions of tourism actors (Mathew & Sreejesh, 2017; Medina, 2005; Mihalic, 2016; Mihalic et al., 2021), the structuration theory implies that regulatory, institutional structures (as sustainable tourism) have the potential to determine individual actions and behaviours (Giddens, 1986; Saarinen, 2021). In this sense, sustainable tourism as a strategy and theory (Mihalic, 2016; Mihalic et al., 2021) manifests in institutional structures that have the potential to enable, limit, shape or constrain actions and behaviours of individuals, who in turn can shape these structures (Giddens, 1986; Saarinen, 2021).

2 An Overview of the Responsible Tourism Concept

2.1 Responsible Tourism Evolution

Responsible tourism emerged during the 1980s, attempting to consider social and, especially, environmental impacts (Krippendorf, 1982). Today, responsible tourism is defined in tourism research as taking responsibility for sustainable tourism development (Bramwell et al., 2008) through responsible tourism actions and behaviours (Mathew & Sreejesh, 2017; Mihalic, 2016; Mihalic et al., 2021). Responsible tourism stakeholder behaviour in this context focuses on protecting natural resources while creating social and economic benefits for local communities (Burrai, Buda & Stanford, 2019; Fang, 2020; Gong, Detchkhajornjaroensri & Knight, 2019; Mathew & Sreejesh, 2017).

Early research on responsible tourism focused specifically on developing an understanding of responsible decision-making by tourists as a result of emerging concerns about the negative impacts of tourism development (Budeanu, 2007; Goodwin & Francis, 2003; Miller, 2003). Starting from a high proportion of studies focusing on tourists' responsible behaviour, in recent years research has evolved to examine the contribution of residents to responsible tourism (Chan, Marzuki & Mohtar, 2021; Gong et al., 2019; Um & Yoon, 2021) as well as engaging tourism businesses in discussions about responsible tourism (Eger, Scarles & Miller, 2019; Moreno-Mendoza, Santana-Talavera & León, 2019; Musavengane, 2019). Thus, previous research on responsible tourism has considered divergent response modes, although the concept's theoretical basis is still considered limited (Mondal & Samaddar, 2021). In line with the already discussed differentiation between responsible and sustainable tourism and the resulting consideration of the concept of responsible tourism as responsible tourism actions and behaviours of tourism stakeholders (Mathew & Sreejesh, 2017; Medina, 2005; Mihalic, 2016; Mihalic et al., 2021), much of the research on responsible tourism approached the theory of planned behaviour to develop an understanding of responsible behaviour by businesses and tourists (Carasuk, Becken & Hughey, 2016; King-Chan, Capistrano & Lopez, 2021; Musavengane, 2019; Tkaczynski, Rundle-Thiele & Truong, 2020).

Although research on responsible tourism has focused on different groups of tourism actors, studies have recognised that all tourism stakeholders are necessary for achieving sustainable tourism development and therefore need to implement sustainable tourism through their responsible actions and behaviours (Blackstock, White, McCrum, Scott & Hunter, 2008; Mathew & Sreejesh, 2017; Mihalic, 2016; Mihalic et al., 2021). Burrai et al. (2019) address the inhomogeneity of tourism stakeholders in this context, despite many of them already acting as advocates for responsible tourism. In addition to these difficulties, the ongoing critique of the theoretical underpinnings of responsible tourism (Mondal & Samaddar, 2021) and the struggle in distinguishing it from sustainable tourism in terms of its limited conceptualisation (Burrai et al., 2019; Mihalic, 2016) requires an in-depth examination of the divergent research perspectives on responsible tourism.

2.2 Responsible Tourism Research Streams

As mentioned previously, research in the field of responsible tourism addresses different perspectives of tourism stakeholders, as shown in the study by Eichelberger (2022), which focuses strongly on (i) the responsibility of tourists (e.g., Hu & Sung, 2022), but also on (ii) the role of residents in responsible tourism (e.g., Mathew & Sreejesh, 2017; Gong et al., 2019), and (iii) the account of responsible tourism by tourism businesses (e.g., Eger et al., 2019; Musavengane, 2019).

Studies on tourists' responsible tourism behaviour evolved from exploring tourists' intentions to behave responsibly (e.g., Lee, Bonn, Reid & Kim, 2017) to quantitative studies that specifically examined responsible tourism behaviour (e.g., Dias, Aldana, Pereira, Da Costa & António, 2021). While these studies include all three pillars of sustainability (Eichelberger, Heigl, Peters & Pikkemaat, 2021), research has also shed light on ecotourism behaviour, focusing on tourists' environmental and social responsibilities (e.g., Lee & Jan, 2018b).

However, focusing on the research stream of "responsible tourist behaviour", it is apparent that studies on their contribution to environmental responsibility with "climate-friendly behaviour", "environmentally responsible behaviour", "eco-friendly behaviour", "environmentally friendly behaviour" and "green behaviour" make up the largest proportion of studies (Han, Lee & Hwang, 2016; Jamal & Smith, 2017; Qiao & Gao, 2017; Yu & Hwang, 2019; Zhang et al., 2019). Researchers have particularly approached the concept of "environmentally responsible behaviour" to explore the role of visitors and tourists, recently examining the impact of the crisis on tourists' adoption of responsible behaviour under the influence of the COVID-19 pandemic (e.g. He, Liu, Song & Li, 2022). While the concept of "climate-friendly behaviour" has been treated as tourists' behaviour to mitigate climate change (Qiao & Gao, 2017), "green tourism behaviour" specifically includes actions to protect the environment (Ibnou-Laaroussi, Rjoub & Wong, 2020).

The more recent research stream on "responsible tourism behaviour of residents" considers the community level of responsible tourism (e.g., Gong et al., 2019; Mathew & Sreejesh, 2017). In this research area, most studies focus on the environmental sustainability pillar, similar to research on responsible tourism behaviour of tourists (e.g., Hu, Xiong, Lv & Pu, 2021; Liu, Qu, Meng & Kou, 2021). In particular, studies on "residents' environmentally responsible behaviour" have been conducted, but they differ from studies on "tourists' environmentally responsible behaviour" as they focus on exploring the impact of community aspects (with community engagement, community attachment or community involvement) (e.g. Safshekan, Ozturen & Ghaedi, 2020; Cheng, Wu, Wang & Wu, 2019). In addition to exploring residents' perceptions of responsible tourism (Chan et al., 2021; Gong et al., 2019; Mathew & Sreejesh, 2017), rather than exploring residents' responsible actions at (i) economic, (ii) environmental, and (iii) social levels as expected, the studies highlighted residents' support for sustainable tourism actions as a necessity and potential contribution to sustainable tourism development (Chamarro, Cobo-Benita & Herrero Amo, 2021; Phuc & Nguyen, 2020).

Previous research also shed light on tourism businesses' accounts of responsible tourism by examining their perceptions of responsible tourism practices, their motivations, as well as their barriers (Carasuk et al., 2016; Eger et al., 2019; Eichelberger & Peters, 2021; Koens & Thomas, 2016; Musavengane, 2019). In addition to these mostly qualitative studies, research has been conducted on examining the corporate social responsibility of tourism businesses (e.g., Chi, Zhang & Liu, 2019; Ferraz & Gallardo-Vázquez, 2016; Luo, Lam, Chau, Shen & Wang, 2017).

3 Responsible Tourism Actions for Sustainable Tourism Development

3.1 Theoretical Implications

The discussion on the distinction between responsible and sustainable tourism, as well as a relation to the prevailing research streams in the field of responsible tourism, makes it possible to derive several implications for theory development. To exploit the synergies between the two concepts of responsible and sustainable tourism, it is important to consider their contextual differences, but also their similarities (Saarinen, 2021). Firstly, both concepts have similar objectives, which overall relate to strengthening the three pillars of sustainable tourism development (Bramwell et al., 2008; Fang, 2020; Higgins-Desbiolles, 2018; Kuščer & Mihalič, 2019) and are thus based on ecological thinking (Dávid, 2011). In addition to the recognition that research on responsible tourism focuses heavily on the environmentally responsible behaviour of tourists and residents (e.g., Lee & Jan, 2018b; Hu et al., 2021; Liu et al., 2021), it can be inferred that responsible tourism should translate in all three levels of sustainable tourism development (economic, social and environmental) and therefore needs to be considered consistently in research.

Following the distinction between sustainable and responsible tourism based on structuration theory (Giddens, 1986; Saarinen, 2021), responsible tourism is regarded as manifesting in the behaviour of individual tourism stakeholders (as in the different research streams that shed light on the responsible behaviour of tourism stakeholders). In this sense, responsible tourism can be considered as an implementation of sustainable tourism development (Mihalic, 2016), contradicting the view that responsible tourism is complementary to the sustainable tourism concept (Pope et al., 2019). Rather, this assumes the need for regulatory, institutional structures to promote individual responsible behaviour (Giddens, 1986).

In line with the assumption that all tourism stakeholders must take responsibility for sustainable tourism development (Blackstock et al., 2008; Mathew & Sreejesh, 2017; Mihalic, 2016; Mihalic et al., 2021), the research streams analysed for responsible tourism showed that studies illustrated (i) the role of tourists (e.g., Lee et al., 2017; Dias et al., 2021), (ii) the role of residents (e.g., Gong et al., 2019; Mathew & Sreejesh, 2017) and (iii) the role of tourism businesses (e.g., Carasuk et al., 2016; Eger et al., 2019) in responsible tourism. Other tourism stakeholders, such as destination management organisations, tourism ministries or governments have not been investigated in responsible tourism research, which corresponds to their potential role in developing institutional regulatory structures for sustainable tourism (Giddens, 1986; Saarinen, 2021).

3.2 Implications for Tourism Practice

Implications for tourism management can be derived from the debate on responsible tourism as the promotion of sustainable tourism development through individual responsible tourism behaviour.

In particular, the distinction between sustainable and responsible tourism based on structuration theory (Giddens, 1986; Saarinen, 2021) highlights the need for regulatory structures to improve individual tourism behaviour and thus achieve sustainable tourism development (Mathew & Sreejesh, 2017; Mihalic, 2016; Mihalic et al., 2021). Furthermore, the role of destination management and government agencies, which has not been examined in research on responsible tourism, suggests that they contribute to the development of regulatory structures as adopted in sustainable tourism (Saarinen, 2021). In this regard, governments, destination management organisations at all levels and tourism ministries are called upon to contribute to sustainable tourism development by creating regulations and institutional structures that foster individual responsible behaviour. These structures can be created through the implementation of promotional programmes for the development of sustainable tourism offers for rewarding responsible tourism behaviour, or also through the establishment of sanctions for non-sustainable behaviour, e.g. in connection with environmentally damaging behaviour.

As shown in the reviews of responsible tourism research (Chamarro et al., 2021; Phuc & Nguyen, 2020), residents' support for sustainable tourism development constitutes a responsible action. For tourism businesses, it can be inferred that their support for tourism development could also drive the establishment of offerings and thus sustainable tourism development.

Research on residents' responsible tourism behaviour has furthermore shown that community aspects are crucial for attributing responsibility. Besides the need to involve the community in tourism planning and development to a greater extent, the inclusion and consideration of all stakeholders could be crucial for sustainable tourism development. Therefore, responsible tourism management needs to understand these stakeholders' concerns regarding tourism development to develop incentives to act responsibly. Destination management organisations have a central role in the stakeholder network and need to educate and train, but also incentivise tourists' and tourism businesses' responsible tourism behaviour.

References

Ahmad, W., Kim, W. G., Anwer, Z., & Zhuang, W. (2020). Schwartz personal values, theory of planned behaviour and environmental consciousness: How tourists' visiting intentions towards eco-friendly destinations are shaped? *Journal of Business Research*, *110*, 228–236. doi:10.1016/j.jbusres.2020.01.040

Becken, S. (2017). Evidence of a low-carbon tourism paradigm? *Journal of Sustainable Tourism*, *25*(6), 832–850. doi:10.1080/09669582.2016.1251446

Becken, S. (2019). Decarbonising tourism: mission impossible? *Tourism Recreation Research*, *44*(4), 419–433. doi:10.1080/02508281.2019.1598042

Bellato, L., Frantzeskaki, N., & Nygaard, C. A. (2022). Regenerative tourism: A conceptual framework leveraging theory and practice. *Tourism Geographies*. doi:10.1080/14616688.2022.2044376

Blackstock, K. L., White, V., McCrum, G., Scott, A., & Hunter, C. (2008). Measuring responsibility: An appraisal of a Scottish national park's sustainable tourism indicators. *Journal of Sustainable Tourism*, *16*(3), 276–297. doi:10.1080/09669580802154090

Bramwell, B., Higham, J., Lane, B., & Miller, G. (2017). Twenty-five years of sustainable tourism and the Journal of Sustainable Tourism: Looking back and moving forward. *Journal of Sustainable Tourism, 25*(1). doi:10.1080/09669582.2017.1251689

Bramwell, B., Lane, B., McCabe, S., Mosedale, J., & Scarles, C. (2008). Research perspectives on responsible tourism. *Journal of Sustainable Tourism, 16*(3), 253–257. doi:10.1080/09669580802208201

Budeanu, A. (2007). Sustainable tourist behaviour – a discussion of opportunities for change. *International Journal of Consumer Studies, 31*(5), 499–508. doi:10.1111/j.1470–6431.2007.00606.x

Burrai, E., Buda, D-M., & Stanford, D. (2019). Rethinking the ideology of responsible tourism. *Journal of Sustainable Tourism, 27*(7), 992–1007. doi:10.1080/09669582.2019.1578365

Butler, R. W. (1996). The concept of carrying capacity for tourism destinations: Dead or merely buried? *Progress in Tourism and Hospitality Research, 2*(3–4), 283–293. doi:10.1002/pth.6070020309

Carasuk, R., Becken, S., & Hughey, K. F. D. (2016). Exploring values, drivers, and barriers as antecedents of implementing responsible tourism. *Journal of Hospitality & Tourism Research, 40*(1), 19–36. doi:10.1177/1096348013491607

Cave, J., & Dredge, D. (2020). Regenerative tourism needs diverse economic practices. *Tourism Geographies, 22*(3), 503–513. doi:10.1080/14616688.2020.1768434

Chamarro, A., Cobo-Benita, J., & Herrero Amo, M. D. (2021). Towards sustainable tourism development in a mature destination: Measuring multi-group invariance between residents and visitors' attitudes with high use of accommodation-sharing platforms. *Journal of Sustainable Tourism*. doi:10.1080/09669582.2020.1870988

Chan, J. K. L., Marzuki, K. M., & Mohtar, T. M. (2021). Local community participation and responsible tourism practices in ecotourism destination: A case of Lower Kinabatangan, Sabah. *Sustainability, 13*(23), 13302. doi:10.3390/su132313302

Cheer, J. M. (2020). Human flourishing, tourism transformation and COVID-19: A conceptual touchstone. *Tourism Geographies, 22*(3), 514–524. doi:10.1080/14616688.2020.1765016

Cheng, J., Chiang, A-H., Yuan, Y., & Huang, M-Y. (2018). Exploring antecedents of green tourism behaviors: A case study in suburban areas of Taipei, Taiwan. *Sustainability, 10*(6), 1928. doi:10.3390/su10061928

Cheng, T. E., Li, S., Zhang, H., & Cao, M. (2021). Examining the antecedents of environmentally responsible behaviour: Relationships among service quality, place attachment and environmentally responsible behaviour. *Sustainability, 13*(18), 10297. doi:10.3390/su131810297

Cheng, T-M., Wu, H. C., Wang, J. T-M., & Wu, M-R. (2019). Community participation as a mediating factor on residents' attitudes towards sustainable tourism development and their personal environmentally responsible behaviour. *Current Issues in Tourism, 22*(14), 1764–1782. doi:10.1080/13683500.2017.1405383

Chi, C. G., Zhang, C., & Liu, Y. (2019). Determinants of corporate social responsibility (CSR) attitudes: Perspective of travel and tourism managers at world heritage sites. *International Journal of Contemporary Hospitality Management, 31*(6), 2253–2269. doi:10.1108/IJCHM-03–2018–0217

Cohen, E. (1984). The sociology of tourism: Approaches, issues, and findings. *Annual Review of Sociology, 10*(1), 373–392. doi:10.1146/annurev.so.10.080184.002105

Dávid, L. (2011). Tourism ecology: Towards the responsible, sustainable tourism future. *Worldwide Hospitality and Tourism Themes, 3*(3), 210–216. doi:10.1108/17554211111142176 de Sausmarez, N. (2007). Crisis management, tourism and sustainability: The role of indicators. *Journal of Sustainable Tourism, 15*(6), 700–714. doi:10.2167/jost653.0

Dias, Á., Aldana, I., Pereira, L., Da Costa, R. L., & António, N. (2021). A measure of tourist responsibility. *Sustainability, 13*(6), 3351. doi:10.3390/su13063351

Dolnicar, S. (2020). Designing for more environmentally friendly tourism. *Annals of Tourism Research, 84*, 102933. doi:10.1016/j.annals.2020.102933

Doxey, G. (1975). A causation theory of visitor-resident irritants: Methodology and research inferences. *The sixth annual TTRA conference proceedings, San Diego*. Travel and Tourism Research Association.

Eger, C., Scarles, C., & Miller, G. (2019). Caring at a distance: A model of business care, trust and displaced responsibility. *Journal of Sustainable Tourism, 27*(1), 34–51. doi:10.1080/09669582.2018.1551403

Eichelberger, S. (2022). Responsible tourism: A literature review on stakeholders' responsible tourism behaviour. In Council for Australasian University Tourism and Hospitality Education (CAUTHE) (Ed.), *CAUTHE 2022 conference online: Shaping the next normal in tourism, hospitality and events* (pp. 91–102). https://search.informit.org/doi/abs/10.3316/informit.408894886697517

Eichelberger, S., Heigl, M., Peters, M., & Pikkemaat, B. (2021). Exploring the role of tourists: Responsible behavior triggered by the COVID-19 pandemic. *Sustainability, 13*(11), 5774. doi:10.3390/su13115774

Eichelberger, S., & Peters, M. (2021). Family firm management in turbulent times. In A. Zehrer, G. Glowka, K. M. Schwaiger, & V. Ranacher-Lackner (Eds.), *Advances in hospitality, tourism, and the services industry: Resiliency models and addressing future risks for family firms in the tourism industry*. IGI Global. doi:10.4018/978-1-7998-7352-5.ch005

Fang, W-T. (2020). Responsible tourism. In W-T. Fang (Ed.), *Tourism in emerging economies* (pp. 131–151). Springer Singapore. doi:10.1007/978–981-15–2463–9_6

Ferraz, F. A. D., & Gallardo-Vázquez, D. (2016). Measurement tool to assess the relationship between corporate social responsibility, training practices and business performance. *Journal of Cleaner Production, 129*, 659–672. doi:10.1016/j.jclepro.2016.03.104

Giddens, A. (1986). The constitution of society: Outline of the theory of structuration (1st paperback ed.). University of California Press.

Gong, J., Detchkhajornjaroensri, P., & Knight, D. W. (2019). Responsible tourism in Bangkok, Thailand: Resident perceptions of Chinese tourist behaviour. *International Journal of Tourism Research, 21*(2), 221–233. doi:10.1002/jtr.2256

Goodwin, H. (2011). *Taking responsibility for tourism: Responsible tourism management*. Goodfellow Publishers.

Goodwin, H., & Francis, J. (2003). Ethical and responsible tourism: Consumer trends in the UK. *Journal of Vacation Marketing, 9*(3), 271–284. doi:10.1177/135676670300900306

Hall, C. M. (2010). Changing paradigms and global change: From sustainable to steady-state tourism. *Tourism Recreation Research, 35*(2), 131–143. doi:10.1080/02508281.2010.11081629

Hall, C. M. (2019). Constructing sustainable tourism development: The 2030 agenda and the managerial ecology of sustainable tourism. *Journal of Sustainable Tourism, 27*(7), 1044–1060. doi:10.1080/09669582.2018.1560456

Hall, C. M., Gossling, S., & Scott, D. (2015). *The Routledge handbook of tourism and sustainability*. Routledge. doi:10.4324/9780203072332

Han, J., Lee, M., & Hwang, Y-S. (2016). Tourists' environmentally responsible behavior in response to climate change and tourist experiences in nature-based tourism. *Sustainability, 8*(7), 644. doi:10.3390/su8070644

Hardy, A., Beeton, R. J. S., & Pearson, L. (2002). Sustainable tourism: An overview of the concept and its position in relation to conceptualisations of tourism. *Journal of Sustainable Tourism*, *10*(6), 475–496. doi:10.1080/09669580208667183

Harrison, D. (2008). Pro-poor tourism: A critique. *Third World Quarterly*, *29*(5), 851–868. doi:10.1080/01436590802105983

He, M., Liu, B., Song, Y., & Li, Y. (2022). Spatial stigma and environmentally responsible behaviors during the pandemic: The moderating role of self-verification. *Tourism Management Perspectives*, *42*, 100959. doi:10.1016/j.tmp. 2022.100959

Higgins-Desbiolles, F. (2018). Sustainable tourism: Sustaining tourism or something more? *Tourism Management Perspectives*, *25*, 157–160. doi:10.1016/j.tmp. 2017. 11.017

Higgins-Desbiolles, F. (2021). The "war over tourism": Challenges to sustainable tourism in the tourism academy after COVID-19. *Journal of Sustainable Tourism*, *29*(4), 551–569. doi:10.1080/09669582.2020.1803334

Hu, H-H., & Sung, Y-K. (2022). Critical influences on responsible tourism behavior and the mediating role of ambivalent emotions. *Sustainability*, *14*(2), 886. doi:10.3390/su14020886

Hu, J., Xiong, L., Lv, X., & Pu, B. (2021). Sustainable rural tourism: Linking residents' environmentally responsible behaviour to tourists' green consumption. *Asia Pacific Journal of Tourism Research*, *26*(8), 879–893. doi:10.1080/10941665.2021.1925316

Hughes, G. (1995). The cultural construction of sustainable tourism. *Tourism Management*, *16*(1), 49–59. doi:10.1016/0261–5177(94)00007-W

Hwang, Y. H., & Moon, H. (2022). When social class and social norms shape word of mouth about eco-friendly tourism businesses. *Journal of Vacation Marketing*. doi:10.1177/13567667221078247

Ibnou-Laaroussi, S., Rjoub, H., & Wong, W-K. (2020). Sustainability of green tourism among international tourists and its influence on the achievement of green environment: Evidence from North Cyprus. *Sustainability*, *12*(14), 5698. doi:10.3390/su12145698

Inskeep, E. (1998). *Tourism planning: An integrated and sustainable development approach*. Wiley.

Jamal, T., & Smith, B. (2017). Tourism pedagogy and visitor responsibilities in destinations of local-global significance: Climate change and social-political action. *Sustainability*, *9*(6), 1082. doi:10.3390/su9061082

King-Chan, M. S. E., Capistrano, R. C. G., & Lopez, E. L. F. (2021). Tourists really do behave responsibly toward the environment in Camiguin Province, Philippines. *Tourism Geographies*, *23*(3), 573–598. doi:10.1080/14616688.2020.1833970

Knight, D. W. (2018). An institutional analysis of local strategies for enhancing pro-poor tourism outcomes in Cuzco, Peru. *Journal of Sustainable Tourism*, *26*(4), 631–648. doi:10.1080/09669582.2017.1377720

Koens, K., & Thomas, R. (2016). "You know that's a rip-off": Policies and practices surrounding micro-enterprises and poverty alleviation in South African township tourism. *Journal of Sustainable Tourism*, *24*(12), 1641–1654. doi:10.1080/09669582.2016.1145230

Krippendorf, J. (1982). Towards new tourism policies. *Tourism Management*, *3*(3), 135–148. doi:10.1016/0261–5177(82)90063-2

Kuščer, K., Eichelberger, S., & Peters, M. (2022). Tourism organizations' responses to the COVID-19 pandemic: An investigation of the lockdown period. *Current Issues in Tourism*, *25*(2), 247–260. doi:10.1080/13683500.2021.1928010

Kuščer, K., & Mihalič, T. (2019). Residents' attitudes towards overtourism from the perspective of tourism impacts and cooperation: The case of Ljubljana. *Sustainability, 11*(6), 1823. doi:10.3390/su11061823

Lee, H. Y., Bonn, M. A., Reid, E. L., & Kim, W. G. (2017). Differences in tourist ethical judgment and responsible tourism intention: An ethical scenario approach. *Tourism Management, 60*, 298–307. doi:10.1016/j.tourman.2016.12.003

Lee, T. H., & Jan, F-H. (2018a). Development and validation of the ecotourism behavior scale. *International Journal of Tourism Research, 20*(2), 191–203. doi:10.1002/jtr.2172

Lee, T. H., & Jan, F-H. (2018b). Ecotourism behavior of nature-based tourists: An integrative framework. *Journal of Travel Research, 57*(6), 792–810. doi:10.1177/0047287517717350

Lee, T. H., & Jan, F-H. (2019). The low-carbon tourism experience: A multidimensional scale development. *Journal of Hospitality & Tourism Research, 43*(6), 890–918. doi:10.1177/1096348019849675

Liu, Y., Qu, Z., Meng, Z., & Kou, Y. (2021). Environmentally responsible behavior of residents in tourist destinations: The mediating role of psychological ownership. *Journal of Sustainable Tourism*. doi:10.1080/09669582.2021.1891238

Liu, Z. (2003). Sustainable tourism development: A critique. *Journal of Sustainable Tourism, 11*(6), 459–475. doi:10.1080/09669580308667216

Luo, J. M., Lam, C. F., Chau, K. Y., Shen, H. W., & Wang, X. (2017). Measuring corporate social responsibility in gambling industry: Multi-items stakeholder based scales. *Sustainability, 9*(11), 2012. doi:10.3390/su9112012

Mathew, P. V., & Sreejesh, S. (2017). Impact of responsible tourism on destination sustainability and quality of life of community in tourism destinations. *Journal of Hospitality and Tourism Management, 31*, 83–89. doi:10.1016/j.jhtm.2016.10.001

Medina, L. K. (2005). Ecotourism and certification: Confronting the principles and pragmatics of socially responsible tourism. *Journal of Sustainable Tourism, 13*(3), 281–295. doi:10.1080/01434630508668557

Mensah, J. (2019). Sustainable development: Meaning, history, principles, pillars, and implications for human action: Literature review. *Cogent Social Sciences, 5*(1), 1653531. doi:10.1080/23311886.2019.1653531

Mihalic, T. (2016). Sustainable-responsible tourism discourse – Towards "responsustable" tourism. *Journal of Cleaner Production, 111*(2), 461–470. doi:10.1016/j.jclepro.2014.12.062

Mihalic, T. (2020). Conceptualising overtourism: A sustainability approach. *Annals of Tourism Research, 84*, 103025. doi:10.1016/j.annals.2020.103025

Mihalic, T., Mohamadi, S., Abbasi, A., & Dávid, L. D. (2021). Mapping a sustainable and responsible tourism paradigm: A bibliometric and citation network analysis. *Sustainability, 13*(2), 853. doi:10.3390/su13020853

Miller, G. A. (2003). Consumerism in sustainable tourism: A survey of UK consumers. *Journal of Sustainable Tourism, 11*(1), 17–39. doi:10.1080/09669580308667191

Mondal, S., & Samaddar, K. (2021). Responsible tourism towards sustainable development: Literature review and research agenda. *Asia Pacific Business Review, 27*(2), 229–266. doi:10.1080/13602381.2021.1857963

Moreno-Mendoza, H., Santana-Talavera, A., & León, C. J. (2019). Stakeholders of cultural heritage as responsible institutional tourism product management agents. *Sustainability, 11*(19), 5192. doi:10.3390/su11195192

Munasinghe, M. (2007). *Making development more sustainable: Sustainomics framework and practical applications*. Munsinghe Institute for Development.

Musavengane, R. (2019). Small hotels and responsible tourism practice: Hoteliers' perspectives. *Journal of Cleaner Production, 220*, 786–799. doi:10.1016/j.jclepro.2019.02.143

Peters, M., Eichelberger, S., & Pikkemaat, B. (2021). Adaptation strategies of destinations in response to the COVID-19 pandemic: Case studies demonstrating the need to change. In P. Callot (Ed.), *Collective TRC publication*: Volume 1. *Tourism post Covid-19: Coping, negotiation, leading change: Original texts in a specific context*. Tourist Research Center Association.

Phuc, H. N., & Nguyen, H. M. (2020). The importance of collaboration and emotional solidarity in residents' support for sustainable urban tourism: Case study Ho Chi Minh City. *Journal of Sustainable Tourism*. doi:10.1080/09669582.2020.1831520

Pope, J., Wessels, J-A., Douglas, A., Hughes, M., & Morrison-Saunders, A. (2019). The potential contribution of environmental impact assessment (EIA) to responsible tourism: The case of the Kruger National Park. *Tourism Management Perspectives, 32*(2), 100557. doi:10.1016/j.tmp. 2019.100557

Qiao, G., & Gao, J. (2017). Chinese tourists' perceptions of climate change and mitigation behavior: An application of norm activation theory. *Sustainability, 9*(8), 1322. doi:10.3390/su9081322

Saarinen, J. (2006). Traditions of sustainability in tourism studies. *Annals of Tourism Research, 33*(4), 1121–1140. doi:10.1016/j.annals.2006.06.007

Saarinen, J. (2021). Is being responsible sustainable in tourism? Connections and critical differences. *Sustainability, 13*(12), 6599. doi:10.3390/su13126599

Safshekan, S., Ozturen, A., & Ghaedi, A. (2020). Residents' environmentally responsible behavior: An insight into sustainable destination development. *Asia Pacific Journal of Tourism Research, 25*(4), 409–423. doi:10.1080/10941665.2020.1737159

Sharpley, R. (2013). Responsible tourism: Whose responsibility? In A. Holden (Ed.), *The Routledge handbook of tourism and the environment*. Routledge.

Tkaczynski, A., Rundle-Thiele, S., & Truong, V. D. (2020). Influencing tourists' pro-environmental behaviours: A social marketing application. *Tourism Management Perspectives, 36*, 100740. doi:10.1016/j.tmp. 2020.100740

Um, J., & Yoon, S. (2021). Evaluating the relationship between perceived value regarding tourism gentrification experience, attitude, and responsible tourism intention. *Journal of Tourism and Cultural Change, 19*(3), 345–361. doi:10.1080/14766825.2019.1707217

Wang, W., Wu, J., Wu, M-Y., & Pearce, P. L. (2018). Shaping tourists' green behavior: The hosts' efforts at rural Chinese B&Bs. *Journal of Destination Marketing & Management, 9*, 194–203. doi:10.1016/j.jdmm.2018.01.006

WCED. (1987). *Our common future*. Oxford University Press.

Weaver, D., Moyle, B. D., Casali, L., & McLennan, C. (2022). Pragmatic engagement with the wicked tourism problem of climate change through "soft" transformative governance. *Tourism Management, 93*, 104573. doi:10.1016/j.tourman.2022.104573

Yu, C., & Hwang, Y. (2019). Do the social responsibility efforts of the destination affect the loyalty of tourists? *Sustainability, 11*(7), 1998. doi:10.3390/su11071998

Zhang, H., Zhang, Y., Song, Z., & Lew, A. A. (2019). Assessment bias of environmental quality (AEQ), consideration of future consequences (CFC), and environmentally responsible behavior (ERB) in tourism. *Journal of Sustainable Tourism, 27*(5), 609–628. doi: 10.1080/09669582.2019.1597102

Excursus

Decarbonising the Tourism Sector First: The Case of the Queenstown Tourism Plan in New Zealand

Karin Malacarne

Overtourism has been affecting many destinations all over the world, creating environmental damage and waste management problems, increasing the carbon footprint, economic dependency and social-cultural pressures, and raising the cost of living (García-Buades et al., 2022; Insch, 2020; Seraphin et al., 2018). Those are just a few of the consequences of overtourism in many places. But what can be done do reverse these effects?

Queenstown, a well-known tourist destination in New Zealand (Aotearoa), has organised a tourism destination plan focused on the visitor economy reaching a zero emissions goal by 2030. Before COVID-19, Queenstown was the most visited city of the South Island in New Zealand (Statista Research Department, 2023). It is possible to draw the following information from the *Discussion Paper* (Destination Queenstown et al., 2023b) created by the collaboration of public and private stakeholders of Queenstown Lakes: visitor expenditure has grown from 1.3 billion (2009) to 3.1 billion (2019) NZ$, reflecting an estimated three million visitors annually; the population itself grew 5.3% annually on average; and the tourism industry accounts for over 41% of the GDP (gross domestic product). The closure of borders due to COVID-19 had a serious effect on the industry as tourists were mainly international.

After detecting a variety of problems caused by the tourism industry, the destination wanted to focus on three main aspects to start the regeneration process (Destination Queenstown et al., 2023a). First, the community (Hapori), which has felt the pressure on public infrastructure and has faced disrespectful behaviour from tourists. Second, the environment (Taiao), which has seen biodiversity loss, greenhouse gas emissions, waste management issues and a decrease in the air and water quality. Finally, the economy (Taiōhaka), understanding the hidden costs for residents such as maintaining and upgrading infrastructure, the problems of a tourist workforce and the related housing challenge, and the lack of availability of data in supporting destination management. Within the many projects presented in the report under the decarbonisation initiatives, the WAO (WAO Climate Action Initiative) shares knowledge and tools with local businesses and schools empowering them to calculate their greenhouse gases in order to be aware and find the solutions to reduce them. An increase of Climate Action Initiatives was seen in 2020, with 38 businesses and nine schools signing up (WAO, 2023).

DOI: 10.4324/9781003365815-16

In conclusion, the Queenstown Lakes Regenerative Tourism Plan 2030 stands on three pillars:

1. Enrich and empower communities and enhance visitor experiences
2. Restore the environment and decarbonise the visitor economy
3. Build economic resilience, capability and productivity

As emphasised in the report, a destination that takes many stakeholders into account is on track to become renowned as a leader in regenerative tourism, green transportation innovation, and environmental solutions. The aim is not only to explore and discover innovative ways to leverage technology but also to guide economic diversification. In this way, collaboration and the networking of varieties of actors will move towards decarbonisation, an improvement in visitor behaviour and a reduction of the economic leakage. Finally, tourism's yield and overall value will take precedence above increased visitor numbers.

References

Destination Queenstown, Queenstown Lakes District Council & Lake Wānaka Tourism (2023a). *Travel to a thriving future: A regenerative tourism plan. Haereka whakamu ki to ao taurikura.* https://assets.simpleviewinc.com/simpleview/image/upload/v1/clients/queenstownnz/Queenstown_Lakes_Regenerative_Tourism_Plan_33b42536-edd1-4086-acc5-708207e134f8.pdf

Destination Queenstown, Queenstown Lakes District Council & Lake Wānaka Tourism (2023b). *Discussion paper: initial findings in the development of a roadmap to carbon zero by 2030.* https://assets.simpleviewinc.com/simpleview/image/upload/v1/clients/queenstownnz/Queenstown_Lakes_Carbon_Zero_Discussion_Paper_3369ba2f-9d38-432a-9255-83de6c18162f.pdf

García-Buades, M. E., García-Sastre, M. A., & Alemany-Hormaeche, M. (2022). Effects of overtourism, local government, and tourist behavior on residents' perceptions in Alcúdia (Majorca, Spain). *Journal of Outdoor Recreation and Tourism, 39*, 100499. https://doi.org/10.1016/j.jort.2022.100499

Insch, A. (2020). The challenges of over-tourism facing New Zealand: Risks and responses. *Journal of Destination Marketing & Management, 15*, 100378. https://doi.org/10.1016/j.jdmm.2019.100378

Seraphin, H., Sheeran, P., & Pilato, M. (2018). Over-tourism and the fall of Venice as a destination. *Journal of Destination Marketing & Management, 9*, 374–376. https://doi.org/10.1016/j.jdmm.2018.01.011

Statista Research Department. (2023). *New Zealand: International visitors by region.* Statista. www.statista.com/statistics/687393/new-zealand-international-visitors-by-region/

WAO. (2023). *WAO annual report 2022: Wao rīpoata ā-tau 2022.* https://static1.squarespace.com/static/62e75a6654804f2488174b3d/t/6457df25afc0820da3663fd2/1683480422620/2022+Wao+Annual+Impact+Report.pdf

Excursus

Regenerative Tourism: A Shift Towards a New Paradigm

Rosa Codina

In recent years, a paradigm shift towards "regenerative tourism" has emerged in tourism practice and research (Cave et al., 2022; Dredge, 2022). This shift occurred partly as a reaction to the perceived failures of sustainable tourism, which has long been criticised for not being able to successfully address the social and environmental issues tourism creates for host communities and natural ecosystems (Cave and Dredge, 2020). At its most basic, the notion of regenerative tourism aims to align the tourism and travel sector towards delivering a net positive benefit to nature, people and places, whilst supporting the longstanding prosperity of social and ecological systems (Dredge, 2022; Hui et al., 2023).

Understanding Regenerative Tourism

Regenerative tourism is underpinned by a living systems perspective, the aim of which is to develop the conditions in which all forms of life can renew and restore themselves (Dredge, 2022; Reed, 2007). In this worldview, humans and nature are connected and intertwined and do not operate as separate entities (Nelson and Shilling, 2021).

Regenerative tourism goes beyond the concept of sustainable tourism in that it seeks to reinvest in the proactive regeneration of the natural environment, communities, cultures, heritage and places (Bellato et al., 2023). In this context, regenerative tourism challenges the dominant scientific paradigms by placing emphasis on Indigenous worldviews which recognise the inherent connection between people and place (Pollock, 2019).

Principles of Regenerative Tourism

Regenerative tourism is underpinned by several core principles (Cave and Dredge, 2020; Hui et al., 2023):

- Net gain for all stakeholders involved in tourism. This includes the environment, host communities, travellers, and future generations.
- Promotes partnerships and collaboration between a range of stakeholders to establish activities that strengthen the social and ecological well-being of places and ensures the long-term feasibility of destinations.

DOI: 10.4324/9781003365815-17

- Recognizes that all living forms are interdependent and interconnected (the complex adaptive systems approach).
- Fosters Indigenous knowledge and holistic ways of being.
- Advances the protection of local cultures and heritage through community-centred, place-based and environment-focused bottom-up approaches.

Examples of Regenerative Tourism in Practice

Tourism practices guided by the principles of regenerative tourism are currently being employed around numerous destinations across the world, from Thailand (Nitsch and Vogels, 2022) to Denmark (Cave and Dredge, 2020). One example is the case of the Blue Penguins Pukekura initiative in New Zealand which offers wildlife tours and is collectively owned and operated by a Maori community (Cave and Dredge, 2020; Blue Penguins, 2023). This regenerative enterprise merges global and Maori Indigenous values to generate fiscal wealth, alongside capacity building, socio-cultural collaboration and enhanced wellbeing for the community and the natural environment. This successful initiative has enabled economic resilience against external shocks for the Maori community, whilst also preserving cultural heritage and customs (Cave and Dredge, 2020; Amoamo et al., 2018).

Challenges and Prospects

Whilst the concept of regenerative tourism holds great promise, it is not without challenges. Traditional tourism management approaches are currently inconsistent with the transition to a regenerative mindset (Cave and Dredge, 2020). For regenerative tourism approaches to succeed, a shift in socio-ecological consciousness must take place. This necessarily involves overcoming the challenges associated with traditional management approaches – such as separation, individualism, reductionism and marketisation – and a shift towards collective thinking and action. Moreover, tourism must be managed as a complex adaptive system approach whereby working with uncertainty and emergence is the norm (Cave and Dredge, 2020).

References

Amoamo, M., Ruckstuhl, K., & Ruwhiu, D. (2018). Balancing indigenous values through diverse economies: A case study of Maori Ecotourism. *Tourism Planning & Development*, *15*(5), 478–495.

Bellato, L., Frantzeskaki, N., & Nygaard, C. A. (2023). Regenerative tourism: A conceptual framework leveraging theory and practice. *Tourism Geographies*, *25*(4), 1026–1046.

Blue Penguins. (2023). *Korero o Takiharuru: Our story*. www.bluepenguins.co.nz/our-story-story-of-takiharuru

Cave, J., & Dredge, D. (2020). Regenerative tourism needs diverse economic practices. *Tourism Geographies*, *22*(3), 503–513.

Cave, J., Dredge, D., Van't Hullenaar, C., Koens Waddilove, A., Lebski, S., Mathieu, O., Mills, M., Parajuli, P., Pecot, M., Peeters, N., & Ricaurte-Quijano, C. (2022). Regenerative tourism: The challenge of transformational leadership. *Journal of Tourism Futures*, *8*(3), 298–311.

Dredge, D. (2022). Regenerative tourism: Transforming mindsets, systems and practices. *Journal of Tourism Futures*, *8*(3), 269–281.

Hui, X., Raza, S. H., Khan, S. W., Zaman, U., & Ogadimma, E. C. (2023). Exploring regenerative tourism using media richness theory: Emerging role of immersive journalism, metaverse-based promotion, eco-literacy, and pro-environmental behavior. *Sustainability*, *15*(6), 5046.

Nelson, M., & Shilling, D. (2021). *Traditional ecological knowledge: Learning from indigenous practices for environmental sustainability*. Cambridge University Press.

Nitsch, B., & Vogels, C. (2022). Gender equality boost for regenerative tourism: The case of Karenni village Huay Pu Keng (Mae Hong Son, Thailand). *Journal of Tourism Futures*, *8*(3), 375–379.

Pollock, A. (2019). *Regenerative tourism: The natural maturation of sustainability*. https://medium.com/activate-the-future/regenerative-tourism-the-natural-maturation-of-sustainability-26e6507d0fcb.

Reed, B. (2007). Shifting from "sustainability" to regeneration. *Building Research and Information*, *35*(6), 674–680.

Part III

Strategies to Deal with Overtourism

12 Last Chance to Behave Badly at Uluṟu, Australia

A Critical Reflection on the Interplay of Overtourism and Last Chance Tourism

Christof Pforr

1 Introduction

Uluṟu-Kata Tjuṯa National Park is a protected area, 1,325 square kilometres (km) in size, in the south-western part of the Northern Territory of Australia. Located 450 km west of the township of Alice Springs, it is commonly referred to as the "heart" (James, 2007: 398) or the symbolic and geographic "centre of the Australian nation" (Everingham et al., 2021: 5) (Figure 12.1). Others refer to it as the "spiritual heart" of Australia's Red Centre, emphasising the great cultural significance Uluṟu holds for the Aṉangu traditional people of the land (Department of Climate Change, Energy, the Environment and Water, 2023a).

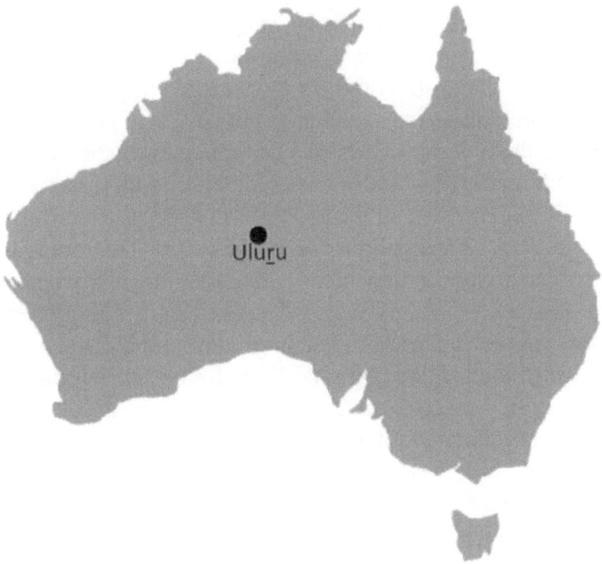

Figure 12.1 Map of Uluṟu/Australia

Source: Author

DOI: 10.4324/9781003365815-19

Figure 12.2 Uluṟu, Australia
Source: Author

The Uluṟu-Kata Tjuṯa National Park is one of Australia's most iconic national parks and home to two famous natural landmarks, Uluṟu (formerly known as Ayers Rock) and Kata Tjuṯa (also known as the Olgas), which are sacred sites for the local Indigenous Aṉangu people Figure 12.2). The National Park is co-managed by Parks Australia in collaboration with its traditional owners and included in the UNESCO World Heritage List as an outstanding cultural site and natural habitat.

However, beyond this rather superficial portrayal as an archetypical Australian landmark and cultural icon, a contested locality emerges that shows deep lines of conflict in its more recent history. Everingham et al. (2021: 6) point out that it is "a site of messy, complex and competing ideologies, practices and performances". In order to better understand these complexities and controversies, including the question of whether or not to climb the *Rock*, which is at the core of this chapter, more cultural and context-specific explanations of the location are needed. Referring to Uluṟu, McKercher et al. (2008: 369/373) highlight, for instance, that "contested places represent social and cultural landscapes in transition [...*and*] are inherently unsustainable".

The debates presented here are primarily based on a comprehensive analysis of secondary data sources, infused by an auto-ethnographic account (Ellis, 2004) of travelling to Uluṟu-Kata Tjuṯa National Park, first as an international tourist in 1992 followed by a visit in 2013 as an Australian citizen and thus as a domestic tourist. However, given the breadth of the subject matter and owing to space limitations, the chapter remains necessarily selective.

To better understand the contested nature of the Uluṟu-Kata Tjuṯa National Park and specifically also Uluṟu itself, important context and background is provided, framed by reference to "overtourism" and "last chance tourism". In the following, attention is directed towards cleavages relating to the recent history of the site since European settlement, its cultural, political and socio-economic implications, as well as environmental realities associated with its remote geography and unique geology.

2 Indigenous Affairs in Australia – a Contested Terrain

Indigenous people have inhabited Australia for at least 60,000 years, but their lives radically changed with the arrival of British settlers in 1788. From the beginning of English colonisation, the relationship between Indigenous and non-Indigenous people was characterised by conflict, struggle and disadvantage; as Storey (2012) argues, a consequence of dispossession of land. To this day, Aboriginal people, who account for about 4% of the Australian population, still represent the country's most disadvantaged cultural group (Australian Bureau of Statistics, 2023; Brueckner et al., 2014).

This history of dispossession and marginalisation and the failure to address the root causes of Indigenous disadvantage led to the "Uluṟu Statement from the Heart" in 2017, a constitutional convention of over 250 Aboriginal and Torres Strait Islander leaders discussing and mapping out an approach to constitutional reform in order to recognise Aboriginal and Torres Strait Islanders as Australia's First Peoples. The Uluṟu Statement from the Heart suggested incorporating a "First Nations Voice", also referred to as an "Indigenous Voice to Parliament", in the Australian Constitution, and proposed the establishment of a "Makarrata Commission" to supervise a process of "agreement-making" (treaty) and "truth-telling" between governments and Aboriginal and Torres Strait Islander peoples (The Uluṟu Dialogue, 2023).

The Uluṟu Statement from the Heart was endorsed by the Commonwealth Government in 2022, followed by a process which will lead to a referendum in the second half of 2023, putting forward, at the time of writing (April 2023), a simple proposition to the Australian people:

> A Proposed Law: to alter the Constitution to recognise the First Peoples of Australia by establishing an Aboriginal and Torres Strait Islander Voice. Do you approve this proposed alteration?

This proposed constitutional amendment was introduced to Commonwealth Parliament through a Constitution Alteration Bill on March 30, 2023 (National Indigenous Australians Agency, 2023; Tingle, 2023) (Figure 12.3).

Shortly after, in April 2023, the Federal Opposition Leader announced that the Liberal Party, following the National Party's decision, would not support the Bill and thus would oppose the suggested Indigenous Voice to Parliament (Butler, 2023). This "No" declaration to a constitutionally enshrined Indigenous Voice to Parliament has once more politicised Indigenous affairs in Australia, denying First Nations People a say in matters that affect them. According to O'Brien, it is a continuation "of institutionalised failure, from generation to generation, from government to government" (Allam, 2023).

This current political battleground reflects a generally highly politicised and contested socio-political environment in dealing with Indigenous affairs in Australia. It brings into sharp relief the difficult relationship between Indigenous and non-Indigenous Australians, a cultural division which is deeply rooted in the country's history.

Figure 12.3 Indigenous Voice to Parliament, community event

Source: City of Belmond (n.d.)

In many respects the question whether or not to climb Uluṟu, or the *Rock*, resonates with this debate as it presents a kaleidoscope of reconciliation, cultural division, cultural respect and also ignorance. As Hueneke (2007: 71) points out, visitors' "decisions to climb or not to climb are deeply situated in wider political contexts and discursive practices that recognise or deny Indigenous claims".

Despite the longstanding wishes of the Aṉangu traditional people not to climb to respect the spiritual significance of Uluṟu, tourists have always travelled to the region with the motivation to climb the *Rock* until the practice, as will be discussed in more detail later, was finally permanently banned on October 26, 2019 (Parks Australia, 2023b). This decision was based on leadership and intervention to initiate change; in that sense it can be seen as a circuit breaker to continued inappropriate tourist behaviour. Although a victory for the traditional owners, it remains to be seen whether the closure of the Uluṟu Climb also represents a step towards reconciliation.

3 Control of and Access to the Park – a Disputed Terrain

To add to the complexity of the site from an administrative point of view, Uluṟu-Kata Tjuṯa National Park, despite being located within the Northern Territory of Australia, it is not part of the Territory's protected area system but is rather managed by the Commonwealth Government through Parks Australia (Department of Environment, Parks and Water Security, 2023).

This peculiar arrangement needs to be seen in the complex context of the Northern Territory's self-government, which sets it apart from other States in Australia's federal system of government. Until 1978, the Northern Territory was administered by the federal government (Commonwealth Government) in Canberra through the Northern Territory (Administration) Act which "remained the core instrument of governance until 1978", when most executive responsibilities were transferred to the Northern Territory (Heatley 1990: 5). Nevertheless, self-governance has remained limited by the fact that the Commonwealth has retained control over certain areas, such as Aboriginal Affairs, National Parks (i.e. Uluṟu-Kata-Tjuṯa and Kakadu) and uranium mining.

This "conflict triangle" of Commonwealth control has also had a major impact on the Northern Territory's tourism sector, which mainly stems from two pieces of Commonwealth legislation, the National Parks and Wildlife Conservation Act, 1975 and the Aboriginal Land Rights (Northern Territory) Act, 1976 (Altman, 1988; Burton, 1994). The former established Kakadu and Uluṟu-Kata Tjuṯa National Parks, which have been managed by the Commonwealth Government through Parks Australia (part of the Department of Agriculture, Water and the Environment), previously the Australian Nature Conservation Agency (ANCA) (Parks Australia, 2023a). Both parks are major cultural and natural attractions, which make them two of the most prominent tourist destinations of the Northern Territory. Their importance as internationally renowned tourist attractions, the associated economic significance, and the fact that the Commonwealth has effectively retained control over a considerable part of the Northern Territory explains why the Northern Territory Government has always fought for the incorporation of these two National Parks into its own parks system managed by the Northern Territory Parks and Wildlife Commission (formerly Conservation Commission).

4 Tourism to the Park – a Growing Terrain

Tourism in the area of today's Uluṟu-Kata Tjuṯa National Park started as early as the 1950s with the establishment of the first guided Uluṟu tour, supported by the construction of a dirt road in 1948. But visitation only became more regular with the establishment of a bore at the end of the 1950s (Davidson & Spearritt, 2000; James, 2007; Parks Australia, 2023b).

From around 100 visitors in the mid-1950s, tourist numbers increased ten-fold within a decade and reached 40,000 at the beginning of the 1970s. Already at that time, a deterioration of the location, its natural environment and its cultural sites, caused through uncontrolled growth of tourism, could be noted. This overtourism situation triggered calls for an intervention to limit visitation, for instance, to a maximum of 5,000 tourists per day (Davidson & Spearritt, 2000). A parliamentary committee issued a report and subsequently a management plan for the National Park was put in place, which in the 1980s saw the moving of accommodation and the site's only airstrip outside the National Park boundaries to a new, custom-built tourist resort, Yulara (Walliss, 2013).

As can be seen in Table 12.1, despite this relocation of critical tourism infrastructure, visitation to Uluṟu-Kata Tjuṯa National Park continued to grow throughout the 1980s and 1990s, further fuelled by its inclusion into the UNESCO World Heritage List, first in 1987 for its natural values and then in 1994 for its cultural values. Tombleson and Wolf (2022: 403) remark that "since the park was listed as a World Heritage Site, annual visitor numbers have steadily risen". More than 100,000 visitors were recorded in the mid-1980s, a number that climbed to almost 400,000 visitors by 2001 and peaked in 2019 with nearly 500,000 (Brown, 1999; McKercher et al., 2008; Parks Australia, 2023d).

Despite this rapid growth in visitor numbers, which made Uluṟu-Kata Tjuṯa one of the most popular national parks in the Northern Territory, the Northern Territory Government has remained sceptical about the opportunity to further grow visitation to the Park. This scepticism is primarily centred around a particular piece of Commonwealth legislation, the Aboriginal Land Rights (Northern Territory) Act, 1976, which has influenced many developments in the Northern Territory tourism sector. It reinstated ownership of land to its traditional people based on ongoing cultural and traditional ties. At present, about half of the Northern Territory has been returned to its traditional owners, who have successfully argued their claim over the land (Brown, 2020). Although not directly dealing with tourism, the Act has had an impact on the tourism system, since it empowers Indigenous people to "control any commercial development on Aboriginal land… and to restrict tourist access", thus, causing uncertainty for the Northern Territory Government's approach to rapid economic development, including tourism development (Altman, 1988: 62; Brown, 2020; Pforr, 2009).

Table 12.1 Visitor numbers to Uluṟu-Kata Tjuṯa National Park

Year	*1984*	*1987*	*1988*	*1990*	*1995*	*2001*	*2005*	*2010*	*2013*	*2015*	*2017*	*2019*	*2020*	*2021*	*2022*
Visitation (000)	100	180	175	214	312	394	249	298	260	277	303	497	93	128	222

Sources: Buckley (2002); Department of Climate Change, Energy, the Environment and Water (2023b); Parks Australia (2023d); Pforr (2009)

In this light the Uluru-Kata Tjuta National Park has in many respects become a symbol of both opposition and also support for Aboriginal Land Rights (Hueneke, 2007). After the local Anangu traditional owners, comprising of the Pitjantjatjara and Yankuntjatjara people, successfully lodged a native title claim over the area in 1979, the Uluru-Kata Tjuta National Park was handed back to the Anangu people in 1985. At the same time, as a pre-condition to the successful claim, the land was leased back to the Australian Government for a period of 99 years in order for it to be managed as a National Park, with a joint management agreement put in place between Parks Australia North (now Parks Australia) and its traditional owners (Bickersteth et al., 2020).

The Commonwealth Aboriginal Land Rights policy has often been seen by subsequent Northern Territory Governments as a major constraint and limitation to develop tourism in the Territory to its full potential, since Aboriginal interest over land created uncertainties for development and was perceived to be a blockade to potential investment. Brown (2020) recently emphasised that it was only after a new government was elected in the Northern Territory that the complex and contested Aboriginal land ownership issues were approached differently. The defeat of the Country Liberal Party by the Northern Territory Labor Party in 2001 marked the end of a 27-year period of conservative rule in the Northern Territory, which constituted a near seismic shift in the Territory's political landscape (Pforr & Brueckner, 2016). "The new government was intent on a new approach to dealings with Aboriginal interests, as well as keen to promote self-determination and Aboriginal employment" (Brown, 2020: 65).

However, on the question of climbing Uluru, the Northern Territory Labor Government made it clear that they would oppose the closure of the "Climb", bringing their position in line with the opposition, the Country Liberal Party (Batty, 2009; Davidson, 2016). This broader battleground over power and control, as well as Indigenous vs. non-Indigenous interests, can also be seen in the contested issue to climb or not to climb Uluru.

5 The Climb – a Controversial Terrain

Until its ban in 2019, climbing Uluru was a key tourist activity and, for many, a prime motivation to visit the National Park. At the same time, as pointed out by Davidson & Spearritt (2000: 200), increased visitation could also be interpreted positively and forward looking as a "tentative step towards trying to understand Aboriginal culture".

From a Western, non-Indigenous perspective, the *Rock* was "discovered" in 1873 by the British explorer William Gosse and named "*Ayers Rock*" after the Chief Secretary of South Australia, Sir Henry Ayers (Bickersteth et al., 2020), before it was re-named *Ayers Rock* / Uluru in 1993 and, later, Uluru / *Ayers Rock* in 2002 (Parks Australia, 2023c).

Visitors began to climb Uluru in the late 1930s (Parks Australia, 2023b) and although, according to Davidson & Spearritt, (2000), only 26 people climbed Uluru between 1931 and 1946, the *Rock* quickly became a site of "conquest" (Everingham

Figure 12.4 Uluṟu climb path
Source: Author

et al., 2021) and its climb a "tourist trophy" (Davidson & Spearritt, 2000). For many years Uluṟu has commonly been perceived as a natural rather than a cultural site (James, 2007). As James (2007: 400) points out, "that visitors are allowed to climb Uluṟu at all is due to the historic connections between National Parks and tourism that constructed it as a natural and (Australian) national site, rather than an Aboriginal cultural landscape". Reflective of a steady increase in visitor numbers climbing Uluṟu, in the mid-1960s a climb chain was installed to ensure tourists' safety (Bickersteth et al., 2020). Whittington and Waterton (2021: 553) remark that "those handrails have since been used by hundreds of thousands of visitors in a ritual that has both re-enacted and reinforced the 'triumph' of settler-colonialism, with each individual 'climb' helping to etch a 1.7 km scar onto the monolith's back, smoothing its vibrant red surface grey" (Figures 12.4 and 12.5).

Over the decades, however, in light of growing opposition from the traditional owners, the Climb became increasingly controversial. While in the early 1990s almost three quarters of visitors mentioned climbing the *Rock* as their main reason for visiting the Park, this figure dropped to about 50% only a decade later (Hueneke, 2007; James, 2007). Nonetheless, despite this decline, about half of the tourists "continue[d] to choose to act in a way that contravenes the wishes of the Aṉangu" (James, 2007: 399).

These changing sentiments can also be seen in the author's diary entries pertaining to his two visits to the National Park, first in the early 1990s and then again in 2013.

Figure 12.5 Uluṟu climb path with chain

Source: Author

Climbing Uluṟu – A Personal Reflection

The first visit was part of a backpacking trip around Australia as an international tourist, which included a visit to the Red Centre of Australia as this was high on my "not to be missed" sites. In Adelaide I took the legendary Ghan train north to Alice Springs, the largest town in Central Australia, from where I continued my journey by bus (Greyhound) to the Yulara Resort, about 450 km away, which is located outside the Uluṟu-Kata Tjuṯa National Park.

The visit to the Park, with Uluṟu as perhaps Australia's most famous landmark, was a highlight of my trip to Australia's Red Centre. My itinerary included the obligatory sunrise and sunset photos (Figure 12.6) at dedicated viewing areas. However, the most important activity for me was to climb Uluṟu, from today's perspective, certainly nothing to be proud of.

Figure 12.6 Sunset at Uluṟu, Australia

Source: Author

Back then in 1992, the evening before the planned ascent, I decided to do an on-site visit of the bottom of the Climb to prepare myself for what was to come the next morning, the Climb in twilight to catch the first glimpse of the rising sun from the top. The next day I got up at 4 a.m. to start the ascent by 5 a.m. at the latest as a means "to beat the crowds" and to reach the *Rock* before the first tourist buses arrived. I even rented a car for this occasion. The 300 (altitude) metres ascent was much steeper and much more physically demanding than expected, proving the installed chain along the ascent to be a useful aid. Despite unexpectedly strong wind awaiting me at the summit, a breathtaking view, just in time for sunrise, made the climb an unforgettable experience…

Finally, at position 12 of the "race to the summit" (after about half an hour), and just in time for sunrise… almost undisturbed, I thoroughly enjoyed the fantastic view of the Olgas (about 50 km away) and Mt Conner (150 km away), and in between just "Nothing"… a truly moving nature experience.

My descent back to the base provided an entirely different perspective, with a never-ending procession of people, mainly visitors from Japan, who struggled up the mountain as the temperature began to rise (Figure 12.7). At about 11 a.m. I was back at the Yulara Resort, concluding my tourist program for the day…

Figure 12.7 Uluṟu carpark

Source: Author

Reading these entries today, I feel ashamed of my naive, ignorant behaviour, even though at that time I was not, at least not fully, aware of the cultural significance of Uluṟu for the Aṉangu traditional owners and their opposition to the Climb. I could also not find any reference to this in the information brochures for visitors, which I kept and read again as I was writing this chapter, so I was one of probably many visitors who experienced Uluṟu primarily as a natural and not necessarily a cultural attraction point.

This personal experience profoundly changed 20 years later, when in 2013, now as an Australian citizen and thus as a domestic tourist, I visited central Australia again, this time with a fundamentally different itinerary – an "ascension" was not even under consideration. Cultural aspects were the prime focus of my trip to Uluṟu-Kata Tjuṯa National Park, which included, for instance, a visit to the Uluṟu-Kat-Tjuṯa Cultural Centre and the 9 km walk around Uluṟu as a cultural learning experience of the history of its traditional owners, the Aṉangu people.

The change in my own tourist behaviour might be a reflection of the growing influence of the local Aṉangu people on the operation of the park, reflected in the information provided to visitors about the cultural significance of the site and appropriate tourist etiquette.

6 You Shouldn't Climb

The traditional owners have long requested that visitors respect their wishes to bar tourists from the Climb (Director of National Parks, 2010). This wish was also affirmed in the Park's 2000 Management Plan (Parks Australia & Uluṟu-Kata Tjuṯa

Board of Management, 2000) with a continuation to discourage climbing and flagging the possibility of introducing stronger measures to deter tourists from the Climb (Hueneke, 2007). This included educational campaigns to inform visitors by creating awareness about the cultural significance of Uluṟu, which included initiatives such as putting signs up asking tourists not to climb (Tombleson & Wolf, 2022).

> This is Aṉangu land and we welcome you. Look around and learn so that you can know something about Aṉangu and understand that Aṉangu culture is strong and really important. We want our visitors to learn about our place and listen to us Aṉangu. Now a lot of visitors are only looking at sunset and climbing Uluṟu. That rock is really important and sacred. You shouldn't climb it! Climbing is not a proper tradition for this place. (Director of National Parks, 2010)

Further, in 2010 a new Park Management Plan mapped out conditions for a possible closure of the Climb. It proposed to develop new products and experiences to provide alternative natural and cultural experiences as critical pull factors to visit the National Park, and on their wider uptake to close the Climb altogether once the share of visitors who climb would fall below a threshold of 20% (Parks Australia, 2023b).

At that time about 38% of visitors climbed Uluṟu, a number which further declined to about 16% in 2015 (Hitch & Hose, 2017). Based on these numbers, the announcement in 2017 to ban the Climb, which was a unanimously made decision by the Uluṟu-Kata Tjuṯa National Park Board and came into effect on October 26, 2019, after a two-year notice period, should not have come as a surprise but should rather have been seen as a logical consequence of the conditions put in place in the 2010 Management Plan. However, the announcement in 2017 triggered a "last chance" and "overtourism" reality, with Everingham et al. (2021: 7) reporting an extra 10,000 visitors every month half a year before the closure, and "on its last day, hundreds of climbers queued amid high tensions, high winds, and scrambles to be the 'last person off Uluṟu'". According to Tombleson and Wolf (2022), mainly domestic visitors in their hundreds climbed Uluṟu day in day out in numbers that had not be seen in a decade, raising questions about the perceived increase in awareness of the cultural significance of the site in the context of decreasing visitor numbers who climbed Uluṟu in the years prior to the ban.

7 "Overtourism" and "Last Chance Tourism" – a "Must See" Terrain

In recent times, the rise in social media motivated tourism, for instance "Insta tourism" or "bucket list tourism", highlights that tourism is increasingly influenced by destination images portrayed on social media platforms such as Instagram. This has created greater visibility of destinations and their attractions but as a consequence has often also generated lists of "must see places" which may not only lead to growing visitation but potentially also to overtourism, a phenomena where too many visitors negatively affect the destinations including the lives of local residents (Dodds & Butler, 2019; Garner, 2020; Kibby, 2020; Mackay et al., 2020; Pechlaner et al., 2020; Tombleson & Wolf, 2022; Wu et al., 2020).

Tombleson and Wolf (2022: 407) described the final rush of people climbing Uluṟu before its closure, which was extensively reported on social media, as bucket list tourism, highlighting that "the month leading up to the closure of Uluṟu to climbing saw the greatest number of comments on social media". This type of bucket list tourism is triggered by increased public attention and a corresponding growing desire to visit these tourist destinations and attractions, as seen in the case of Uluṟu, where a spike in visitation and also in climbs was triggered by the limited time left to experience the ascent after its ban was announced in 2017. These phenomena of changed tourist flows are associated with the concept of "last chance tourism" (Lemelin et al., 2013; Mackay et al., 2020; Piggott-McKellar & McNamara, 2017). The less time left to visit a place, the greater the incentive to travel to this locality. Last chance tourism can lead to increased, unsustainable visitor numbers to destinations that are under threat of disappearing due to climate change, natural disasters or human activities. It can contribute to the problem it aims to address, as it might increase demand for unsustainable tourism activities and place additional pressure on already vulnerable destinations and attraction points. This demand pattern highlights the need to change tourist behaviour, challenging Destination Management Organisations to show leadership in managing destinations more sustainably, with, for instance, demarketing initiatives of certain attraction points (Hall & Wood, 2021).

These developments can also be seen in the Uluṟu-Kata Tjuṯa National Park. The urgency associated with a last chance to climb the *Rock* (Everingham et al., 2021) and the resulting boom in tourist numbers intensified pressure on the destination and, ironically, stands in direct contrast to the cultural reason for closing the Climb and the changes in visitor behaviour seen in the preceding years.

This shows that the effects of last chance tourism are often similar to the impacts associated with overtourism, both intensifying pressures on destinations associated with increased visitation. The issue of overtourism, with associated problems such as overcrowding, environmental degradation and cultural disruption, has been a concern for many destinations around the world, including Uluṟu-Kata Tjuṯa National Park, which has seen a significant increase in visitor numbers over the years (see Table 12.1).

According to Tombleson and Wolf (2022: 402), reflective of what has been observed in the Uluṟu-Kata Tjuṯa National Park, many "UNESCO World Heritage sites face increasing pressure due to rapidly growing tourist numbers, leading in some cases to overcrowding". Therefore, next to cultural reasons, the decision to ban climbing Uluṟu was also made due to concerns about overtourism and the impact that the activity had on the environment. The constant foot traffic from climbers, for instance, caused significant erosion and damage to the *Rock*, with long-term implications for the future sustainability of the site.

In many respects, these developments are a continuation of the early days of tourism expansion at Uluṟu-Kata Tjuṯa National Park in the 1970s and 1980s, when negative consequences of too much and unmanaged visitation were already apparent, prompting a call for management intervention and a cap on visitor numbers in line with the sites' carrying capacities (Davidson & Spearritt, 2000).

8 Conclusion

The discussed interplay between last chance tourism and overtourism, with Uluṟu as a case in point, has illustrated not only the issue of too many visitors and their impact on the destination's environment, but has also highlighted possible solutions to respond to these tourism phenomena which are embedded in a complex destination reality. In the context of the Uluṟu case example, synergies appeared from the literature pointing to the importance of leadership, public debate and education to increase awareness, the right marketing message and, most importantly, the need to listen to and empower Indigenous people.

Tombleson and Wolf (2022: 402) for instance see ripple effects from the Uluṟu case as a landmark model to manage access to culturally significant sites and even for destinations under threat from overtourism. In addition to educational campaigns aiming to create greater awareness for environmental and cultural vulnerabilities of destinations, which should lead to more respectful tourist behaviours, destination management and marketing organisations will also need to lead the way with supporting initiatives such as the active demarketing of certain attraction points (Hall & Wood, 2021).

Prior to the ban on climbing Uluṟu, climbing was a key tourist activity and for many, including the author in the early 1990s, a prime motivation to visit the Uluṟu-Kata Tjuṯa National Park. These situations highlight that the Climb was mainly an adventure and physical challenge, but for some the activity was also a matter of free choice, displaying a "right to climb" attitude. Both motivations are in diametrical contrast and thus disrespectful to the culture and beliefs of the traditional owners.

Tombleson and Wolf (2022) identified three narratives in their qualitative research which have framed positive and negative arguments surrounding the closure: (1) an economic perspective, which highlights the economic significance of the site as a key destination for income and job generation which might be undermined by banning the Climb and an associated decline in visitation (Davidson, 2016); (2) a notion of "entitlement", with Uluṟu being seen as a national icon which should be accessible to all Australians, thus implying a right to climb and social acceptability; and (3) an "inclusive sustainability" narrative, referring to notions of "respect" and "empowerment" to enable a strong culture for future generations.

These facets to the question whether to climb or not to climb Uluṟu have played out in the National Park for many decades, in particular also in the context of the last chance tourism rush after the ban was announced in 2017. They illustrate broader, historically entrenched cultural divisions between Indigenous and non-Indigenous Australians. It is argued here that this complex and challenging socio-political environment resonates in the Uluṟu case example, which brings into sharp relief not only the difficult road towards reconciliation but also the highly politicised arena of Indigenous affairs in Australia. In that sense Uluṟu can also be seen as a kaleidoscope of the difficult relationship between Indigenous and non-Indigenous Australians.

Therefore, James' (2007: 406) observation that Uluṟu "is perhaps not a site of either division or reconciliation, but a site of continuing dynamic complex negotiations of post-colonial relations between Indigenous and non-Indigenous Australians in relation to place", remains relevant. Thus, in the long term, there remains hope that the unsustainable nature of contested places (McKercher et al., 2008), such as Uluṟu, will be transitioned into more sustainable pathways. This view is supported by Whittington & Waterton (2021: 567) through their fieldwork conducted in 2012 and then again in 2019. They concluded that "a clear majority, on both occasions [...] supported the closure of the Climb", which reflect in their view "an increasing trend to support Aboriginal and Torres Strait Islander self-determination and rights and a desire to establish right relations with Aboriginal peoples".

Since the closure of the Climb at the end of October 2019, according to Parks Australia, visitation to Uluṟu-Kata Tjuṯa National Park declined in the first year (October 2020) by 60%.

However, in the context of the onset of the COVID-19 pandemic it has been difficult to determine the impact of the ban on climbing the *Rock* on this decline. In 2021, visitation to the National Park increased again to almost 128,000 visitors with a further increase to more than 220,000 visitors in 2022 (although this figure is still 45% below pre-pandemic numbers) (Department of Climate Change, Energy, the Environment and Water, 2023b; Parks Australia, 2023d).

Maybe the COVID-19 pandemic has created a buffer, with the Uluṟu Climb fading over time from the lived experience of travellers. Visitors today will have to accept the realities created back in 2019 and instead enjoy the diverse cultural and natural experiences on offer in Uluṟu-Kata Tjuṯa National Park. The notice below, put in place by Parks Australia (2023b), may therefore, as time passes, become obsolete.

Notes

Visitors are advised that climbing Uluṟu is a breach of the Environmental Protection and Biodiversity (EPBC) Act, and penalties will be issued to visitors attempting to do so.

References

Allam, L. (2023, 14 April). Kerry O'Brien gives impassioned defence of Indigenous voice to parliament at hearing. *The Guardian*. www.theguardian.com/australia-news/2023/apr/14/kerry-obrien-gives-impassioned-defence-of-indigenous-voice-to-parliament-at-hearing

Altman, J. C. (1988). *Aborigines, tourism and development: The Northern Territory experience*. North Australia Research Unit (NARU).

Australian Bureau of Statistics (ABS). (2023). *Estimates of Aboriginal and Torres Strait Islander Australians*. www.abs.gov.au/statistics/people/aboriginal-and-torres-strait-islander-peoples/estimates-aboriginal-and-torres-strait-islander-australians

Bickersteth, J., West, D., & Wallis, D. (2020). Returning Uluṟu. *Studies in Conservation, 65*, 1–9.

Batty, D. (2009, 8 July). Uluṟu visitors face climbing ban. *The Guardian*. www.theguardian.com/world/2009/jul/08/uluru-climbing-ban-plan

Brown, P. R. (2020). Framing, agency and multiple streams – a case study of parks policy in the Northern Territory. *Australian Journal of Political Science*, *55*(1): 55–71.

Brown, T. (1999). Antecedents of culturally significant tourist behaviour. *Annals of Tourism Research*, *26*, 676–700.

Brueckner, M., Spencer, R., Wise, G., & Marika, B. (2014). Indigenous entrepreneurship: Closing the gap on local terms. *Journal of Australian Indigenous Issues*, *17*(2), 2–24.

Buckley, R. (2002). *World heritage icon value: Contribution of World Heritage branding to nature tourism*. Australian Heritage Commission.

Burton, R. C. J. (1994). Making sustainable tourism a reality. Tourism management in the national parks of Australia's Top End. Expert Meeting on Sustainable Tourism and Leisure (8–10 December). Department of Leisure Studies, Tilburg University, The Netherlands.

Butler, J. (2023, 7 April). Peter Dutton and the voice: What the Liberal party has got wrong about Indigenous recognition. *The Guardian*. www.theguardian.com/australia-news/2023/apr/07/peter-dutton-and-the-voice-what-the-liberal-party-has-got-wrong-about-indigenous-recognition

City of Belmond. (n.d.). Home [Facebook page]. Facebook. Retrieved May 2, 2024, from https://www.facebook.com/BelmontCouncilWA/

Davidson, H. (2016, 20 April). Uluṟu: Northern territory chief minister opposes climbing ban. *The Guardian*. www. theguardian.com/australia-news/2016/apr/20/dont-stop-tourists-climbing-uluru-says-northern-territory-chief-minister

Davidson, J., & Spearritt, P. (2000). *Holiday business: Tourism in Australia since 1870*. Melbourne University Press.

Department of Climate Change, Energy, the Environment and Water. (2023a). *Welcome to Uluṟu-Kata Tjuṯa National Park*. www.dcceew.gov.au/parks-heritage/national-parks/uluru-kata-tjuta-national-park

Department of Climate Change, Energy, the Environment and Water. (2023b). *Visitors and tourism: Covid-19 impacts, numbers, revenue, closures*. www.dcceew.gov.au/sites/default/files/documents/72878.pdf

Department of Environment, Parks and Water Security. (2023). *NT Parks masterplan 2023–2053*. Department of Environment, Parks and Water Security.

Director of National Parks. (2010). *Uluṟu-Kata Tjuṯa National Park: Tourism directions: Stage 1 September 2010*. www.dcceew.gov.au/sites/default/files/documents/tourismdirections.pdf

Dodds, R., & Butler, R. (Eds.). (2019). *Overtourism: Issues, realities and solutions*. De Gruyter.

Ellis, C. (2004). *The ethnographic I: A methodological novel about autoethnography*. AltaMira Press.

Everingham, P., Peters, A., & Higgins-Desbiolles, F. (2021). The (im)possibilities of doing tourism otherwise: The case of settler colonial Australia and the closure of the climb at Uluṟu. *Annals of Tourism Research*, 88, 1–11.

Garner, A. O. (2020). I came, I saw, I selfied. Travelling in the age of Instagram. In M. Månsson, A. Buchmann, C. Cassinger & L. Eskilsson (Eds.), *The Routledge Companion to Media and Tourism*. Routledge.

Hall, C. M., & Wood, K. (2021). Demarketing tourism for sustainability: Degrowing tourism or moving the deckchairs on the Titanic? *Sustainability*, *13*, 1585.

Heatley, A. (1990). *Almost Australians: The politics of Northern Territory self-government*. North Australia Research Unit (NARU).

Hitch, G., & Hose. N. (2017, 1 November). Uluṟu climbs banned from October 2019 after unanimous board decision to "close the playground". *ABC News*. www.abc.net.au/news/2017-11-01/uluru-climbs-banned-after-unanimous-board-decision/9103512

Hueneke, H. (2007). *To climb or not to climb?* ANU.

James, S. (2007). Constructing the climb: Visitor decisionmaking at Uluṟu. *Geographical Research, 45*, 398–407.

Kibby, M. (2020). Instafamous: Social media influencers and Australian beaches. In E. Ellison & D. Brien (Eds.), *Writing the Australian Beach*. Palgrave Macmillan.

Lemelin, H., Dawson, J., & Stewart, E. J. (Eds.). (2013). *Last chance tourism: Adapting tourism opportunities in a changing world*. Taylor & Francis.

Mackay, R. M., Minunno, R., & Morrison, G. M. (2020). Strategic decisions for sustainable management at significant tourist sites. *Sustainability, 12*(21), 1–22.

McKercher, B., Weber, K., & du Cros, H. (2008). Rationalising inappropriate tourism behaviour at contested sites. *Journal of Sustainable Tourism, 16*(4), 369–385.

National Indigenous Australians Agency. (2023). *Referendum on an Aboriginal and Torres Strait Islander Voice*. www.niaa.gov.au/indigenous-affairs/referendum-aboriginal-and-torres-strait-islander-voice

Parks Australia. (2023a). *Joint management*. https://parksaustralia.gov.au/uluru/about/joint-management

Parks Australia. (2023b). *Uluṟu climb close*. https://parksaustralia.gov.au/uluru/discover/culture/uluru-climb/

Parks Australia. (2023c). *Ayers Rock or Uluṟu?* https://parksaustralia.gov.au/uluru/about/ayers-rock-or-uluru/

Parks Australia. (2023d). *Visitor numbers*. www.google.com/url?sa=t&rct=j&q=&esrc=s&source=web&cd=&ved=2ahUKEwjV4NnTg43-AhXM8DgGHW2JA1wQFnoECDUQAQ&url=https%3A%2F%2Fwww.aph.gov.au%2F-%2Fmedia%2FEstimates%2Fec%2Fadd2122%2Ftabled_docs%2FParks_Australia_visitor_numbers.pdf%3Fla%3Den%26hash%3DE28ECC2FE731A2DE6A268CE408251E3AEAB2598B&usg=AOvVaw1HKKMYYPFgNBSFOSEs4YuP

Parks Australia & Uluṟu-Kata Tjuṯa Board of Management. (2000). *Fourth Uluṟu-Kata Tjuṯa National Park Plan of Management*. Commonwealth of Australia.

Pechlaner, H., Innerhofer, E., & Erschbamer, G. (Eds.). (2020). *Overtourism: Tourism management and solutions*. Routledge.

Pforr, C. (2009). *Tourism public policy in the Northern Territory of Australia*. LAP.

Pforr, C., & Brueckner, M. (2016). The quagmire of stakeholder engagement in tourism planning: A case example from Australia. *Tourism Analysis, 21*(1), 61–76.

Piggott-McKellar, A. E., & McNamara, K. E. (2017). Last chance tourism and the Great Barrier Reef. *Journal of Sustainable Tourism, 25*(3), 397–415.

Storey, M. (2012). 20 years after Mabo v Commonwealth. *Alternative Law Journal, 37*(3), 190–191.

Tingle, L. (2023, 25 March). The Voice debate turned a corner this week but it's splintered into so many shards that confuse and dazzle. *ABC News*. www.abc.net.au/news/2023-03-25/wording-voice-referendum-question-unsure-how-make-a-difference/102142918

The Uluṟu Dialogue. (2023). *The Uluṟu statement from the heart*. https://ulurustatement.org/the-statement/view-the-statement/

Tombleson, B. & Wolf, K. (2022). Sustainable tourism and public opinion: Examining the language surrounding the closure of Uluru to climbers. In M. Sigala, A. Yeark, R.

Presbury, M. Fang, & K.A. Smith (Eds.), *Case based research in tourism, travel, hospitality and events*. Springer.
Walliss, J. (2013). Transformative landscapes: Postcolonial representations of Uluṟu-Kata Tjuṯa and Tongariro National Parks. *Space and Culture*, *17*, 280–296.
Whittington, V., & Waterton, E. (2021). Closing the climb: Refusal or reconciliation in Uluṟu-Kata Tjuṯa National Park? *Settler Colonial Studies*, *11*(4), 553–572.
Wu, H-C., Cheng, C-C., Ai, C-H., & Wu, T-P. (2020). Fast-disappearing destinations: The relationships among experiential authenticity, last-chance attachment and experiential relationship quality. *Journal of Sustainable Tourism*, *28*(7), 956–977.

13 Spotting Overtourism and Identifying Actions to Address It with the Help of the Visitor Flow Approach

Pietro Beritelli

1 Introduction

1.1 Problem Statement and Objective

Although not new as a phenomenon, overtourism[1] has increasingly become an area that practice and research find a hard nut to crack. There is an increasing amount of studies and concepts that attempt on the one hand to observe and measure overtourism, and on the other hand to control and steer it (for a current overview see Séraphin, Gladkikh & Vo Thanh, 2020). Above all, spotting and exploring the phenomenon in places, cities and regions is a challenge in itself. An extensive study has dared to develop and communicate metrics and key figures of overtourism (Weber, Stettler, Crameri, Eggli & Barth, 2019). Today, we continue to face a conundrum of how to properly capture overtourism, and even more so, how to then manage it best.

At this point it should be emphasised that overtourism is a collective problem both on the side of demand, that is, from the visitors themselves, and on the side of supply. Not only does it usually take place in the public space, but it also directly affects all stakeholders who find themselves there in terms of time and space, while enjoying their free time, working or doing business. In this sense, overtourism is a special case of the tragedy of the commons (Hardin, 1968) that must consider visitor's behaviour and their dynamics (Goodwin, 2017). As such, not only the solution can be found in the community. The origin must also be considered in a differentiated way, too. Here, the visitor flow approach comes into play. This article, aware of the plethora of extant theoretical and practical contributions, attempts to provide a new perspective on understanding overtourism. To do so, it makes use of the visitor flow approach that has proven itself in recent years, first in practice (e.g. Beritelli, 2020; Beritelli, Crescini, Reinhold & Schanderl, 2019; Beritelli & Laesser, 2017), and then scientifically, described as a conceptual framework (Beritelli, Reinhold & Laesser, 2020).

1.2 Structure of the Article

Overtourism will be characterised in terms of visitor flows and, thus, from a specific perspective. In doing so, it will be shown how, thanks to recent research results, it is possible to capture the variety and diversity of visitor flows and, at

DOI: 10.4324/9781003365815-20

the same time, to obtain a differentiated and very specific reconstruction of the emergence and development of these movements in time and space. Not surprisingly, it will demonstrate that even locals and inhabitants of the surrounding region are part of the multitude of visitor flows. In the third section, the article will show how possible solutions can be derived thanks to this differentiated view. This takes place, on the one hand, by focusing on individual tourist attractions and thus on critical areas. On the other hand, it is precisely along the reconstructed visitor flows that the tangle of local and temporal emergence and concentration that produce overtourism can be unravelled and examined more closely. In this way, promising interventions in space and time can be derived. The article concludes with an overview of the fundamental implications of this approach that not only help manage current problems but also offer new paths for development in tourist places.

2 A Specific View on Overtourism

2.1 Visitor Flows

People move along flows, both during normal working hours (e.g. as commuters) and during their leisure time (MacCannell, 2001; Sheller & Urry, 2006). When following travellers, we learn not only when and where they go or what they do, but also what they purchase and which services they use. This allows us to reconstruct the service chain from the perspectives of supply and demand simultaneously. Since the visitor is the main actor of their experiences (Smith, 1994), because they decide with whom, what, when and where they create moments of experience together with other travellers, by studying them we can understand how travel and ultimately tourism works in detail. Indeed, travel is a form of household production (Ironmonger, 2000; Maggi, 2014; Muth, 1966). Attractions are first created in the eye of the beholder – when many individuals value an activity or a place as being meaningful and important, it becomes a tourist activity/attraction in the collective perception. Many of today's tourist attractions were not originally intended for leisure travel (e.g. The Colosseum in Rome, The Eiffel Tower in Paris). Also, in the case of highly visited places in nature, it becomes clear that travellers themselves have turned them into tourist attractions (e.g. Ayers Rock in Australia). Consequently, visitors are the main actors in their very own play, with places providing the stage for their performance. Visitors shape their individual experiences through their own decisions and actions, drawing on selected resources. Travellers make tourist destinations.

In the past 30 years or so, science has made great progress in understanding the phenomenon of travel and studying it more closely. What used to be conceptually described and supported mostly anecdotally by tourism sociologists (e.g. Urry, 1990) can now be measured and analysed precisely on the basis of travellers' data traces thanks to modern information technology. Arguably, one of the most important findings in recent years has been reconstructing visitor flows. Geo- and time-tagged data extracted, for instance, from user-generated

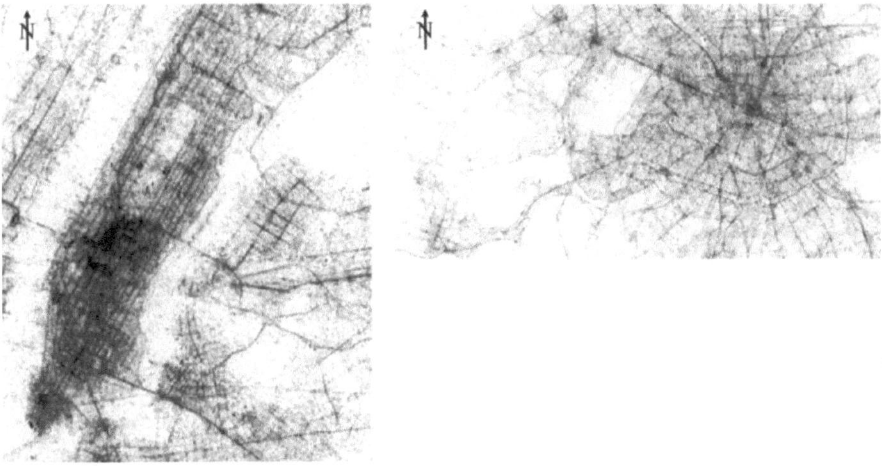

Figure 13.1 Areas and visitor presence as well as their movements. Locating visitors and locals in Manhattan and Paris

Source: Beritelli et al. (2020)

content of social media (e.g. Kádár & Gede, 2021; Paulino, Lozano & Prats, 2021), from credit card transactions (e.g. Aparicio, Hernández Martín-Caro, García-Palomares & Gutiérrez, 2021), from smartphones with their GPS locations (e.g. Baggio & Scaglione, 2018; Hardy et al., 2017; Raun, Shoval & Tiru, 2020) or captured from tourist cards (e.g. Steiner, Baggio, Scaglione & Favre, 2016) show that visitors are more likely to be at landmarks, attractions and tourist sites than at other "ordinary" places. Thus, visitors connect tourist attractions or sites by moving through other, non-tourist sites (Liu, Dong & Chen, 2017; Scaglione, Baggio & Doctor, 2021), creating a dense web of individually produced threads.

Figure 13.1 visualises the data traces captured from visitors. It can be clearly seen that:

- there are tourist attractions/localities and non-tourist localities
- tourist attractions/locations are connected to each other
- most travellers prefer certain tourist attractions/locations and corresponding paths.

2.2 *A Different Origin, a Different Geometry*

Tourism is first and foremost a social phenomenon, which is only subsequently turned into a business. Economic mechanisms come into play mostly when tourists have needs during their trip (food, accommodation, information and intermediation, etc.). It is also clear that day-trippers and locals are also included when they have leisure time available and want to do something in their own place of

Figure 13.2 Travel is not areas/territories for tourists, but points and lines/trajectories

Source: Own illustration, adapted from Laesser et al. (2023, p. 53)

residence or in their own region (see the current definition of the UNWTO, 2022). Over the past 50 years or so, research and practice on locals' perceptions and attitudes towards tourism (for an overview see Williams & Lawson, 2001) has generally assumed a separation between visitors and local people. It should be made clear that (1) locals should not in principle be regarded as antagonists of visitors or tourism businesses; they themselves are often part of tourism and leisure, and (2) there is no such thing as the local; the boundaries are fluid, the degree of concern varies, and interests are diverse. The local population must be considered as an essential because original part of local tourism. Indeed, without the appreciation of the locals for various places and attractions, many of today's tourist destinations would not have become such.

Mobility is a daily occurrence for everyone. Travel and tourism permeate our daily lives more than we may realise. The following sketch shows the logical consequence of what constitutes a journey for the tourist but also for the providers of transport and tourist services (see Figure 13.2). It is a sequence of movements and stays in space. A destination must geometrically be understood as a network of points and lines or trajectories, not an area or surface.

Thus, there are no tourist cities, regions or countries; rather, cities, regions or countries have different, specific, attractive tourist places, which are connected by other places in between with tourists, or no tourists at all.

3 Learnings from Practice

3.1 Working with Maps and the Bottom-Up Process

In the last ten years or so, we have worked with maps and a simple legend to understand the details of what happens when travellers move in and through places. The maps invite the various actors in a place to intuitively draw the visitor flows. On the map one sketches where the visitors come from, what they do, and where they continue to. Each individual visitor flow represents approximately one day's sequence. Studies have confirmed that visitors roughly plan and engage in a new sequence of activities each day (Moore, Smallman, Wilson & Simmons, 2012; Smallman & Moore, 2010), pointing to on-trip decisions as an important determinant of how trips unfold (Choi, Lehto & Oleary, 2007; DiPietro, Wang, Rompf & Severt, 2007; Fesenmaier & Jeng, 2000). Other reported phenomena include visitors meeting each other at different places during longer stays, and visitors meeting acquaintances and relatives unexpectedly, by chance, during their travel (Beritelli & Reinhold, 2018).

The visitor flows are additionally characterised on the corresponding legend. The selected maps in Figure 13.3 show this as an example. Tripographics in the legend help to describe the visitor flows in more detail. Of particular interest in the legend – in addition to describing the visitors, what they do and how much they consume along a daily sequence – is information about seasonality (when during the year?), the number of visitors (how many?), and what phase of the life cycle it is in (increasing, stagnant, decreasing number of visitors?). In Figure 13.3, the image top left is an example of an empty map with legend (notice seasonality, number of visitors and life cycle phase at the upper line at the right). The images top right show a selection of single flows. Thanks to specific hypotheses[2] the individually selected visitor flows can be overlaid (producing a "variable geometry"), and the legend contents listed in a separate table. In this way, the temporal and local coincidence of several visitor flows and finally the complex tangle of a mass of tourists, which can be perceived and observed (i.e., "overcrowding", "overtourism"), is revealed. In the lower part of the figure, two examples of overlaid flows with tables are shown.

The visitor flows are drawn and described by so-called informants. Informants are frontline employees of tourist companies. They can observe visitors and have conversations with them every day, and thus know where they come from, what they do or will do, and why. Other informants are locals who, like frontline staff, have good knowledge about some aspects of the visitors on site. An informant does not need to know everything about any or all of the visitor flows. Nevertheless, they can contribute to complete the overall picture of existing visitor flows. This makes the maps method a kind of comprehensive analysis, evaluation and decision technique. In the final phase of each workshop, the informants are asked to draw and describe new visitor flows as well. In doing so, they hypothesise new visitor flows based on their place's very specific context, given the unique resources of the place as well as the current visitor flows. In so doing, innovation is contextualised and suggested in a more realistic framework.

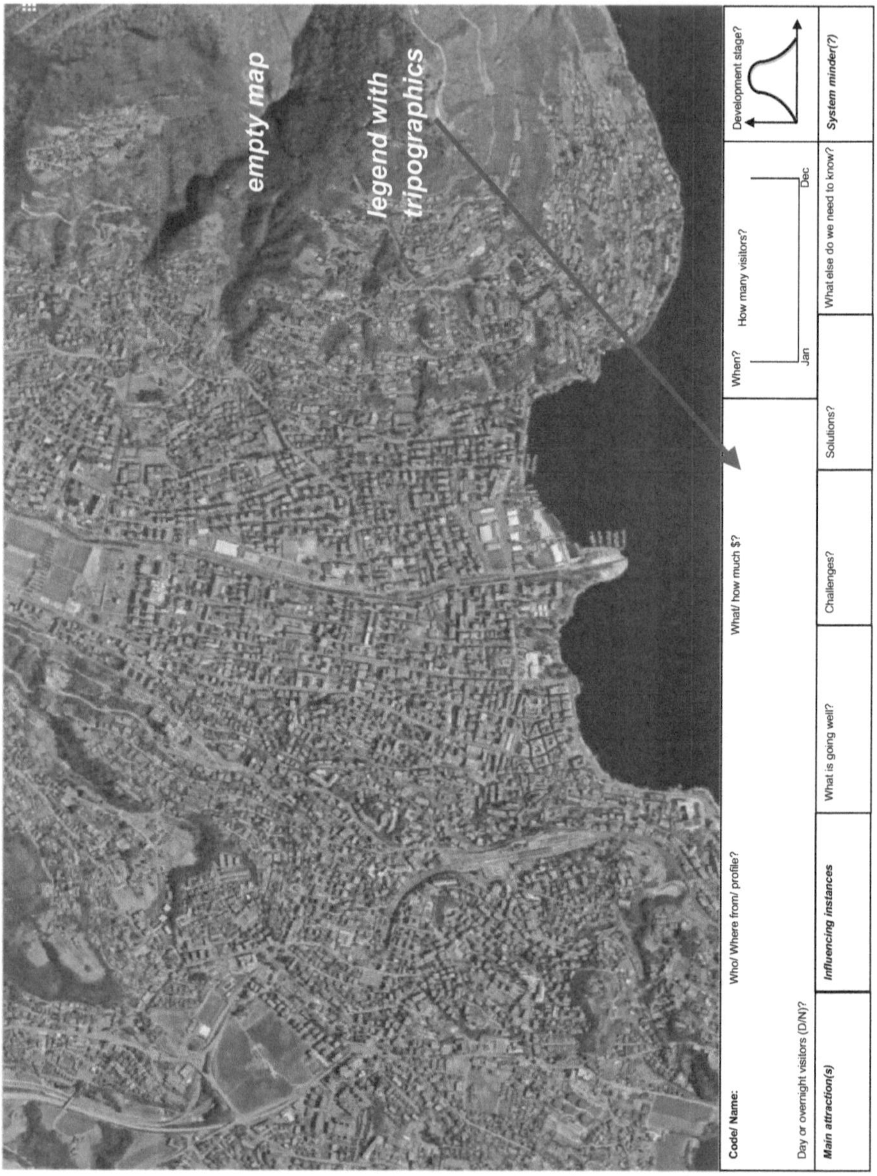

Figure 13.3 Empty map with legend, single flows, overlaid flows with table

Source: Own illustrations

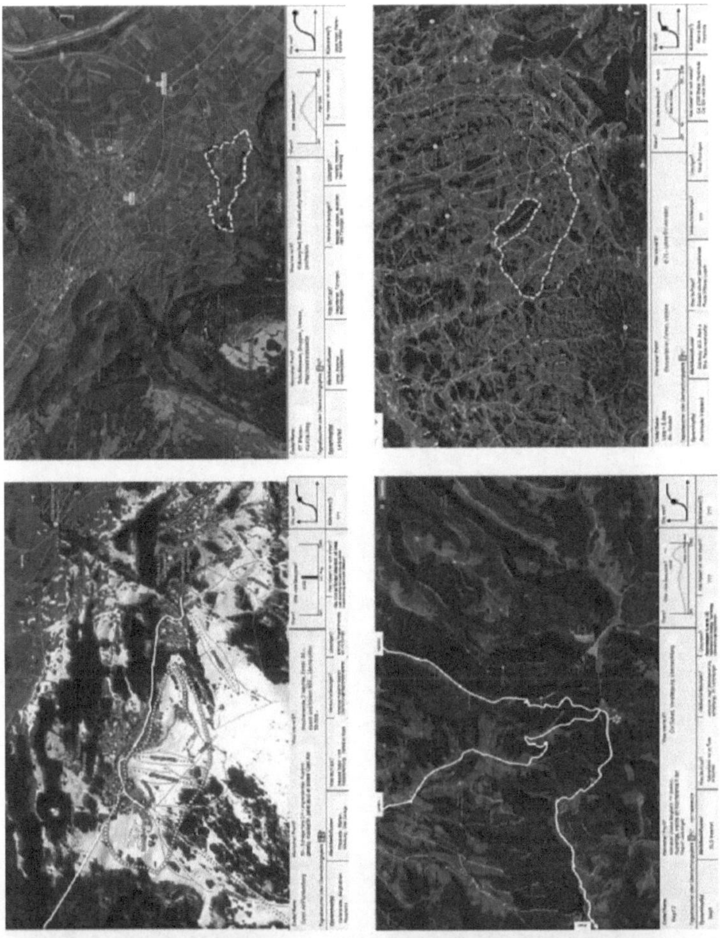

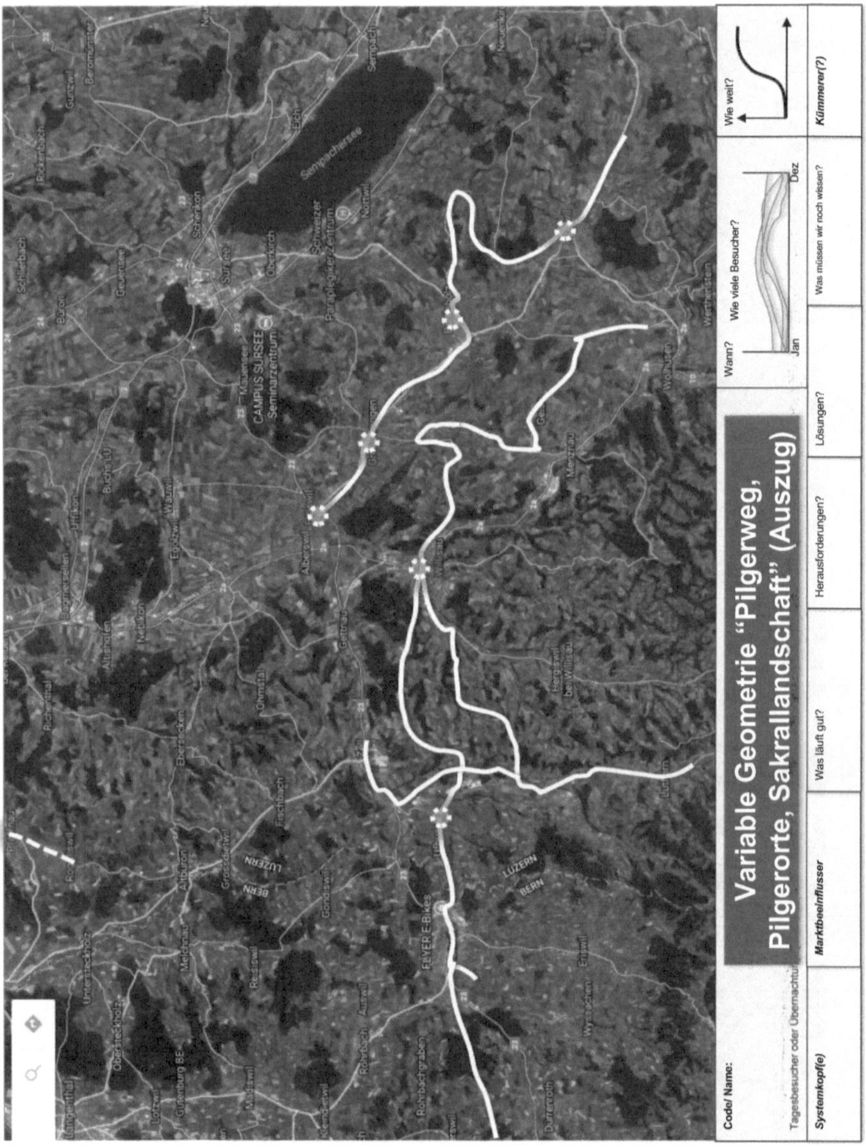
Variable Geometrie "Pilgerweg, Pilgerorte, Sakrallandschaft" (Auszug)
Code/ Name:
Wann?
Wie viele Besucher?
Wie weit?
Jan
Dez
Systemkopf(e)
Marktbeeinflusser
Was läuft gut?
Herausforderungen?
Lösungen?
Was müssen wir noch wissen?
Kümmerer(?)
Sempachersee
CAMPUS SURSEE
Seminarzentrum
LUZERN
BERN

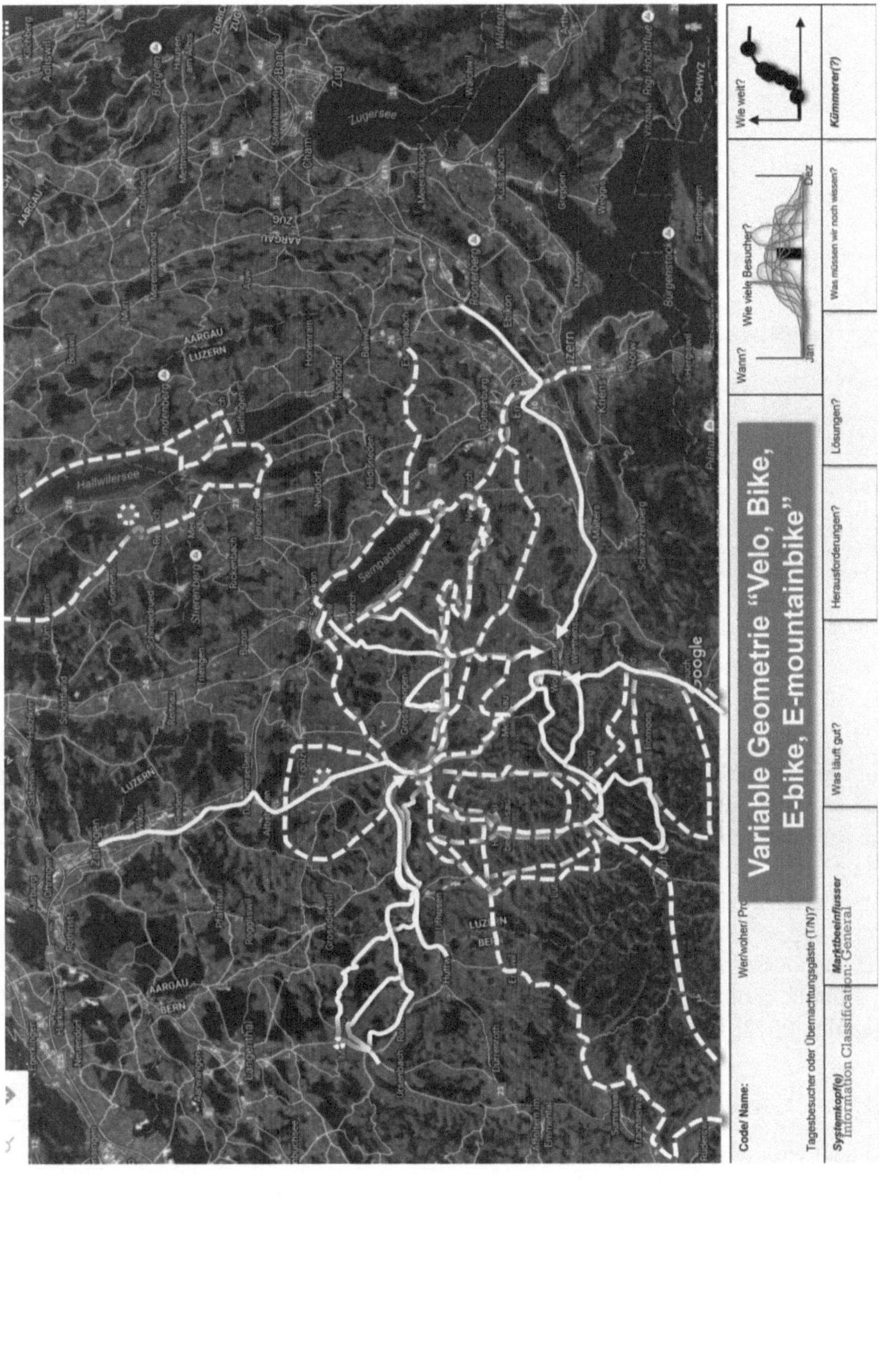
Variable Geometrie “Velo, Bike, E-bike, E-mountainbike”
Code/ Name:
Tagesbesucher oder Übernachtungsgäste (T/N)?
Systemkopf(e)
Marktbeeinflusser
Was läuft gut?
Herausforderungen?
Lösungen?
Wann?
Wie viele Besucher?
Was müssen wir noch wissen?
Wie weit?
Kümmerer(?)
Information Classification: General
Sempachersee
Hallwilersee
Zugersee
Luzern

Figure 13.4 Working with the visitor flow approach, selected impressions

Source: Own pictures

Finally, the individual visitor flows as well as the variable geometries are laid out or hung on walls to allow all actors at different levels to identify challenges and to concretely identify and discuss possible solutions. Not only the quality of the services provided to the visitors or their general satisfaction, but also the interaction in space and time with different visitor flows lead the actors to gain a differentiated view of phenomena that today may fall under the term "overtourism". Figure 13.4 presents a selection of moments of interaction between actors in various tourist places in the world as they work with the help of visitor flow maps in a public room, an instance called "destinorama" (Beritelli, Laesser, Reinhold & Kappler, 2013; Beritelli, Reinhold, Laesser & Bieger, 2015).

3.2 *Possible Solutions*

To deal with overtourism, extant literature presents a rich list of approaches at different levels (e.g. Dodds & Butler, 2019; Koens & Postma, 2017; Smeral, 2019). This section is not intended to list and assess them all, but to present only those that are directly and immediately (specifically in space and time) identified and decided by the actors, based on the visitor flows. In fact, visitor management must go beyond simply closing a tourist area or attraction (Innerhofer, Erschbamer & Pechlaner, 2019). The different approaches have been observed directly from practical work in numerous places and regions around the world.

3.2.1 On Points (Elements Along "Corridors")

a. Locate points and make new points possible

The more different visitor flows exist in an area, the more likely the tourism economy in that area will be resilient. A rich portfolio of visitor flows at different times of the year, in different places, with different types of visitors, guarantees that an area will be popular in the long term. In this sense, destinations could be compared to complex and dynamic ecosystems that are constantly renewing themselves, such as forests (Beritelli & Reinhold, 2021). Simply creating new stopping or meeting places can influence the movement of visitors. Potentially attractive places can be designed and offered along corridors, which include connecting streets, passages in shopping centres, alleyways, etc. Sketch A in Figure 13.5 at the end of this section shows how stakeholders can think about how to create potential places to stay and at the same time relieve pressure in the vicinity of a highly frequented attraction point.

b. Reduce or increase capacity

In certain cases, it is worthwhile to increase the absorption capacity of a particular zone or attraction. Where there is enough space, increasing the capacity of a place can provide a more pleasant atmosphere and enliven the area. In many cases, however, access to certain attractions must be limited. Quotas, usually coupled with raising access fees or increasing existing entrance prices, are the typical solutions. The attraction in green (Sketch B, Figure 13.5) increases or decreases in size.

c. Close or cancel points

In some cases, if the attraction is so heavily visited that its preservation for the community is endangered or even threatens to collapse, there is also the possibility of closing it (see Sketch C, Figure 13.5). Historical buildings worthy of protection are then only visited, for instance, by researchers and under strict conditions. The same can also happen in nature, where, for example, a heavily visited glacier retreats and geological risks (e.g. mud- or landslides) mean that it is no longer possible to walk on it.

d. Select and differentiate visitors

The price differentiation described in point B results in certain visitor flows being given preferential treatment, for example when admission costs a lot or when groups receive a discount. However, visitor flows can also be managed in a differentiated manner, such as staggering admission times. School classes, for example, could be admitted to a museum at an earlier or later time. This is shown in Sketch D, Figure 13.5. The dashed blue line represents a staggered or

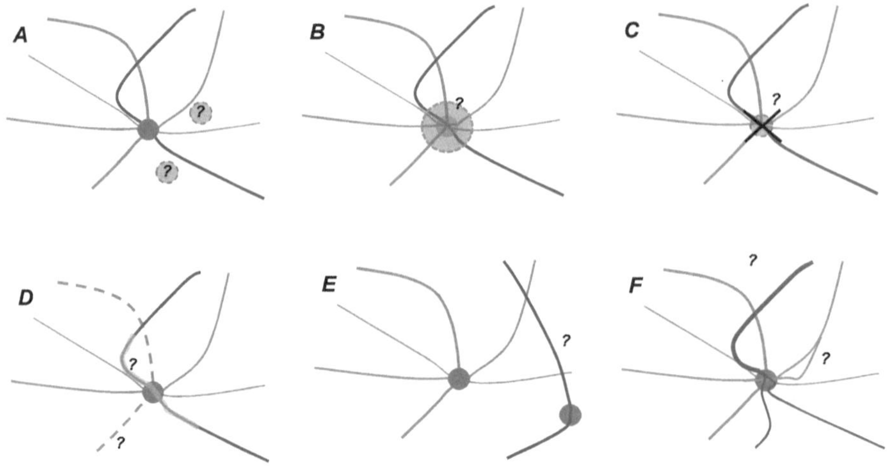

Figure 13.5 Selection of solutions A–F with the help of the visitor flow approach

Source: Own illustration

intermittent flow of visitors. The thick red line highlighted in green refers to a specially escorted flow that moves in narrow lanes and is guided through the attraction at a faster pace, for example.

3.2.2 On Lines ("Trajectories and Flows")

e. Deviate flows

Particularly for guided groups, but also for travellers under specific conditions and with specific needs, such as families with young children, alternative locations can result in individual visitor flows moving in other directions. This may be done by providing information as part of (pre-)trip planning and by directing visitors on site. In the latter case, the tourist gaze plays a decisive role in the popularity of a spot. Visitors will then spontaneously decide to move to another place. This is visualised in Sketch E, Figure 13.5.

f. Ramify flows

Ramifying particularly prominent flows along their path allows diluting them and spreading the people across different zones. Alternative routes such as streets, alleys, but also attractively designed over- or underpasses play a central role. In Sketch F, Figure 13.5, the red visitor flow is split into two flows after they have visited the central attraction. Each branch will continue to a different place and possibly reunite at a later point and time.

4 Overview and Conclusion

This article is not intended to provide an exhaustive list of methods and solutions to address overtourism. In the previous section, some proven and well-documented solutions were presented. Yet, it is amazing how a wide repertoire of ideas is intuitively and purposefully discussed and decided upon when working with a simple tool like visitor flows on maps (i.e. the destinorama). The common learning platform enables stakeholders at different levels to find solutions. This takes place by installing a bottom-up process of joint viewing, learning and decision-making. In this sense, this approach also demonstrates a new method of governance in communities.

Even for companies in which the ability to plan and manage exists to a large extent, there are more emergent strategies than intentionally intended and correspondingly realised strategies (Mintzberg & Waters, 1985). This fact is evident for the public sector (Jørgensen & Mintzberg, 1987) and consequently also in public spaces. Places or regions have generally been developed following emergent rather than written strategies.

Therefore, the "success" (or "failure") of a tourism destination is not the result of one strategy, but of several decisions made with varying degrees of clarity and intention. Purposefully intended and formulated comprehensive strategies of a tourism destination have very little if any impact. The reason why tourists come to, for example, a well-visited mountain location throughout the year is the result of several different tourism forms (e.g. spa guests in summer, winter sports guests, congress visitors), all of which have emerged, multiplied and decayed in a historical context on the basis of decisions made by different actors at different levels over several generations (e.g. construction of the congress centre, expansion of the cable car capacity, expansion of a camping site, renovation and expansion of a hotel). It is precisely these decisions and actions that produce the green dots in Figure 13.2. In sum, the actions may have contributed together to the collective "success" or the current state of development of a particular place. However, it is obvious that there could not have been an intended plan behind it according to which the actors decided and acted. The development was characterised over a longer period by different decisions, which resulted in smaller or larger impulses for the tourism destination.

For this reason, research on evolutionary economic geography and on path dependence has also gained importance in destination management in recent years (cf. Beritelli & Laesser, 2017; Ma & Hassink, 2013; Sanz-Ibáñez & Anton Clavé, 2014). This assumes that:

- basic resources (natural and social, cultural) determine the unique contextual factors of an area
- historical development, including the past and current forms of tourism, explains the current state and shapes future development
- the development itself, in different aspects, localities, times, can be completed by numerous independent decisions and actions, somewhat, or partially coordinated with each other (co-evolution).

The development of a destination is a process that arises in the longer term through different actors and organisations and their decisions, sometimes coherent, sometimes conflicting. This process is neither dependent on one individual or one organisation alone, nor can it be intentionally planned and implemented. Therefore, a differentiated bottom-up approach that recalls every actor's own responsibility while knowing and respecting the other co-existing realities of other actors in the same place is promising for practice and research.

The rise of concepts like tourism forecasting, destination marketing or management, carrying capacity, sustainable travel and overtourism unite our desire for the predictability or at least the plannability of a social phenomenon over which we will have only very limited control because we basically have still not understood the underlying mechanisms of its emergence and how it evolves. We are left with a common-sense approach that generates our curiosity, inquiry and will to learn: namely, trial and error.

Notes

1 Let's stick to the general definition for the term overtourism suggested by Milano, Cheer & Novelli who describe it as "the excessive growth of visitors leading to overcrowding in areas where residents suffer the consequences of temporary and seasonal tourism peaks, which have caused permanent changes to their lifestyles, denied access to amenities and damaged their general well-being" (2019, p. 1).

2 For example, what is concentrated at a certain place? What time of the week or season do which flows unfold? Which similar activities take place when and where?

References

Aparicio, D., Hernández Martín-Caro, M. S., García-Palomares, J. C., & Gutiérrez, J. (2021). Exploring the spatial patterns of visitor expenditure in cities using bank card transactions data. *Current Issues in Tourism*. https://doi.org/10.1080/13683500.2021.1991898

Baggio, R., & Scaglione, M. (2018). Destination attractions system and strategic visitor flows. Paper presented at the International Conference on Web Engineering, Cáceres, Spain, 5–8 June.

Beritelli, P. (2020). From flow analysis to shared insight to planning for impact: The development campaign of Altdorf (CH). In M. Volgger & D. Pfister (Eds.), *Atmospheric turn in culture and tourism: Place, design and process impacts on customer behaviour, marketing and branding*, Vol. 16. Emerald.

Beritelli, P., Crescini, G., Reinhold, S., & Schanderl, V. (2019). How flow-based destination management blends theory and method for practical impact. In N. Kozak & M. Kozak (Eds.), *Tourist destination management*. Springer.

Beritelli, P., & Laesser, C. (2017). The dynamics of destinations and tourism development. In D. R. Fesenmaier & Z. Xiang (Eds.), *Design science in tourism*. Springer.

Beritelli, P., Laesser, C., Reinhold, S., & Kappler, A. (2013). *Das St. Galler Modell für Destinationsmanagement – Geschäftsfeldinnovation in Netzwerken*. IMP-HSG.

Beritelli, P., & Reinhold, S. (2018). Chance meetings, the destination paradox, and the social origins of travel: Predicting traveler's whereabouts? *Tourist Studies*, *18*(4), 417–441.

Beritelli, P., & Reinhold, S. (2021). Sustainable destination management – What can we learn from forestry? In M. Valeri, A. Scuttari, & H. Pechlaner (Eds.), *Resilienza e sostenibilità: dinamiche globali e risposte locali*. Giappichelli.

Beritelli, P., Reinhold, S., & Laesser, C. (2020). Visitor flows, trajectories and corridors: Planning and designing places from the traveler's point of view. *Annals of Tourism Research*, *82*. https://doi.org/10.1016/j.annals.2020.102936

Beritelli, P., Reinhold, S., Laesser, C., & Bieger, T. (2015). *The St. Gallen model for destination management*. IMP-HSG.

Choi, S., Lehto, X. Y., & Oleary, J. T. (2007). What does the consumer want from a DMO website? A study of US and Canadian tourists' perspectives. *International Journal of Tourism Research, 9*(2), 59–72.

DiPietro, R. B., Wang, Y., Rompf, P., & Severt, D. (2007). At-destination visitor information search and venue decision strategies. *International Journal of Tourism Research, 9*(3), 175–188.

Dodds, R., & Butler, R. (2019). *Overtourism: Issues, realities and solutions* (Vol. 1). Walter de Gruyter GmbH & Co KG.

Fesenmaier, D. R., & Jeng, J. (2000). Assessing structure in the pleasure trip planning process. *Tourism Analysis, 5*(1), 13–27.

Goodwin, H. (2017). The challenge of overtourism. *Responsible Tourism Partnership, 4*, 1–19.

Hardin, G. (1968). The tragedy of the commons. *Science, 162*(3859), 1243–1248.

Hardy, A., Hyslop, S., Booth, K., Robards, B., Aryal, J., Gretzel, U., & Eccleston, R. (2017). Tracking tourists' travel with smartphone-based GPS technology: A methodological discussion. *Information Technology & Tourism, 17*(3), 255–274.

Innerhofer, E., Erschbamer, G., & Pechlaner, H. (2019). Overtourism: The challenge of managing the limits. In H. Pechlaner, E. Innerhofer & G. Erschbamer (Eds.), *Overtourism: Tourism management and solutions*. Routledge.

Ironmonger, D. (2000). *Household production and the household economy*. Melbourne.

Jørgensen, J., & Mintzberg, H. (1987). Emergent strategy for public policy. *Canadian Public Administration, 30*(2), 214–229.

Kádár, B., & Gede, M. (2021). Tourism flows in large-scale destination systems. *Annals of Tourism Research, 87*, 103113.

Koens, K., & Postma, A. (2017). *Understanding and managing visitor pressure in urban tourism: A study to into the nature of and tools used to manage visitor pressure in six major European cities*. CELTH.

Laesser, C., Küng, B., Beritelli, P., Boetsch, T., & Weilenmann, T. (2023). *Tourismus-Destinationen: Strukturen und Aufgaben sowie Herausforderungen und Perspektiven. Bericht im Auftrag des Staatssekretariats für Wirtschaft SECO*. [Tourism destinations: Structures and tasks as well as challenges and perspectives. Report on behalf of the State Secretary for Economy SECO] SECO.

Liu, W., Dong, C., & Chen, W. (2017). Mapping and quantifying spatial and temporal dynamics and bundles of travel flows of residents visiting urban parks. *Sustainability, 9*. doi:10.3390/su9081296

Ma, M., & Hassink, R. (2013). An evolutionary perspective on tourism area development. *Annals of Tourism Research, 41*, 89–109.

MacCannell, D. (2001). Tourist agency. *Tourist Studies, 1*(1), 23–37.

Maggi, R. (2014). *"Get there", "Stay there", "Live there" – Household production of the tourist experience and its implications for destinations*. Paper presented at the 2nd Biennial Forum "Advances in Destination Management", 10–13 June, St. Gallen, Switzerland.

Milano, C., Cheer, J., & Novelli, M. (2019). Overtourism: an evolving phenomenon. In C. Milano, J. Cheer & M. Novelli (Eds.), *Overtourism: Excesses, discontents and measures in travel and tourism*, CABI.

Mintzberg, H., & Waters, J. A. (1985). Of strategies, deliberate and emergent. *Strategic Management Journal, 6*(3), 257–272.

Moore, K., Smallman, C., Wilson, J., & Simmons, D. (2012). Dynamic in-destination decision-making: An adjustment model. *Tourism Management, 33*(3), 635–645.

Muth, R. F. (1966). Household production and consumer demand functions. *Econometrica, 34*(3), 699–708.

Paulino, I., Lozano, S., & Prats, L. (2021). Identifying tourism destinations from tourists' travel patterns. *Journal of Destination Marketing & Management, 19*, 100508.

Raun, J., Shoval, N., & Tiru, M. (2020). Gateways for intra-national tourism flows: Measured using two types of tracking technologies. *International Journal of Tourism Cities, 6*(2), 261–278.

Sanz-Ibáñez, C., & Anton Clavé, S. (2014). The evolution of destinations: Towards an evolutionary and relational economic geography approach. *Tourism Geographies, 16*(4), 563–579.

Scaglione, M., Baggio, R., & Doctor, M. (2021). *Evidence of the impact of weather conditions on visitor flows in urban destinations: The case of the "Geneva City pass"*. Paper presented at the 6th World Research Summit for Hospitality and Tourism, 14–15 December, online.

Séraphin, H., Gladkikh, T., & Vo Thanh, T. (2020). *Overtourism*. Springer.

Sheller, M., & Urry, J. (2006). The new mobilities paradigm. *Environment and Planning A, 38*(2), 207–226.

Smallman, C., & Moore, K. (2010). Process studies of tourist's decision-making. *Annals of Tourism Research, 37*(2), 397–422.

Smeral, E. (2019). Overcrowding of tourism destinations: Some suggestions for a solution. In H. Pechlaner, E. Innerhofer & G. Erschbamer (Eds.), *Overtourism: Tourism management and solutions*. Routledge.

Smith, S. L. J. (1994). The tourism product. *Annals of Tourism Research, 21*(3), 582–595.

Steiner, T., Baggio, R., Scaglione, M., & Favre, P. (2016). *Implementing lean destination management with strategic visitor flow (SVF) analysis*. Paper presented at the AIEST, Malta, 28–31 August.

UNWTO. (2022). *Glossary of tourism terms*. www.unwto.org/glossary-tourism-terms

Urry, J. (1990). *The tourist gaze: Leisure and travel in contemporary societies, theory, culture & society*. Sage.

Weber, F., Stettler, J., Crameri, U., Eggli, F., & Barth, M. (2019). *Measuring overtourism – Indicators for overtourism: Challenges and opportunities*. Lucerne University of Applied Sciences and Arts.

Williams, J., & Lawson, R. (2001). Community issues and resident opinions of tourism. *Annals of Tourism Research, 28*(2), 269–290.

14 Calling for a Transformative Destination Development

Narrative, Mindset and System

Harald Pechlaner, Hannes Thees and Julian Philipp

1 Crises and Transformation: The Lack of Responsibility of the Tourism System

The current manifold crises in the world – the political, economic, migration and demographic crises in particular – along with the multi-layered discussions on climate change are driving forces for transformation in economy, society and politics. Accordingly, they also challenge the tourism system and it is necessary to specifically address conflicts surrounding sustainability in order to put the transformation on an ethical path by reflecting on the conflicts (Philipp & Pechlaner, 2023).

In times of political instability and fragile mobility chains, issues around income, societal value shifts, and an increasingly important corporate crisis management provoke several questions that we will address subsequently in relation to the future viability of the tourism system. In this context, the "system" refers to the ability of self-controlling and self-monitoring of the tourism sector and tourism policy. Tourism stakeholders act in accordance to their interests which are based on specific topics – a systemic context helps to reduce conflicts and fragmentation, and build commitment and coalitions (Pechlaner et al., 2024). Thus, the elaborations in this chapter follow a conceptual approach.

The global COVID-19 pandemic showed once again that tourism actors do not interact as a coherent system. Instead, the many separate tourism segments called for political support independent of each other. In such an uncoordinated crisis mode, political measures were implemented in an uncoordinated way, without the possibility of distributing subsidies fairly (Dupeyras et al., 2020). It became clear that in many countries and regions the tourism system failed to communicate both the interconnectedness and network-relatedness of the sector (with its small and medium-sized companies) to tourism politics, and the vulnerability of the network in the event of a global crisis. This led to questions such as: Were political actors aware of the importance, structure and vulnerability of tourism? And in relation to the tourism system: What efforts did tourism actors and associations make in recent years to strengthen the industry's resilience in exchange with political actors?

Both resilience in times of crisis and a transformation towards sustainable tourism require a high degree of cooperation, which should be immanent in tourism (Meriläinen & Lemmetyinen, 2011). However, these recent events showed that thinking in systemic contexts is still lacking, which also reduces their power when

DOI: 10.4324/9781003365815-21

dealing with political actors. Destinations, tour operators, and accommodation and gastronomy businesses cooperate – if necessary – on the operational level (joint offers, networking), but they do not think in a macro and systemic context (strategic thinking, future prospects for the tourism value chain as a whole), which is in line with the cross-sectoral character of tourism and its societal relevance.

The tourism economy emphasises its role as an economic factor (Cárdenas-García & Pulido-Fernández, 2019), but merely as a societal phenomenon. The perception of tourism politics being social politics and not exclusively economic politics is only slowly gaining ground. Discussions on overtourism and the defensive, faltering dealing with it by the tourism system prove this. That climate change is still being underestimated, for example in the context of mega events in winter tourism, indicates business models and power that are still oriented at skimming off the demand as long as possible. Even after the COVID-19 pandemic, KPIs (key performance indicators) and reporting focus on pre-pandemic scales: many destinations and tour operators celebrate achieving their pre-COVID-19 levels. Thinking only in terms of arrivals or overnight stays prevents seeing the big picture around the social importance and responsibility of the global tourism industry. These socio-economic tensions could be summarised as follows:

> Tourism helps wide social classes with distraction, but it also represses the responsibility for shaping societal transformations itself.

Against this background, critical questions about the tourism system emerge:

- How can tourism as a system work as an agent of change?
- Although it is a socio-political function, is it enough to focus on providing a societal distraction from everyday problems and crises? How can a comprehensive participation of citizens in tourism development be enabled?

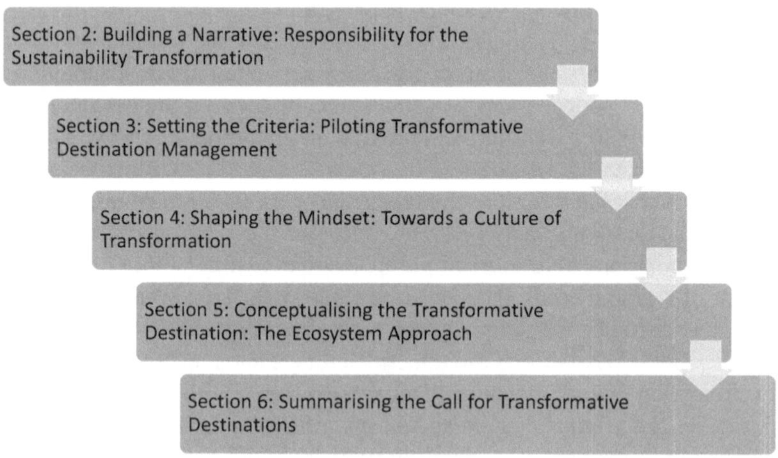

Figure 14.1 Structure of the chapter

Source: Own elaboration

- Is tourism primarily focusing on incremental adaptations to economic and societal changes, and is incremental adaption sufficient for substantial innovation and transformation of the underlying system itself?
- Is it possible that tourism – as a key global economic sector – does not contribute substantially to the ubiquitous socio-ecological transformation and to central questions of the future, as implied by its economic relevance?

Looking forward to a transformation where tourism is a proactive and responsible agent of change requires not only a new transformation mindset, but also a holistic and integrative development and management framework that focuses not only on economic goals, but also on societal values, environmental protection and future issues. This will be explored in the following sections that build upon each other (see Figure 14.1).

2 Building a Narrative: Responsibility for the Sustainability Transformation

In relation to sustainability, some tourism practices, such as the current level of air travel, are becoming increasingly controversial in society. This may be just the beginning of questioning levels of tourism that have been taken for granted before (Goessling et al., 2020). There are no system-led answers to the big questions indicated above; instead, particularistic areas of interest dominate the public debate. It is increasingly becoming apparent that the Sustainable Development Goals of the United Nations cause a variety of conflicts, especially social conflicts, around the question of how economic and social inequalities – for instance in terms of income (SDGs 1 and 8) or food security (SDGs 2 and 6) – can be reconciled with climate protection (Wong & van der Heijden, 2019). In addition, the COVID-19 pandemic and the respective debates on social media have considerably challenged or even harmed social structures.

The tourism system takes care of people while traveling; it is about hospitality, which only occurs through encounters. The creation and development of spaces of encounters is a central aspect of social interaction. Tourism has always done this. That also makes it extremely important. Nevertheless, the tourism system is not a driver for the development of new forms of social interaction; it certainly follows trends, but is it future-oriented enough to support and promote lifestyles and living environments that contribute to future transformations?

It should be tourism – given that it perceives itself as a coherent system more strongly – that leads the debate on future lifestyles actively and responsibly and thus inspires young people in particular (who seek meaningfulness in their actions) to work in tourism (Robinson & Schänzel, 2019). Lifestyles, in this context, refer to blurring boundaries between work and leisure, and, as such, to "an individual's personal and social behaviour, including consumption behaviour, leisure, work or civic activities" (Thees & Philipp, 2022, p. 68). The attractiveness of an industry for workers and jobseekers also relates to how it deals with change and whether it is possible to guarantee fairness in working relationships. If fairness is possible, the social situation and upgrading of the many professions in tourism will also be achieved (cf. Santos, 2023). However, the most important question is whether the tourism system adopts a leadership in this transformation towards climate neutrality. Related to this is the question of what contribution tourism can make to the preservation of biodiversity, as the IPBES (2019) is stressing the critical situation with regard to pollution, climate change, invasive alien species, exploitation of organisms or changes in land and sea use.

Global environmental changes inevitably require a central responsibility of the tourism system. The dramatic reduction in biodiversity and the increasingly visible consequences of climate change make it clear that tourism has to change in order to be fit for the future. Changes in land use or intensive agriculture and consumption of wild animals accelerate the loss of biodiversity and climate change globally. In the climate context, this requires a comprehensive monitoring system for companies and destinations in order to be able to determine evidence-based decisions.

Against this background, tourism could build upon its societal relevance and follow a new narrative:

> The tourism system has a decent culture of transformation and is a driver for change. It is also recognised as such, both inside and outside of the sector. Tourism stakeholders are aware of the broad cross-sectoral and transnational impacts and opportunities of their economy, which go far beyond economic aspects and encompass ecological, societal, and cultural aspects as well. Accordingly, tourism stakeholders understand their responsibility as societal actors and act as agents of change.

To ensure that their actions and goals are in line with the interests, goals and values in the destination as well as those of their employees and customers, a close exchange with their diverse stakeholders – customers, politics and the broader society – throughout the entire value chain takes place. Following such a responsible, sustainable and inclusive approach makes the upcoming and inevitable transformation the entire tourism economy's greatest strength.

3 Setting the Criteria: Piloting Transformative Destination Development

Against the previously sketched narrative on transformation in tourism, destinations are in a central position as places of encounters between guests and locals, and places where numerous and diverse hospitality services and lifestyle offers are available. How can a destination react quickly to external disruptions and new internal needs? What roles do destination management organisations (DMOs), service providers and the public have? To achieve even the double transformation (sustainability and digital transformation) of the tourism industry, destinations and DMOs face *inter alia* a re-orientation of processes and actions, collaboration or technologies (Reinhold et al., 2023).

In this regard, transformative destination development, on the one hand, means inspiring people about questions of the future and making appropriate suggestions for lifestyles in the destination that do not hide the conflicts, but actually address them. On the other hand, it means that guests need to contribute to sustainable and responsible development as well. This can be facilitated by companies going beyond the necessary, for example, by actively addressing human rights issues (and the difficulties of the global tourism system in dealing with them) instead of just communicating them. Doing so requires a certain knowledge of action, and an implementation strategy where responsibility is shared with guests.

Transformative action in tourism is aimed at connecting innovation in the economic system with social innovation, understanding civic engagement as an enrichment for the successful combination of holiday and living environments, and creating living environments that enable encounters but also allow for more social accommodation and innovations avoiding overcrowding (see Chapter 5). Corporate appreciation is one thing, social appreciation is another. Instead of "Corporate Social Responsibility", the term should rather be "Corporate towards Social Responsibility". It is about developing transformative skills in the destination system that foster future-shaping learning processes of guests and providers, with particular consideration of local society.

A detailed analysis of current crises and their possible effects is, at best, a necessary starting point to enable transformative learning. Realigning meaning structures also refers to the ability to access the appropriate mindset that enables an examination of one's own value and meaning foundations (Graupe & Bäuerle, 2022). The combination of mindset and current crises leads to new narratives.

Transformation is always associated with a paradigm shift. It can only work if humans are recognized not only as part of the economy and society, but participants instead (Razavi, 2022). The resulting ideas about possible future scenarios also create initial transition paths that need to be experimented with and piloted. Transformative destination management is necessary for "pilot destinations" that make experimentation with new lifestyles their mission.

Destination management is not only an enabler of transformative forces, i.e. those forces that want transformation, but also and above all should be understood as a way of shaping the future in a rapidly changing world. Dealing professionally with distraction and alleged "impossibilities" is just as important as enduring failed experiments; both must enable learning, ideally accessible to the entire destination network. Transformation competence means developing a high level of reflection on the tourism system, taking into account the potential of contributions to social transformation, and in particular developing specific action steps for the destination. Accordingly, destination management organizations must become sustainable tourism organizations (Philipp & Pechlaner, 2023).

In sum, the criteria for a transformative destination development are (among others):

- Observing, accompanying, anticipating and integrating trends (cf. Álvarez Jaramillo et al., 2018)
- Thinking and strategising holistically at the interface of resilience, sustainability, values and learning (cf. Thees et al., 2022)

- Strengthening regional responsibility and civic participation (cf. Philipp & Pechlaner, 2023)
- Understanding processes of transformation and the role of learning, experimenting and failing (cf. Zacher et al., 2021)
- Reflecting and discussing fundamental elements of values and meaning (cf. Seeler et al., 2021)
- Extending the boundaries of the tourism system (cf. Pechlaner et al., 2022a)

4 Shaping the Mindset: Towards a Culture of Transformation

The culture of transformation is central to the evolution of transformative destination development, with questions such as: What skills are needed to support transformation in an organisation or region? Is there a mindset for transformation and if so, what does it look like? Against the background of the issues of the tourism system in crisis situations described above, the culture of transformation emerges as a critical factor in enabling organisations and networks to navigate change successfully. Therefore, this section explores essential criteria for its development from literature and other branches.

Searching for the roots of a culture of transformation leads to a diverse set of organisational theories under the umbrella of the similar-sounding "cultural transformation". Related terms and theories are learning organisations (including fostering open communication, reflection and experimentation), agility (including iterative work processes, self-organisation and rapid feedback), change management (including planning, execution and control of processes), innovation culture (including creativity or new perspectives) and Positive Organizational Scholarship (focusing on strengths and engagement). These theories have different emphases that mainly refer to the transformation of culture, which is an *enabler* for the culture of transformation, but not the *same* as the culture of transformation. The same logic applies to organisational transformation (e.g. Levy & Merry, 1986) that aims to develop an organisation itself rather than shaping its socio-economic transformation.

As an enabler, culture has many facets: it promotes learning on an individual level, explains the behaviour of whole organisations, or even represents societies (McCalman, 2015, p. 4). Culture Change Theories are central to understanding a change in society. Those theories can be adopted for the transformation of the tourism sector towards more sustainable development. However, such transformation requires "A fundamental change in the meanings that cultural members attribute to their values and assumptions, which leads to a shift in the nature of cultural themes in use and the expressive content of the cultural paradigm" (McCalman, 2015, p. 4). Sketching the pathway towards cultural change includes several steps (see Figure 14.2): After *cultural reproduction* (repetition of established and known processes) and *cultural adaption* (changing form or tools, but not the meaning and goals of processes), *cultural transformation* is the most advanced type of cultural change, and aims to identify elements of organisational culture that are deemed redundant, thereby changing organisational form and the overall meaning and attitudes (McCalman, 2015).

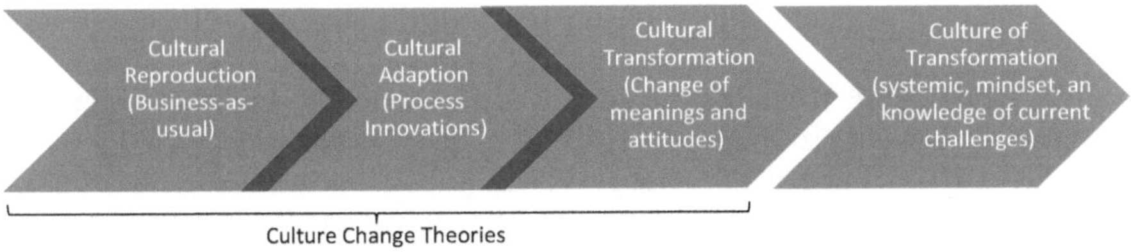

Figure 14.2 The emergence of cultural change theories

Source: Own elaboration

However, a more advanced *culture of transformation* is seldom discussed in the literature (cf. Lugtu, 2022). Which mindsets do we need to achieve the transformation of culture towards a culture of transformation? A first **definition** could be:

> The culture of transformation refers to a collective and holistic mindset, shared values, and behaviours within a group (organisation, region) or individual that foster adaptability, innovation and resilience in the light of multiple crises and a required transformation. This culture provides the foundation for embracing change, enabling organisations to seize opportunities and respond effectively to disruptions.

Reflection as a competence might be a central component for the sustainability transformation on an individual level (Seeler et al., 2021) but also in reference to the system. The stronger the ability to reflect, the better the system is prepared to take responsibility for the transformation. Such a transformation of culture exceeds the first-order change with established thinking and cultural sense-making towards a second-order change that impacts the cultural DNA and the paradigms of an organisation. Therefore, it is multidimensional, qualitative, discontinuous and radical (Levy & Merry, 1986).

Research has shown that a strong culture of transformation contributes to improved organisational performance, enhanced employee engagement, and a sustained competitive advantage (Quinn & Cameron, 2019). Although culture was long underrated in sustainability concepts, today, it gains importance in implementing a sustainability mindset. So far, culture in sustainability is isolated to issues of local implementation and regional differences, art and elites, education for sustainable development, lifestyles and sustainable consumption (Parodi, 2015). Parodi (2015) identified four deficits that include a lack of sensitivity for other cultures, handling sustainability primarily as a collective interest, formulating sustainability as an ethical rather than an aesthetic and practice-oriented approach, and neglecting non-material aspects of sustainability. However, other voices increasingly stress creativity and transformation: "Sustainability is only attainable if we regard it as a culture-transforming, creative project for the entire society" (Packalén, 2010, p. 118). This quote also highlights the close relationship between sustainability and transformation. Moreover, "Culture provides the necessary transformative dimension that ensures the sustainability of development processes", as UNESCO (2023) expressed it.

The culture of transformation is relevant to different groups of actors:

- **Organisations** often discuss the culture of sustainability across all functions, from management processes to the role of employees and accountability (Galpin et al., 2015). For example, Google is renowned for its culture of transformation, which encourages employees to experiment, take risks and challenge the status quo.
- **Individuals** often report positive, transformative changes in response to adversity. Cognitive transformation involves a turning point in a person's life characterised by (1) the recognition that coping with adversity can result in new opportunities and (2) the re-evaluation of the experience from one that was primarily traumatic or threatening to one that is growth-promoting. Research findings strongly supported the hypothesis that transformation predicts resilience (Tebes et al., 2004).
- **Regions** are seldom discussed while building a culture of transformation. Of course, there are similarities to community development (Mann et al., 2017), regional resilience (Thees et al., 2022) and urban transformation (Pechlaner et al., 2022a). However, tourism destinations are familiar with local networks, cooperation and interdependencies, and provide research on change and disruption, as seen especially during the COVID-19 pandemic.

Assuming that the culture of transformation requires the cooperation and integration of the mentioned actors, several obstacles emerge in practice:

- A lack of knowledge and experience in system change and turning away from growth thinking
- Conflicting goals and priorities
- Inadequate supporting data for evidence-based decision-making
- A lack of motivation and incentives
- Insufficient time and money to engage in transformation and make appropriate investments
- Failure to involve the community and relevant stakeholders
- Misunderstanding and mismanaging the cultural dimensions of change (Álvarez Jaramillo et al., 2018; Komatsu et al., 2019; McCalman, 2015; Stewart et al., 2016)

These obstacles can also be found in theory, e.g. discussing the Culture of Change or Resistance to Change. It is difficult to sketch a fast way to implement such a culture of transformation. McCalman (2015) is even calling for an overnight change of the company culture towards a more social capitalism. Besides such disruption, there are five critical elements for implementing the Culture of Transformation:

1. Leadership Commitment
2. Learning Orientation
3. Transparent Communication

4. Agile Structures and Processes
5. Collaboration and Empowerment

The implementation of the culture of transformation is a process of multiple stages that is characterised by transdisciplinarity, including *inter alia* behavioural science, organisational theories and human–nature interaction. In practice, there might be supportive guidelines or values, such as the Agile Manifesto, which defines, for example, socio-ecological restoration over economic justification or transformative system change over small steps to keep business as usual (Beck et al., 2001). However, in order for such transdisciplinary approaches to be implemented successfully and allow for transformative thinking, traditional theories and practices must be rethought and further developed. They need to become more agile, flexible and holistic in order to consider the interests of various actors and stakeholder groups, and thereby meet the demands of an ever more complex and diverse global society.

5 Conceptualising the Transformative Destination: The Ecosystem Approach

Against the background of the current transformation in destinations, the calls for thinking in systemic contexts, and possible paths through the culture of transformation, the tourism system lacks a conceptual framework that addresses those issues, as described in Section 1. What are the necessary factors for establishing such a framework and facilitating the tourism system to take up its cross-sectoral and interdisciplinary responsibility? With respect to the aforementioned changing framework conditions in the context of socio-ecological values, public sustainability debates, and new, blended forms of work, life and leisure, respective spatial and tourism frameworks need to be increasingly flexible (Bieger & Klumbies, 2022; Pechlaner, 2022). In spatial or societal contexts, **flexibility** allows for the ability to adapt to changing circumstances and adjust processes, strategies or initiatives where needed. A high flexibility also helps individuals, organisations and networks to both prepare for and react to crisis situations and, thus, increase their resilience (Pechlaner et al., 2022b). Accordingly, flexibility can be seen as a fundamental prerequisite for fostering and strengthening experimentation and innovation, which, in turn, facilitate the development of new paths and directions (Brouder, 2020). An **openness to innovation** not only helps solve problems or tackle challenges, but also allows for the creation of new ideas, technologies and processes. This may ultimately lead to a set of products and services that are different from those of competitors and, thus, result in a competitive advantage (Eckert & Pechlaner, 2019; Tessarin & Azzoni, 2022). Keeping in mind the increasing integrated consideration of destination and living spaces and the respective synergies and overlaps of destination development and living space development, **cooperation** of all stakeholders is a basis for transformational processes (Zacher et al., 2021). This allows for the exchange of knowledge and resources and, thereby, reduces risks and conflicts and enhances synergies (Nielsen, 2005).

Those criteria mentioned are represented by the sum of different approaches that shape current destination management. A more flexible and holistic approach is the Ecosystem of Hospitality (EoH; see Figure 14.3), representing a spatial adoption of the ecosystem approach. In recent years, the ecosystem term has been increasingly adopted by a variety of disciplines and contexts, particularly by the business and economics environment. Stam and Spigel (2017, p. 1), for example, describe an entrepreneurial ecosystem as a "set of independent actors and factors coordinated in such a way that they enable productive entrepreneurship within a particular territory". It has both spatial and organisational dimensions and consists of formal, physical, market, cultural and systemic elements. Accordingly, an entrepreneurial ecosystem with its focus on actors, interactions and relationships has the potential to foster local and regional development (Bachinger et al., 2020). Adopted to the spatial development context, the ecosystem approach can help to connect the different and diverse spatial, social and economic networks and integrate them into a larger spatial ecosystem.

The EoH mentioned above aims to integrate various geographical layers such as the business location, the tourist destination, and the living and leisure space of locals and residents (Pechlaner et al., 2022a; Philipp et al., 2022). It allows for not only a rethinking of traditional and well-established structures within destinations as well as the entire tourism system, but also for a particular focus on sustaining and expanding stakeholder networks. Following the earlier elaborations on culture as an enabler for change and transformation, the EoH can be an approach to further develop tourism to a more sustainable tourism culture. To better understand the transformative potential of the EoH, it will subsequently be illustrated by focusing on four key characteristics:

1. **Actors and competencies**: The focus of the EoH is not on organisations, but rather on "the individual and the opportunity for encounters between individuals [as well as] on issues surrounding quality of life, resilience, culture, mobility and connectivity" (Pechlaner et al., 2022a, p. 12). The holistic approach of the EoH, both in terms of spatial settings and target groups, allows not only for a consideration of different opinions and discussions, but also for an integration of the various local and regional stakeholder groups – politicians, residents, visitors, businesses, institutions or entrepreneurs alike – and their quite diverse lifestyles, experiences and visions, which were often influenced by the COVID-19 pandemic (Philipp et al., 2023). Transformative actors can actively integrate their competencies, responsibilities and self-efficacy, making spatial development a shared responsibility.
2. **Organisational models, participation and leadership**: Customer and demand-side needs are in a constant state of change (cf. Thorns, 2002) and, thus, require urban and rural spaces to adapt accordingly and focus on public interest and sustainable, long-term spatial development rather than short-term interests or economically driven business models (Ferguson, 2019). The participation and involvement of all stakeholders through bottom-up approaches (cf. Thees et al., 2020) is crucial to ensure that all needs are met and that meaning structures can be adapted. New organisational models beyond growth- and success-oriented

models need to be developed. Clear leadership that defines the rights, roles, relationships and responsibilities of actors and stakeholders is essential for integrated approaches to work (cf. Aitken & Campelo, 2011).

3. **Processes**: Key processes that help achieve a transformative destination are manifold. Of particular importance is strategic planning, as it helps to develop a clear and understandable vision and plan long-term. Community engagement, including regular feedback and consultations, is vital to ensure that the transformation reflects the actual needs. This may include changing guest structures. A collaborative governance, encompassing public–private partnerships or stakeholder committees, can help with this (cf. Pechlaner & Philipp, 2024). These processes, among others, can be supported by policy development, integrating innovation, inclusivity and sustainability and adjusting regulations, where needed. However, this requires a certain openness to the deconstruction of paradigms, experimentation with new paths of transition, and the integration of customers along the entire process, built on visions and strategies.
4. **Tasks and fields of action**: The tasks and fields of action illustrated in the EoH are diverse and linked to many ideas and terms mentioned above. Fostering the start-up scene is important to attract talent and businesses, including their innovative ideas and networks, to the respective place. Engagement and involvement of various stakeholders needs to happen at the earliest possible stage (cf. Philipp et al., 2023). The boundaries between urban and rural spaces need to be overcome through investments in digitalisation, mobility and others (Pechlaner, 2022). The well-being of stakeholders and modern leisure options need to be integrated. Culture and creativity play an increasingly important role, allowing for individual and authentic experiences and, thereby, enhancing the quality of life of individuals. Furthermore, culture constitutes identity and strengthens the coordination and collaboration of individuals, making it an essential component of integrated approaches.

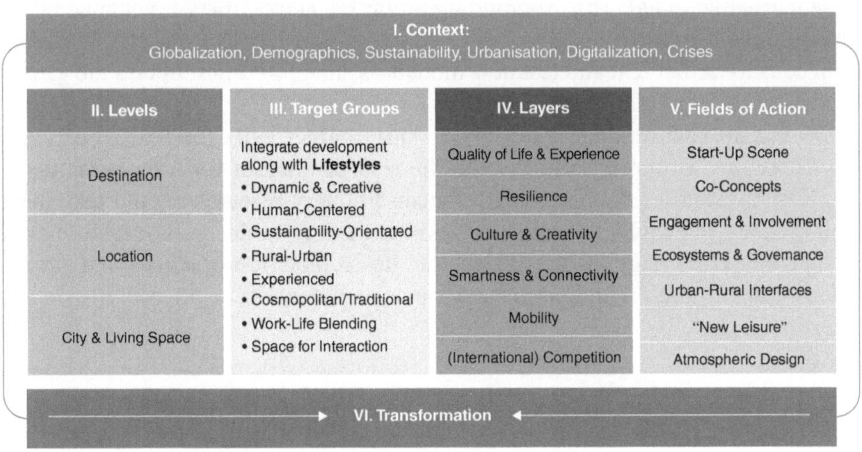

Figure 14.3 The Ecosystem of Hospitality

Source: Pechlaner et al., 2022a

As shown in Figure 14.3, the EoH and its holistic integration of the location, destination and living space facilitates the creation, development and implementation of new concepts and respective fields of action that meet the requirements for a spatial transformation. This fosters the development of a transformative destination by focusing on more individualistic and target-group-oriented discussions.

6 Summarising the Call for Transformative Destinations

Numerous crises, challenges and trends affect the ability of the tourism system to act as an enabler of change and transformation. At the same time, tourism with its cross-sectoral nature has the potential to actively contribute to discussions on future lifestyles and questions of economic, ecological and social sustainability and transformation. Guest flows enable the activation of supply or offer systems, each with their own dynamics. This complexity must be taken into account in order to recognise the right balance between transformation and sustainability (St. Gallen Model for Destination Management; Reinhold et al., 2023).

From our conceptual perspective, transformative destination development goes beyond the mere management of tourism destinations – it is a result of different criteria, measures and actions that actively shape the future through observation, adaptation, experimentation and learning. Such a new understanding of tourism builds on the Culture of Transformation in a systemic view. The stronger this system is able to reflect on itself, the better it can take over responsibility. The EoH is a holistic and integrated tool that can support achieving this vision by empowering transformative actors with their competencies and responsibilities through adjustable and participative organisational models and processes. To support this transformation and take over a leadership responsibility, DMOs should become SDMOs (sustainable destination management organizations) or SMOs (sustainability management organizations). This re-branding might be an example of a change which then requires new structures, tasks and narratives. The tourism industry should be courageous to implement and test new models as such piloting supports the gathering of stakeholders behind new narratives.

Opening up such holistic questions in this chapter calls for new research as well. Therefore, we propose to discuss the role of tourism for transformation at first on an abstract level. Afterwards, transdisciplinary research should take place that integrates stakeholders, travellers and the society to design a transformative destination. At this stage, it is also important to include best practices that already piloted a transformation in their destination or organization.

References

Aitken, R., & Campelo, A. (2011). The four Rs of place branding. *Journal of Place Marketing Management*, *27*(9–10), 913–933.

Álvarez Jaramillo, J., Zartha Sossa, J. W., & Orozco Mendoza, G. L. (2018). Barriers to sustainability for small and medium enterprises in the framework of sustainable development – Literature review. *Business Strategy and the Environment*. https://doi.org/10.1002/bse.2261

Bachinger, M., Kofler, I., & Pechlaner, H. (2020). Sustainable instead of high-growth? Entrepreneurial ecosystems in tourism. *Journal of Hospitality and Tourism Management, 44*, 238–244.

Beck, K., Beedle, M., van Bennekum, A., Cockburn, A., Cunningham, W., Fowler, M., Grenning, J., Highsmith, J., Hunt, A., & Jeffries, R. (2001). *The agile manifesto*. The Agile Alliance. www.agilemanifesto.org.

Bieger, T., & Klumbies, A. (2022). From destination management to integrated development of places – enabling personal networks instead of management and control. In H. Pechlaner, N. Olbrich, J. Philipp & H. Thees (Eds.), *Towards an ecosystem of hospitality – Location:City:Destination*. Graffeg.

Brouder, P. (2020). Reset redux: Possible evolutionary pathways towards the transformation of tourism in a COVID-19 world. *Tourism Geographies, 22*(3), 484–490.

Cárdenas-García, P. J., & Pulido-Fernández, J. I. (2019). Tourism as an economic development tool: Key factors. *Current Issues in Tourism, 22*(17), 2082–2108.

Dupeyras, A., Haxton, P., & Stacey, J. (2020). The Covid-19 crisis and tourism: Response and recovery measures to support the tourism sector in OECD countries. *G20 Insights*. www.global-solutions-initiative.org/policy_brief/the-covid-19-crisis-and-tourism-response-and-recovery-measures-to-support-the-tourism-sector-in-oecd-countries/

Eckert, C., & Pechlaner, H. (2019). Alternative product development as strategy towards sustainability in tourism: The case of Lanzarote. *Sustainability, 11*(13), 3588.

Ferguson, F. (2019). *Make city: A compendium of urban alternatives*. Jovis.

Galpin, T., Whittington, J. L., & Bell, G. (2015). Is your sustainability strategy sustainable? Creating a culture of sustainability. *Corporate Governance, 15*(1), 1–17. https://doi.org/10.1108/CG-01-2013-0004

Goessling, S., Humpe, A., & Bausch, T. (2020). Does "flight shame" affect social norms? Changing perspectives on the desirability of air travel in Germany. *Journal of Cleaner Production, 266*, 122015.

Graupe, S. & Bäuerle, L. (2022). Bildung in fragilen Zeiten: Die Spirale transformativen Lernens [Education in fragile times: The spiral of transformative learning]. Working Paper Series, No. 70, Cusanus University of Society Design. www.econstor.eu/handle/10419/261476

IPBES. (2019). *Global assessment report on biodiversity and ecosystem services of the Intergovernmental Science-Policy Platform on Biodiversity and Ecosystem Services*. IPBES Secretariat. https://doi.org/10.5281/zenodo.3831673

Komatsu, H., Rappleye, J., & Silova, I. (2019). Culture and the independent self: Obstacles to environmental sustainability? *Anthropocene, 26*, 100198. https://doi.org/10.1016/j.ancene.2019.100198

Levy, A., & Merry, U. (1986). *Organizational transformation: Approaches, strategies, theories*. Praeger.

Lugtu, R. J. (2022). The culture of transformation. In F. Granito (Ed.), *Digital transformation demystified*. World Scientific.

Mann, S., Eden-Mann, P., Smith, L., Ker, G., Osborne, P., & Crawford, P. A. (2017). A transformation mindset as the basis for sustainable community development. *Community Development, 15*, 59.

McCalman, J. (2015). *Leading cultural change: The theory and practice of successful organizational transformation*. Kogan Page.

Meriläinen, K., & Lemmetyinen, A. (2011). Destination network management: A conceptual analysis. *Tourism Review, 66*(3), 25–31.

Nielsen, B. B. (2005). The role of knowledge embeddedness in the creation of synergies in strategic alliances. *Journal of Business Research, 58*(9), 1194–1204.

Packalén, S. (2010). *Culture and sustainability: Corporate social responsibility and environmental management.* https://doi.org/10.1002/csr.236

Parodi, O. (2015). The missing aspect of culture in sustainability concepts. In J. C. Enders & M. Remig (Eds.), *Routledge studies in sustainable development. Theories of sustainable development.* Routledge.

Pechlaner, H. (2022). From urban and rural places to integrative and transformative spaces – Ecosystems of hospitality as an expression of future-oriented living together. In H. Pechlaner, N. Olbrich, J. Philipp & H. Thees (Eds.), *Towards an ecosystem of hospitality – Location:City:Destination.* Graffeg.

Pechlaner, H., Olbrich, N., Philipp, J., & Thees, H. (Eds.). (2022a). *Towards an ecosystem of hospitality – Location:City:Destination.* Graffeg.

Pechlaner, H., & Philipp, J. (2024). Addressing wicked problems through integrated policy-making: An ecosystem-based approach. In C. Pforr, M. Pillmayer, M. Joppe, N. Scherle & H. Pechlaner (Eds.), *Tourism policy-making in the context of contested wicked problems.* Emerald.

Pechlaner, H., Philipp, J., & Olbrich, N. (2024). Destination governance: The new role of destination management, stakeholder networks and sustainability. In J. Saarinen & C. M. Hall (Eds.), *The handbook of destination governance.* Edward Elgar.

Pechlaner, H., Störmann, E., & Zacher, D. (Eds.) (2022b). *Resilienz als Strategie in Region, Destination und Unternehmen [Resilience as strategy in region, destination and business].* Springer.

Philipp, J., & Pechlaner, H. (2023). Towards places and ecosystems: The integrated management of locations, destinations, and the living space. In H. Y. Arias Gomez & G. Antošová (Eds.), *Considerations of territorial planning, space, and economic activity in the global economy.* IGI Global.

Philipp, J., Thees, H., Olbrich, N., & Pechlaner, H. (2022). Towards an ecosystem of hospitality: The dynamic future of destinations. *Sustainability, 14*(2), 821.

Philipp, J., Volgger, M., Pforr, C., Pechlaner, H., & Gon, M. (2023). *Transformations for a more resilient tourism system: Learnings from COVID-19 in Australia, Germany and Italy.* CAUTHE 2023 Conference, Perth, Australia, February 7–9.

Quinn, R. E., & Cameron, K. S. (2019). Positive Organizational Scholarship and agents of change. In A. B. Shami & D. A. Noumair (Eds.), *Research in organizational change and development* (Vol. 27). Emerald Publishing.

Razavi, R. (2022). *Die Magie der Transformation: Wie wir in Zukunft in Wirtschaft und Gesellschaft gemeinsam gestalten* [*The magic of transformation: How we shape economy and society together in the future*]. Haufe.

Reinhold, S., Beritelli, P., Fyall, A., Choi, H.-S. C., Laesser, C., & Joppe, M. (2023). State-of-the-art review on destination marketing and destination management. *Tourism and Hospitality, 4*(4), 584–603. https://doi.org/10.3390/tourhosp4040036

Robinson, V. M., & Schänzel, H. A. (2019). A tourism inflex: Generation Z travel experiences. *Journal of Tourism Futures, 5*(2), 127–141.

Santos, E. (2023). From neglect to progress: Assessing social sustainability and decent work in the tourism sector. *Sustainability, 15*(13), 10329.

Seeler, S., Zacher, D., Pechlaner, H., & Thees, H. (2021). Tourists as reflexive agents of change: Proposing a conceptual framework towards sustainable consumption. *Scandinavian Journal of Tourism and Hospitality, 21*(3), 1–19.

Stam, E., & Spigel, B. (2017). Entrepreneurial ecosystems. In R. Blackburn, D. De Clercq, J. Heinonen & Z. Wang (Eds.), *Handbook of entrepreneurship and small business.* Sage.

Stewart, R., Bey, N., & Boks, C. (2016). Exploration of the barriers to implementing different types of sustainability approaches. *Procedia CIRP*, *48*, 22–27.

Tebes, J. K., Irish, J. T., Puglisi Vasquez, M. J., & Perkins, D. V. (2004). Cognitive transformation as a marker of resilience. *Substance Use & Misuse*, *39*(5), 769–788. https://doi.org/10.1081/ja-120034015

Tessarin, M. S., & Azzoni, C. R. (2022). Innovation and competitiveness: The regional dimension. In E. Amann & P. N. Figueiredo (Eds.), *Policy for innovation, competitiveness and development in Latin America: Lessons from the past and perspectives for the future*. Oxford University Press.

Thees, H., Pechlaner, H., Olbrich, N., & Schuhbert, A. (2020). The living lab as a tool to promote residents' participation in destination governance. *Sustainability*, *12*(3), 1120.

Thees, H., & Philipp, J. (2022). Maximum flexibility and experience! Lifestyles as the new customer approach? In H. Pechlaner, N. Olbrich, J. Philipp & H. Thees (Eds.), *Towards an ecosystem of hospitality – Location:City:Destination*. Graffeg.

Thees, H., Störmann, E., & Pechlaner, H. (2022). Business modeling for resilient destination development: A multi-method approach for the case of destination Franconia, Germany. *Tourism Planning & Development*. https://doi.org/10.1080/21568316.2022.2121313

Thorns, D. C. (2002). *The transformation of cities: Urban theory and urban life*. Palgrave Macmillan.

UNESCO. (Ed.). (2023, 21 June). *Culture & sustainable development*. www.unesco.org/en/sustainable-development/culture

Wong, R., & van der Heijden, J. (2019). Avoidance of conflicts and trade-offs: A challenge for the policy integration of the United Nations Sustainable Development Goals. *Sustainable Development*, *27*(5). https://doi.org/10.1002/sd.1944

Zacher, D., Thees, H., & Herbold, V. (2021). The Tourism Lab: A place for change, participation, and future destination development. In A. Farmaki & N. Pappas (Eds.), *Emerging Transformations in Tourism and Hospitality*. Routledge.

15 Destination Design for Sustainable Tourism

Considerations for Bridging the Gap Between Design Thinking and Critical Perspectives

Greta Erschbamer and Natalie Olbrich

1.0 Introduction

Tourism has experienced exponential growth in recent years and has become a major global industry (World Tourism Organization [UNWTO], 2023). While this growth brought economic benefits to many destinations, it also resulted in a number of negative impacts. These undesirable consequences provoked a discussion on overtourism as a phenomenon, including concepts of quantitative measurement such as carrying capacity. As an impacting factor, overtourism can lead to stress on infrastructure, destruction of cultural heritage and natural resources, and conflict between tourists and locals. Popular tourist destinations such as Venice, Barcelona and Bali have become symbols of the challenges associated with overtourism (Pechlaner et al., 2018; Zacher et al., 2020). These issues led to the recognition of the need for change. Tourism stakeholders, including governments, destination management organisations, local communities and businesses, have embraced sustainable tourism policies as a solution. This expanded approach aims to balance the economic benefits of tourism with its impact on the environment, culture and society (Dávid, 2011; Olbrich et al., 2022).

At the same time, the transformation towards sustainability in connection with challenges for the environment and society are central topics of current discussions in tourism research (Budeanu et al., 2016). To date, various approaches from management, leadership and governance research have been used to discuss key issues in tourism. Many of these considerations have already found their way into tourism research and the tourism industry, influencing political measures and decisions. Changing political, social, technological and economic developments increase the complexity of processes in destinations and require approaches that contribute to a better understanding of destinations in terms of their future development, governance and management, particularly in relation to the quality of life of inhabitants and experiences of guests (Butler, 2023).

In this context, the fusion of research in the domains of tourism and design has made its mark within the field of destination research in the form of destination design. This is an approach to tourism that aims to create sustainable and attractive destinations. It involves developing a holistic approach to destination

DOI: 10.4324/9781003365815-22

development that considers the needs of both visitors and locals and focuses on long-term economic, social and environmental impacts (Erschbamer et al., 2023; Koens et al., 2021). It can be seen as a structured process that includes steps to make a destination attractive, sustainable and competitive: (1) strategic planning, (2) market research and target group analysis, (3) infrastructure and supply design, (4) sustainability and environmental protection, (5) local community involvement, and (6) marketing and promotion (Tussyadiah, 2014).

2.0 Advancing a Holistic View on Destinations

The concept of destination design enables a holistic view of a destination in order to analyse previously separate components and processes as a unified element. This holistic view of destinations is achieved by including design as an instrument for visualising processes, as a disruptive element for creating new perspectives, as a vehicle for trans- and interdisciplinarity, and as a tool that enables participation. Destination design is characterised by its blurred boundaries, as challenges are approached from a non-disciplinary perspective and solutions are sketched and prototyped at the interface of existing disciplines. Therefore, it should support the transition from multidisciplinarity to transdisciplinarity in tourism research (Scuttari et al., 2020). Undoubtedly, the discussion around destination design is still in its infancy and appears somewhat unstructured, and theoretical concepts and foundations need to be refined. Above all, an expansive combination with other topics and perspectives from different disciplines should be cultivated, despite some of the imprecision it entails (Volgger et al., 2021).

Design and design research harbour great potential for the innovative development of sustainable and liveable destinations. The focus is primarily on the various starting points for design interventions, as well as processes and methods. Design – whether as "silent design" in the early phase of destination development or, as is currently the case, in a transdisciplinary network – can make important contributions to the very different challenges facing a destination (Steffen, 2023).

Likewise, other authors discussed design approaches in tourism: Fesenmaier and Xiang introduced the concept of design science as a framework which can guide both the theoretical foundations and applications in tourism design. Their advancements in theoretical frameworks, research methodologies and practical applications lay the groundwork for a novel paradigm known as Design Science in Tourism (DST). DST introduces a comprehensive framework for crafting systems and artifacts aimed at enhancing the daily lives and travel experiences of individuals. It should be emphasised that DST transcends the mere creation of events or locations to enhance the traveller's experience; rather, it serves as a fundamental framework for conducting research and designing tourism destinations. Furthermore, DST places explicit emphasis on the creation of innovative artifacts, providing a solid basis for tourism managers to devise pioneering processes, systems and destinations. Consequently, DST can be effectively employed to inform tourism research by amalgamating the principles of design thinking and the science of design, the intrinsic nature of the visitor's experience, and the potential artifacts that can be devised to govern and enhance these experiences (Fesenmaier & Xiang, 2017).

In this context, the interweaving of tourism research and design research has found its way into destination research. The utilisation of design thinking and participatory design approaches in the context of tourism involves the application of methodologies to enhance and innovate various aspects of the tourism industry. Design thinking, a problem-solving approach, is employed to develop creative solutions that cater to the needs and desires of tourists. It emphasises empathy, ideation and prototyping to create user-centred experiences and services in tourism. Taking a broader view, participatory design encourages the active involvement of all relevant stakeholders in the design process, including tourists, local communities and businesses. It aims to ensure that the outcomes of tourism development initiatives are more inclusive, sustainable and reflective of the collective aspirations and values of the involved parties. Together, design thinking and participatory design provide a comprehensive framework for designing tourism experiences and destinations that are both innovative and responsive to the diverse needs and perspectives of the tourism ecosystem.

3.0 Design Thinking for Participation as "Cure-All"?

Isolated approaches from design thinking and participatory design seem to have been particularly successful in their application in destinations, where design approaches in tourism research are predominantly governed by productive business agendas (i.e. fiscal gain and/or competitive advantages of a destination), choosing certain methodical orientations such as design thinking and other user- and customer-oriented approaches (Boedker, 2023).

Nevertheless, critics of design thinking are not new and have voiced several reservations and objections regarding their application. Some of the key criticisms encompass, for example, the tendency of design thinking to prioritise commercial interests and profitability over broader social and environmental concerns.

The author Jonas (1998) outlines the traditional notion of design as problem-solving suitable at a cognitive level but becoming misleading when applied to societal contexts. Unlike in mathematics, where solutions eliminate problems and establish secure knowledge, the dynamics of the market reveal a different reality, where also research on tourism and destinations would be accounted to. The market often generates "solutions" for existing "problems" or even invents "problems" to fit new "solutions", challenging the conventional problem–solution dichotomy. This commercial orientation can lead to the over-commercialisation of tourist destinations, potentially eroding the authenticity of local cultures and exacerbating negative socio-economic impacts. Other design scholars such as Nigel Cross and Charles Burnett were among the early voices to critique the overly normative interpretation of the concept of design thinking. Jonas (2010), while recognising that the use of design thinking in decision-making could broaden the range of challenges that design could tackle significantly, voiced concerns about its limited capacity to encourage reflection on the unique characteristics of a particular problem and its ability to support a contextually grounded design process effectively (Jonas, 2010). Moreover, design thinking is often promoted as the new medium/method/tool to advance what is sometimes fuzzily labelled the "great transformation" towards the better (Jonas, 2010). Also, Ackermann (2023) shows a multitude of challenges still unresolved nearly two decades after the ascent of design thinking. Design processes themselves need to evolve beyond design thinking. A fundamental critique of design thinking is that it has corporate origins, which firmly embed the methodology within a capitalist framework. More recently, a strand has been developed in design research that is primarily characterised by the term "critical design", where an alternative viewpoint on the design process has emerged, one that is less driven by the generation of economic value in the marketplace (such as design thinking) and more concerned with fostering reflection, addressing political dimensions, upholding values and confronting the intrinsic challenges within the field of tourism (Boedker, 2023). In this context, adopting a justice-oriented perspective could facilitate collaboration and creativity on a broader scale, transcending existing power structures. Contemplating and recognising that capitalism does not constitute an immutable or intrinsic foundational law of nature can serve as a catalyst for critical design within the concept of destination design, which seeks to redefine societal structures and promote sustainable, participative, equitable systems in tourism as a sector and destinations as physical spaces while welcoming guests and being living spaces for inhabitants.

References

Ackermann, R. (2023). Design thinking was supposed to fix the world. Where did it go wrong? An approach that promised to democratize design may have done the opposite. *MIT Technology Review*. www.technologyreview.com/2023/02/09/1067821/design-thinking-retrospective-what-went-wrong

Boedker, M. (2023). Design as inquiry. Critical design in tourism. In H. Pechlaner, G. Erschbamer, & N. Olbrich (Eds.). *Destination design: Neue Ansätze und Perspektiven aus der Designforschung für die Entwicklung von Regionen und Destinationen*. Springer Fachmedien Wiesbaden; Imprint Springer Gabler. https://doi.org/10.1007/978-3-658-39879-8_8

Budeanu, A., Miller, G., Moscardo, G., & Ooi, C. (2016). Sustainable tourism, progress, challenges and opportunities: An introduction. *Journal of Cleaner Production, 111*, 285–294. https://doi.org/10.1016/j.jclepro.2015.10.027.

Butler, R. (2023). Rethinking tourism: Why and who? *Worldwide Hospitality and Tourism Themes*. https://doi.org/10.1108/WHATT-09-2023-0108

Dávid, L. (2011). Tourism ecology: Towards the responsible, sustainable tourism future. *Worldwide Hospitality and Tourism Themes*, *3*(3), 210–216. https://doi.org/10.1108/17554211111142176

Erschbamer, G., Pechlaner, H., & Olbrich, N. (2023). Destination design – Evolution und Revolution des Wandels [Destination design – evolution and revolution of change]. In H. Pechlaner, G. Erschbamer, & N. Olbrich (Eds.). *Destination design: Neue Ansätze und Perspektiven aus der Designforschung für die Entwicklung von Regionen und Destinationen*. Springer Fachmedien Wiesbaden; Imprint Springer Gabler.

Fesenmaier, D., & Xiang, Z. (2017). Introduction to tourism design and Design Science in Tourism. In D. Fesenmaier & Z. Xiang (Eds.), *Design Science in Tourism: Tourism on the verge*. Springer. https://doi.org/10.1007/978-3-319-42773-7_1

Jonas, W. (1998). Design as problem-solving? or: Here is the solution – what was the problem? *Design Studies*, *14*(2), 157–170. doi:10.1016/0142–694X(93)80045-E

Jonas, W. (2010). *A sense of vertigo: Design thinking as general problem solver.* Institute for Transportation Design. http://8149.website.snafu.de/wordpress/wp-content/uploads/2011/06/EAD09.Jonas_.pdf

Koens, K., Smit, B., & Melissen, F. (2021). Designing destinations for good: Using design roadmapping to support pro-active destination development. *Annals of Tourism Research*, *89*, 103233. https://doi.org/10.1016/j.annals.2021.103233

Olbrich, N., Philipp, J., & Thees, H. (2022). Tourism and tourism services in transition: What are the challenges and how should tourism change? In H. Pechlaner, N. Olbrich, J. Philipp, & H. Thees (Eds.), *Towards an ecosystem of hospitality – Location:City:Destination.* Graffeg.

Pechlaner, H., Eckert, C., & Olbrich, N. (2018). Zu viel Tourismus?: Lösungsansätze zu Over-Crowding und Overtourism [Too much tourism? Solutions for overcrowding and overtourism]. *Tourismus Wissen – Quarterly: Wissenschaftliches Magazin Für Touristisches Know-How*, *14*, 291–297.

Scuttari, A., Pechlaner, H., & Erschbamer, G. (2020). Destination design: A heuristic case study approach to sustainability-oriented innovation. *Annals of Tourism Research, 86.* https://doi.org/10.1016/j.annals.2020.103068.

Steffen, D. (2023). Designtheoretische Zugänge für die nachhaltige Gestaltung von Destinationen [Design-theoretical approaches for a sustainable design of destinations]. In H. Pechlaner, G. Erschbamer & N. Olbrich (Eds.), *Destination design: Neue Ansätze und Perspektiven aus der Designforschung für die Entwicklung von Regionen und Destinationen.* Springer Fachmedien Wiesbaden; Imprint Springer Gabler. https://doi.org/10.1007/978-3-658-39879-8_2

Tussyadiah, I. P. (2014). Toward a theoretical foundation for experience design in tourism. *Journal of Travel Research*, *53*(5), 543–564. https://doi.org/10.1177/0047287513513172

Volgger, M., Erschbamer, G., & Pechlaner, H. (2021). Destination design: New perspectives for tourism destination development. *Journal of Destination Marketing & Management*, *1*. https://doi.org/10.1016/j.jdmm.2021.100561

World Tourism Organization (UNWTO). (2023). *World Tourism Organization: A UN specialized agency*. www.unwto.org/

Zacher, D., Pechlaner, H., & Olbrich, N. (2020). Strategy is the art of combining short- and long-term measures – Empirical evidence on overtourism from European cities and regions. In H. Pechlaner, E. Innerhofer, & G. Erschbamer (Eds.), *Contemporary geographies of leisure, tourism and mobility. Overtourism: Tourism management and solutions.* Routledge

16 Don't Write Cheques You Cannot Cash

Challenges and Struggles with Participatory Governance

Donagh Horgan and Ko Koens

1 Introduction

Prior to the global pandemic, a growing chorus of critical voices raised the alarm with regards to overtourism, particularly in urban destinations (Milano et al., 2019). The negative impacts of tourism are not a new phenomenon and their origins and drivers, including the relationship of dependency of local communities, have been debated by scholars since the 1970s (Pizam, 1978; Boissevain, 1979). Whereas in the 1980s the discussion focused mostly on the number of visitors and carrying capacity of an area, recent approaches highlight how negative impacts are, to a large extent, also based on how these impacts are perceived by different stakeholders, including local communities (Lindberg et al., 1997; Pechlaner et al., 2019). These perceptions can be influenced by wider societal issues, such as the quality of places, general well-being, and the extent to which certain stakeholders have a direct or indirect involvement in tourism (Koens et al., 2021). The negative issues related to tourism commonly encompass a plurality of sectors – with varying degrees of influence within the wider urban ecosystem. Tourism's interconnectedness with urban development thus necessitates a collective response, which also includes the participation of stakeholders from outside of the tourism domain (Koens & Milano, 2023).

Participatory governance is increasingly being discussed as a route to developing a collective response to the challenges presented by overtourism. This type of governance assumes that the engagement of different stakeholders with tourism will contribute to a more collectively shared sense of responsibility for tourism, and reduce conflicts (Moscardo, 2011; Phi & Dredge, 2019). At the same time, participation of residents has been problematised by critical researchers who note that participation is hindered by a lack of time, access, awareness, knowledge and opportunities (Erdmenger, 2022). In addition, conflicting worldviews of stakeholders can limit opportunities for collective solutions (Boom et al., 2021). The global shutdown explicitly emphasised the political nature of managing tourism. Whereas prior to the pandemic tourism growth was not questioned – and degrowth was not seen as a viable outcome of participatory processes – the pandemic showed how, in a situation where tourism is deprioritised in favour of public health, there are possibilities to enact degrowth policies that may favour the quality of life of local

DOI: 10.4324/9781003365815-23

populations over tourism growth. However, political will is needed to enact these kinds of policies (Horgan and Dimitrijević, 2020).

This chapter discusses the debate around participatory governance in tourism. Following a short description of the literature, cases in Amsterdam, Barcelona and Lisbon are used to highlight challenges and opportunities that come with this kind of governance. Finally, pathways towards using participatory governance to stimulate a sustainable visitor economy are provided.

2 The Complexity of Participatory Governance

The subject of tourism management attracted considerable interest in the years immediately preceding the pandemic, but the extent to which that attention has helped overcome issues related to overtourism and (un)sustainability remains unclear. This may be because the impacts of tourism are determined by many contextual factors and local conditions that are in constant flux, which manifest differently in each setting (Nilsson, 2020). As in the case of other complex or wicked problems facing society, standard one-size-fits-all solutions are not possible. New alliances between diverse and interdisciplinary actors are required to support social innovation in the built environment and to address underlying socio-spatial inequalities (Horgan & Dimitrijević, 2020). Participatory governance may be a way to help with this process. The concept refers to participatory forms of political decision-making that involve organised and non-organised citizens and other stakeholders and aim to improve the quality of open democracy and outcomes for citizens (Geissel, 2009). It has even been argued that participatory governance will deepen citizens' democratic engagement in the governmental processes by placing them at the centre of grass-root social innovation – where they can actively contribute to discussions on the balance of social, cultural, political and weigh up environmental concerns with financial gain (Voorberg et al., 2014).

Following the definition offered by García (2006: 745), governance can be understood as a "negotiation mechanism for formulating and implementing policy that actively seeks the involvement of stakeholders and civil society organisations besides government bodies and experts". It is a model of decision-making that emphasises consensus, and thus allows for the integration of perspectives of historically excluded groups. In a contribution looking at urban tourism as a source of contention and social mobilisations, Novy and Colomb (2021) note recent scholarship integrating tourism within broader urban struggles in order to encourage genuine participation in decision-making, orientating policymaking around rights-based arguments. This is necessary to eschew a logic of dualism – tourists versus locals – and to shift the focus of dialogue away from protecting the city from tourism (Arias-Sans and Russo, 2016). A seminal article by Galuszka (2019) on what makes urban governance co-productive provides clarity on definitions and contradictions on co-production in planning. In a comprehensive review of the literature, the author cites a variety of studies that link participatory governance to emerging forms of co-production and invited spaces of participation in policymaking. These different types of activities are characterised by including the involvement of citizens in the co-planning, co-design, co-prioritising, co-management, co-financing and co-assessment of interventions (Bovaird and Loeffler, 2013; Bovaird et al., 2019). Galuzka (2019: 156) found that "institutional change is highly unlikely to take place without an active civil-society sector that is able to build up its own knowledge and resource bases, which includes a capacity to operate in conflict spaces", emphasising the advantage of structures that maintain a degree of independence from specific legal frameworks or institutional settings. Seeing a resurgence in the planning discourse, Arnstein's Ladder of Participation (1969; see Figure 16.1) has been embraced by many tourism scholars seeking to categorise various types of public participation in planning processes. A number of studies use this model to propose an alternative public participation framework for sustainable

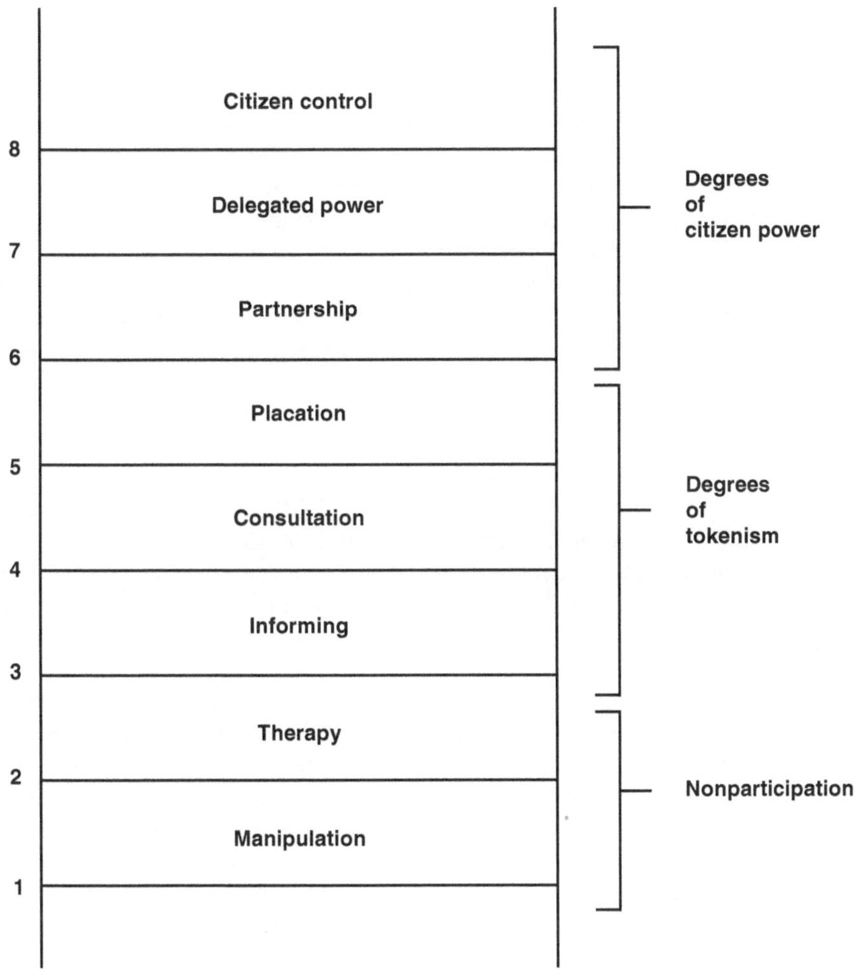

Figure 16.1 Arnstein's original Ladder of Participation

Source: Arnstein (1969)

tourism planning (Marzuki and Hay, 2013; Mak et al., 2017), as it is useful when applied to (tourism) governance given that it identifies challenges to participatory dialogue – including ideological and political prejudices among powerful stakeholders on one side, and knowledge and skills deficits among citizen groups on the other (Horgan and Dimitrijević, 2020). Ranging across manipulation, therapy, informing, consultation, placation, partnership, delegated power and citizen control, the model is helpful in locating participation – given the power vacuum that can seem to befall communities faced with tourism pressures.

The eight levels of community participation climb towards increased devolvement of power to communities to give them greater ownership of the planning development process. The model allows for comparison of levels of impactful participation across destinations seeking to gain consensus for tourism policy across a diverse ecosystem of actors. In the case of Hong Kong, Mak et al. (2017) used it to explore the differences between local residents' expected and actual participation in public consultation activities. Findings identified shortfalls in the public consultation process; in particular, local residents were not well enough informed, and called for more effective strategies in promoting public consultation activities. This demonstrates a type of participatory governance that lies towards the manipulation end of the scale rather than as an exercise in delegated power. We can use this model to describe the forms of participatory governance observed within the content of this chapter. Without adequately addressing some of the political conflicts that exist within urban governance, even aspirational participatory processes (such as those observed for this research) stay within the realms of tokenism – according to the Ladder (Horgan & Dimitrijević, 2020).

In an interdisciplinary review, Phi and Dredge (2019) emphasise how co-creation in tourism can help break down traditional tensions inherent in the literature that separate locals and visitors in destination management. The authors refer to the destination management organisation (DMO) of Copenhagen, which in 2016 heralded the end of tourism as we know it, positioning the tourist as a temporary local, seeking authenticity and locally embedded shared experiences (Wonderful Copenhagen, 2017). In their conceptualisation, *localhood* involves a more intimate relationship with local residents, and collaboration between public and private actors – who can initiate shared value propositions that incorporate features of social, economic, environmental and political innovation. This is then supposed to lead to a form of tourism that is more in tune with local needs and more appreciated by residents than more negative forms. Such a perspective is increasingly supported in tourism academia. Based on a review of the literature on tourism governance, Bichler (2021: 1) argues that "local residents' participation is an essential aspect of effective tourism governance" and that they should be better facilitated to engage with tourism management. For example, by means of collaborative or innovation platforms (Lalicic, 2018), design-oriented tools that improve communication between stakeholders (Koens et al., 2022).

In practice, however, it is proving difficult to implement participatory tourism development processes (Bichler, 2021), which suggests certain participatory tourism governance approaches may not be without challenge (Pechlaner et al., 2015).

Indeed, several issues have been recognised. To start with, stakeholders commonly have different interests that are not compatible with each other. Local enterprises are not likely to want to see their income reduced to accommodate a reduction in visitors, nor will management organisations want to see the image of their destination tarnished by policies seen as unwelcoming to tourists, which could halt rather than reduce the flow of visitors (Butler & Dodds, 2022). Government actors have also been noted to do too little to implement appropriate actions and limits on tourism, with tourism policies sometimes merely paying lip service to sustainable development principles (Mihalic, 2016). At the same time residents are often concerned with regards to tourism development, and particularly those who feel strongly attached to the place where they live can oppose further tourism growth (Lalicic & Garaus, 2022). These diverging interests make participatory governance very difficult. To understand why this is the case, it is useful to appreciate that the potential success of participatory arrangements strongly depends on the institutional logic and underlying governance structures of the destination (Beaumont & Dredge, 2010). Historically, tourism is seen primarily as an economic activity driven by a logic of growth, and this appears set to continue post-COVID (Milano & Koens, 2021). Even participative strategies that are supposed to emphasise *localhood*, like that of the city of Copenhagen, explicitly or tacitly accept tourism growth as an underlying premise.

It is unclear to what extent participatory governance practices can help overcome these issues unless they engage more with questions relating to the lack of decision-making powers among certain stakeholders (Bichler, 2021). While outcomes associated with co-creation in the tourism literature have mostly been discussed in the context of inclusivity and democracy, other bodies of literature (e.g. planning and governance) highlight how the very act of co-creation is a political act as it may allow certain actors to exercise their agency (Phi & Dredge, 2019). Seen from this perspective, it is key to appreciate the extent to which stakeholders can engage, activate agency and wield their political, social or economic power. At this point in time residents may fail to engage in a meaningful way due to a variety of barriers. These may be due to structural reasons, such as the way these processes are set up (e.g. gatekeepers discouraging participation of residents), but may also include more mundane causes like a lack of time or interest in participation (Presenza et al., 2013; Erdmenger, 2022). Eagles (2009) notes the significance of involving local community champions on this matter. They can help make sense of different community perspectives and can act as ambassadors to promote and encourage the participation of local residents (McGehee et al., 2015). Other authors have stressed the broader importance of leadership for successful participatory governance in tourism and of designing participatory structures around the specific and contextual needs of different stakeholders (McGehee et al., 2015; Cross, 2011). Setting long-term values and directions, based on frameworks for bottom-up processes, can help stimulate actors to become and remain involved (Valente et al., 2015; Pechlaner et al., 2014). This may help stimulate clarity and transparency regarding the roles and responsibilities of stakeholders in developing participatory governance arrangements. This is particularly important since, as with other governance arrangements, there will be winners and losers in participative processes. If these are built on unbalanced power relations, they will not lead to valuable co-creation, but rather to value co-destruction (Echeverri & Skålén, 2011).

3 Participatory Governance in Practice; Experiences from Amsterdam

To better appreciate the political nature of participatory governance, this section discusses experiences from Amsterdam, based on literature and findings from a Horizon 2020 project (SMARTDEST) that ran from 2020 until 2023. Amsterdam is among the major tourism destinations in Europe. Whereas the city is proud of its reputation for tolerance and openness, the discourse towards tourism and its excesses has become increasingly negative since the early 2010s. Since then, discussions about overcrowding and the negative impact of tourism on the city have regularly featured in the local newspapers. Grassroots movements have been active to keep the topic high on the policy agenda. For the city, tourism represents a complex challenge due to the competing interests of different stakeholders. The city sought to deal with tourism through a special *Stad in Balans (City in Balance)* programme, which included multiple experiments on a local level to stimulate liveability of the city in relation to tourism (Gemeente Amsterdam, 2018). This involved participatory activities with different stakeholders – which could be understood as Level 2 on Arnstein's Ladder (Therapy). The programme gained reasonable traction in the city, although it was criticised for not achieving enough results. It was abruptly stopped in 2020, with some activities being transferred to a new work programme that focuses on the city centre, and other activities placed under the responsibility of other policy departments for whom tourism often is not a priority (Amsterdam, 2023).

When the COVID-19 pandemic struck, entrepreneurs and those working in tourism were hit hard. However, for many residents it was the first time that they saw the city without crowds. This may have been a driver for the public petition *Amsterdam heeft een keuze (Amsterdam has a choice)*, which demanded a cap on the number of visitors of 12 million overnight stays and which was signed 30,000 times. This "uninvited" form of participation had a big impact, maybe even more so than the years of collaboration between stakeholders leading up the pandemic. As a result of the petition, the council accepted a proposal to cap overnight stays at 20 million, which is more or less the same as in 2019 (12 million being deemed unrealistic) (Boonstra, 2021). Whilst the expectation was that it would take several years for tourism to again reach this number, in 2022 the number of overnight stays was already so high that the city put forward several new proposals. Many of these focused on reducing and restricting tourism using more top-down oriented measures (Gemeente Amsterdam, 2022; anon, 2022), and included the much maligned *Stay Away* campaign, which actually increased interest in the city among "party tourists" – having the opposite effect than intended (Boyd, 2023). This type of engagement can be placed higher on the Ladder, at Level 5 (Placation).

For the SMARTDEST project, we engaged with a wide range of key stakeholders within the tourism governance ecosystem. Research into same revealed a lack of central coordination, given that tourism in Amsterdam is entangled with other issues such as housing and urban planning. At the same time, local politics are rather myopic, focusing mostly on highly visible hyperlocal problems like waste and disturbance in urban neighbourhoods (for example, in the Red-Light district). Within such a fragmented landscape of policymaking, it is very difficult to come to a long-term holistic vision and to show clear leadership. Stakeholders noted that the policy ecosystem was fuzzy, in that it was unclear which department or organisation could effectuate change. This could also be observed in discussions with stakeholders, as the DMO was more often mentioned as central to tourism development than the municipality. More generally, both stakeholders and residents commonly felt they were somewhat on the periphery of the policy discourse or did not feel represented. This is despite multiple participatory and co-creative activities undertaken by different municipality departments on multiple levels (e.g. *Stad in Balans* programme, discussions with residents and representative organisations, support for activities of entrepreneurial grassroots movements such as the Reinvent Tourism Festival).

In addition, there is a lack of clarity among stakeholders with regards to the expected outcomes of participation, which can result in inflated expectations. Stakeholders mentioned being discouraged when their opinions were not visibly incorporated in policy, while others note they did not have a "seat" at the decision-making "table" (Stompff & Gerritsma, in revision). Given the complexity of tourism and the highly differing interests, such expectations are simply not realistic. Participative governance does not mean direct involvement in decision-making. When different departments need to collaborate, inevitably trade-offs need to be made, and decisions also may be influenced by politics. If this is not clarified sufficiently, participatory processes may even act as a form of de-politicisation. Furthermore, many things which have a major impact on tourism are not even decided in Amsterdam (e.g. flight numbers to and from Schiphol). All of these issues expose the limits of participation, which need to be managed to prevent participants becoming cynical or losing faith in the process. The extent to which this has happened in Amsterdam is not clear, but participatory governance does not seem to have increased understanding or support for tourism. In fact, recent research suggests that residents and stakeholder organisations alike suffer from participation fatigue and are increasingly unwilling to voluntarily invest time and/or resources to participate in collaborative processes (Stompff & Gerritsma, in revision).

These findings from the Amsterdam case mirrored those from the other cities, partner case studies on the SMARTDEST project. Romão et al. (2021) found that while in Barcelona, the Tourism Council is understood to be highly representative of the interests involved in the city, there are enduring difficulties in achieving consensual positions for the definition and implementation of tourism policies. The authors emphasised the process of airing different positions involved, which contributed to more informed decision-making on the part of the local authority – allowing public opinion to be included. The participatory Tourism Council includes a plenary with a diversity of representatives of tourism, retail and hospitality, culture and sports, unions, environmental associations, social organisations, special interest groups and local residents. All political parties are represented alongside subject matter experts and the official DMO – representing Level 4 on the Ladder (Consultation) (Tourism of Barcelona). Romão et al. (2021) point to deep conflicts of interest among different stakeholder groups, necessitating important participatory mechanisms of engagement. They caution, however, that the effectiveness of consultation is dependent on the motivation of the stakeholders involved (Font et al., 2019). In Barcelona, the establishment of the Tourism Council has been effective in setting an agenda for sustainable tourism but limited in programme and policy implementation (Romão et al., 2021).

Comprehensive analysis from Ivars-Baidal et al. (2023) for SMARTDEST raises questions about the association of smart cities (or destinations) with participatory governance. Their analysis found that holistic integrated planning was being challenged by entrepreneurial approaches categorised by an emphasis on experimentation. While these can, in theory, open up participation through innovative collaboration and partnerships, they can also have diverse implications in terms of scalability, cost–benefit analysis of public resources, and stakeholder representation and benefits. This activity can best be understood to sit at Level 2 (Therapy). Elsewhere critics have associated these approaches (engagement platforms, city labs and hackathons) with other "smart initiatives", which are often used to obscure exercises in opaque public procurement, and neoliberal relationships with technology providers that lack accountability (Bria, 2019; Horgan and Dimitrijević, 2019). From a participatory governance perspective, these concerns resonate with those of Angelidou (2017), whose analysis of smart city strategies noticed low or no participation. The authors report that the development of smart initiatives (at the destination level) is framed within a logic of experimental governance, which seems to favour bottom-up engagement – at least in theory (Cardullo, 2020). With limited feedback loops in practice, and sometimes fuzzy levels of accountability, the smart approach tends to lead to the consolidation of a triple helix model with selective stakeholder participation (Ivars-Baidal et al., 2023).

In a related upcoming publication on smart tourism ecosystems and urban governance in Barcelona, Pastor et al. (2024) share findings from qualitative research which identify numerous opportunities for enhancing synergies between smart initiatives and tourism governance in practice. Presently, the city falls short in fully leveraging the potential benefits of coordination between smart city and tourism governance initiatives. Based on their analysis of the Barcelona case, the authors are positive about the direction of tourism governance which has progressively incorporated cross-cutting perspectives – seeking a better balance between tourism management and promotion – widely demanded in both theoretical and practical terms. Pastor et al. (2024) found that while the importance of tourism in the city is acknowledged, the negative impacts of tourism and monoculture are evident. Cooperation and co-creation relationships are complicated by the absence of shared strategies, despite the use of the *decidim.barcelona* platform and the Tourism Council to stimulate engagement, limiting the practical impact.

The authors note the urgency in this context to clarify the scope of the smart city projects in Barcelona and the role reserved for tourism. As the smart city concept is replaced by other data-led initiatives in (neoliberal) urban governance, the need for genuine participation of community actors is increasingly apparent – as is the need for better links between the city's innovative start-up and the tourism sector. Pastor et al. (2024) propose a robust multilevel governance structure to integrate policymaking at different levels – metropolitan, regional, national and continental. The exponential growth of tourism in Lisbon, another SMARTDEST case study, has turned the Portuguese city into a popular destination for digital nomads (Buhr, forthcoming). The influence of mobile populations is noticeable especially in spaces that mix consumption and work practices, such as coffee shops, coworking and coliving spaces. Buhr (forthcoming) notes how coffee shops facilitate the maintenance of these lifestyles, and how they are embedded in broader processes of transnational gentrification. Alongside the other cases of overtourism, this case necessitates deep engagement with stakeholders in order to protect the attrition of local services – giving agency to overlooked residents.

4 Pathways and Future Suggestions

The idea of participatory governance is enticing. Participatory design exercises can help generate consensus on development among competing agendas, but they can only do this within existing structures of governance (Horgan, 2022). Looking at the experiences in Amsterdam, a relatively open and collaborative city, this is far from easy. A perceived lack of connected policymaking in the city appears to be reducing the trust of locals and professional stakeholders in governance processes, as well as their willingness to participate in engagement-led co-design exercises. Challenges to meaningful participation – and the difficulty of including citizens in co-design and decision-making – are visible in other European cities as well (Bua & Bussu, 2021). Whilst these findings are not particularly positive, this does not mean that participatory governance is undesirable. In fact, we believe participatory processes have great potential value for tourism policymaking, if only because the alternative – top-down oriented policymaking – is potentially more problematic (see e.g. González-Reverté, 2019; Wan & Pinheiro, 2014). However, it is necessary to acknowledge the limitations of participatory governance approaches and look for ways to overcome their weaknesses and maximise their potential within a wider governance structure. To assist with this, this section contains suggestions that deal with the overarching structure of participatory governance.

An important finding in the literature and in Amsterdam is that of clear leadership on different levels. The absence of clear roles, responsibilities and feedback loops between citizens and decision-makers makes it harder to engage stakeholders in policy innovation (Pechlaner et al., 2014). Whilst a range of visions have arisen from different destinations (e.g. Gemeente Amsterdam, 2022; Wonderful Copenhagen, 2017, 2021; Ministry of Industry, Trade and Tourism, 2021), it is not common for these to be accompanied by practice-oriented action plans that provide clarity on who could be involved or take ownership over outcomes. Even when such documents do exist, they lack clarity and information on how to provide resources for the work that engaged stakeholders need to do (European Commission, 2021). This makes it difficult to appreciate if and how stakeholders can engage. For participatory governance to create impact, decision-makers need to develop sufficient processes – channels and feedback loops – to support a more honest and open discussion in our planning systems.

This also includes acknowledging the limitations of participative approaches from the start, designing appropriate mitigations, and including discussion on how (not) to implement outcomes. As it is, participatory processes create great expectations that cannot be met and that, with a cynical hat on, can be seen as an elite-led (top-down) way to address the legitimacy crisis of current policymaking (Bua & Bussu, 2021). To prevent this, it is essential to be honest about the extent to which processes are truly participative, even if this means that many interventions that are currently described as being participative have to be relabelled as purely consultative in nature. This may appear a minor difference, but it is significantly different with regards to both expectations and outcomes. Moving towards more participation means making a concerted effort to include opportunities for participation in the design – at all stages of tourism governance. This means involving local people in problem definition and narrative building as well as in the testing and implementation of new interventions, policy initiatives and programmes. Sectoral partners need to be upskilled in methods to better moderate the processes so that the proposals become more realistic and solution-oriented – having been tested in a robust setting through platforms such as the urban living lab. Although participative processes often take place on a local level, the current system struggles to accommodate a wide variety of participative processes on which open urban governance should depend. As a result, debates tend to focus on the most problematic conflicted parts of the city, where emotions are likely to run high (Gemeente Amsterdam, 2023). To overcome this, new ways of engagement may be required. Thees et al. (2020) positions the living lab as a tool to promote residents' participation in destination governance, pointing to experimental interventions that empower local communities. As a platform (and site) to facilitate joint problematisation and participatory experimentation with tourism partners and community stakeholders, the urban living lab has emerged as a concept to support more locally owned solutions and therefore innovative forms of governance. Mahmoud et al. (2021) propose a new concept of the living lab – aimed at remodelling the face of policy making and participatory governance tools for sustainable urban development – based on their experience examining strategies for co-creation across a set of cases.

Co-creation processes produce multiple benefits as well, if correctly tailored for their community and embedded into public decision-making. In order to support sustainable transitions in tourism and other areas of urban governance, ongoing participatory governance mechanisms need to be cross-sectoral and intersectional (Mahmoud et al., 2021). Living labs can act as participatory breeding grounds by situating residents' needs within a holistic city perspective and including public and administrative decision-makers in the discussion. Within the container of the living lab, actors within the tourism ecosystem can test and develop new propositions to tackle overtourism and break traditional hierarchies and unequal power relationships. As a tool for participatory governance, the living lab can help to secure long-term engagement of the community and collective ownership over decision-making, policy innovation and behavioural change. To make this possible, it is extremely important that structures of decision-making are present that are open to take the outcomes of such labs seriously (Thees et al., 2020). This is particularly crucial for tourism, as this is an emotive topic where local public perceptions may differ from "desired" policy outcomes on a destination-wide level (Lalicic & Garaus, 2022). If decisions are taken where short-term political or economic advantage comes at the expense of longer-term sustainability and the resilience of a destination (Horgan, 2022), the use of participatory approaches is likely to lead to protest and social discord.

References

Amsterdam. (2023). Volg het beleid: toerisme [Follow the policy: Tourism]. Stad Amsterdam. www.amsterdam.nl/bestuur-organisatie/volg-beleid/toerisme/

Angelidou, M. (2017). Smart city planning and development shortcomings. *TeMA – Journal of Land Use, Mobility and Environment*, *10*(1), 77–94. https://doi.org/10.6092/1970-9870/4032

Anon. (2022). Gemeente Amsterdam neemt maatregelen om groei en overlast toerisme te beperken [Municipality of Amsterdam takes measures to limit growth and nuisance tourism]. Amsterdams Dagblad. www.amsterdamsdagblad.nl/gemeente/gemeente-amsterdam-neemt-maatregelen-om-groei-en-overlast-toerisme-te-beperken

Arias-Sans, A., & Russo, A. P. (2016). The right to Gaudí: What can we learn from the commoning of Park Güell, Barcelona? In C. Colomb & J. Novy (Eds.), *Protest and Resistance in the Tourist City*. Routledge.

Arnstein, S. R. (1969). A ladder of citizen participation. *Journal of the American Institute of Planners*, *35*(4), 216–224.

Beaumont, N., & Dredge, D. (2010). Local tourism governance: A comparison of three network approaches. *Journal of Sustainable Tourism*, *18*(1), 7–28.

Bichler, B. F. (2021). Designing tourism governance: The role of local residents. *Journal of Destination Marketing & Management*, *19*, 100389.

Boissevain, J. (1979). The impact of tourism on a dependent island: Gozo, Malta. *Annals of Tourism Research*, *6*(1), 76–90.

Boom, S., Weijschede, J., Melissen, F., Koens, K., & Mayer, I. (2021). Identifying stakeholder perspectives and worldviews on sustainable urban tourism development using a Q-sort methodology. *Current Issues in Tourism*, *24*(4), 520–535.

Boonstra, W. (2021). Amsterdamse raad neemt toerismeverordening aan [Amsterdam Council adopts tourism ordinance]. Binneland Bestuur. www.binnenlandsbestuur.nl/ruimte-en-milieu/raad-stemt-met-maximum-toeristenovernachtingen

Bovaird, T., Flemig, S., Loeffler, E., & Osborne, S. P. (2019). How far have we come with co-production – and what's next? *Public Money & Management*, *39*(4), 229–232.

Bovaird, T., & Loeffler, E. (2013). We're all in this together: Harnessing user and community co-production of public outcomes. *Birmingham: Institute of Local Government Studies: University of Birmingham*, *1*, 15.

Boyd, M. (2023). Stag-do bookings to Amsterdam triple after city warned Brits to "stay away". *Daily Mirror*. www.mirror.co.uk/travel/europe/stag-bookings-amsterdam-triple-after-29591746

Bria, F. (2019). *Building digital cities from the ground up based around data sovereignty and participatory democracy: The case of Barcelona.* Barcelona Centre for International Affairs.

Bua, A., & Bussu, S. (2021). Between governance-driven democratisation and democracy-driven governance: Explaining changes in participatory governance in the case of Barcelona. *European Journal of Political Research, 60*(3), 716–737.

Buhr, F. (forthcoming). Estilos de vida móveis e suas infraestruturas: notas sobre Lisboa. [Mobile lifestyles and their infrastructures: Notes about Lisbon]. *Revista Brasileira de Sociologia.*

Butler, R. W., & Dodds, R. (2022). Overcoming overtourism: A review of failure. *Tourism Review, 77*(1), 35–53.

Cardullo, P. (2020). *Citizens in the 'smart city': Participation, co-production, governance.* Routledge.

Cross, N. (2011). *Design thinking: Understanding how designers think and work.* Berg Publishers.

Eagles, P. J. F. (2009). Governance of recreation and tourism partnerships in parks and protected areas. *Journal of Sustainable Tourism, 17*(2), 231–248. https://doi.org/10.1080/09669580802495725

Echeverri, P., & Skålén, P. (2011). Co-creation and co-destruction: A practice-theory based study of interactive value formation. *Marketing Theory, 11*(3), 351–373.

Erdmenger, E. C. (2022). The end of participatory destination governance as we thought to know it. *Tourism Geographies, 25*(4), 1104–1126.

European Commission. (2021). *Scenarios towards co-creation of transition pathway for tourism for a more resilient, innovative and sustainable ecosystem.* European Commission.

Font, J., Pasadas, S., & Fernández-Martínez, J. L. (2019). Participatory motivations in advisory councils: Exploring different reasons to participate. *Journal of Representative Democracy, 57*, 1–19. https://doi.org/10.1080/00344893.2019.1643774

Galuszka, J. (2019). What makes urban governance co-productive? Contradictions in the current debate on co-production. *Planning Theory, 18*(1), 143–160. https://doi.org/10.1177/1473095218780535

García, M. (2006). Citizenship practices and urban governance in European cities. *Urban Studies, 43*, 745–765. doi:10.1080/00420980600597491

Geissel, B. (2009). Participatory governance: Hope or danger for democracy? A case study of local agenda 21. *Local Government Studies, 35*(4), 401–414.

Gemeente Amsterdam. (2018). *Stad in Balans 2018–2022; Naar een nieuw evenwicht tussen leefbaarheid en gastvrijheid.* Gemeente Amsterdam.

Gemeente Amsterdam. (2022). *Visie bezoekerseconomie in Amsterdam 2035* [*Vision of the visitor economy in Amsterdam 2035 economy*]. https://openresearch.amsterdam/nl/page/90775/visie-bezoekerseconomie-amsterdam-2035

Gemeente Amsterdam. (2023). *Aanpak Binnenstad* [*City Centre Approach*]. www.amsterdam.nl/stadsdelen/centrum/aanpak-binnenstad/

González-Reverté, F. (2019). Building sustainable smart destinations: An approach based on the development of Spanish smart tourism plans. *Sustainability, 11*(23), 6874.

Horgan, D. (2022). Devolution of decision-making: Tools and technologies towards equitable place-based participation in planning. In J. Gilbert (Ed.), *Urban agglomeration.* https://statesacademicpress.com/book/267

Horgan, D., & Dimitrijević, B. (2019). Frameworks for citizens participation in planning: From conversational to smart tools. *Sustainable Cities and Society, 48*, 101550.

Horgan, D., & Dimitrijević, B. (2020). Social innovation in the built environment: The challenges presented by the politics of space. *Urban Science*, *5*(1), 1.

Ivars-Baidal, J. A., Celdrán-Bernabeu, M. A., Femenia-Serra, F., Perles-Ribes, J. F., & Vera-Rebollo, J. F. (2023). Smart city and smart destination planning: Examining instruments and perceived impacts in Spain. *Cities*, *137*, 104266. https://doi.org/10.1016/j.cities.2023.104266

Koens, K., Klijs, J., Weber-Sabil, J., Melissen, F., Lalicic, L., Mayer, I., Önder, I., & Aall, C. (2022). Serious gaming to stimulate participatory urban tourism planning. *Journal of Sustainable Tourism*, *30*(9), 2167–2186.

Koens, K., Melissen, F., Mayer, I., & Aall, C. (2021). The Smart City Hospitality Framework: Creating a foundation for collaborative reflections on overtourism that support destination design. *Journal of Destination Marketing & Management*, *19*, 100376.

Koens, K., & Milano, C. (2024). Urban tourism studies: A transversal research agenda. *Tourism, Culture & Communication*. https://doi.org/10.3727/109830423X16999785101653

Lalicic, L. (2018). Open innovation platforms in tourism: How do stakeholders engage and reach consensus? *International Journal of Contemporary Hospitality Management*, *30*(6), 2517–2536.

Lalicic, L., & Garaus, M. (2022). Tourism-induced place change: The role of place attachment, emotions, and tourism concern in predicting supportive or oppositional behavioral responses. *Journal of Travel Research*, *61*(1), 202–213.

Lindberg, K., McCool, S., & Stankey, G. (1997). Rethinking carrying capacity. *Annals of Tourism Research*, *24*(2), 461–465.

Mahmoud, I. H., Morello, E., Ludlow, D., & Salvia, G. (2021). Co-creation pathways to inform shared governance of urban living labs in practice: Lessons from three European projects. *Frontiers in Sustainable Cities*, 3, 690458. doi:10.3389/frsc.2021.690458

Mak, B. K., Cheung, L. T., & Hui, D. L. (2017). Community participation in the decision-making process for sustainable tourism development in rural areas of Hong Kong, China. *Sustainability*, *9*(10), 1695.

Marzuki, A., & Hay, I. (2013). Towards a public participation framework in tourism planning. *Tourism Planning & Development*, *10*(4), 494–512.

McGehee, N. G., Knollenberg, W., & Komorowski, A. (2015). The central role of leadership in rural tourism development: A theoretical framework and case studies. *Journal of Sustainable Tourism*, *23*(8–9), 1277–1297.

Mihalic, T. (2016). Sustainable-responsible tourism discourse – Towards "responsustable" tourism. *Journal of Cleaner Production*, *111*, 461–470.

Milano, C., & Koens, K. (2021). The paradox of tourism extremes: Excesses and restraints in times of COVID-19. *Current Issues in Tourism*. https://doi.org/10.1080/13683500.2021.1908967

Milano, C., Novelli, M., & Cheer, J. M. (2019). Overtourism and tourismphobia: A journey through four decades of tourism development, planning and local concerns. *Tourism Planning & Development*, *16*(4), 353–357.

Ministry of Industry, Trade and Tourism. (2021). *General Guidelines of the Sustainable Tourism Strategy of Spain 2030*. Madrid, Ministry of Industry, Trade and Tourism. https://turismo.gob.es/en-us/estrategia-turismo-sostenible/paginas/index.aspx

Moscardo, G. (2011). Exploring social representations of tourism planning: Issues for governance. *Journal of Sustainable Tourism*, *19*(4–5), 423–436.

Nilsson, J. H. (2020). Conceptualizing and contextualizing overtourism: The dynamics of accelerating urban tourism. *International Journal of Tourism Cities*, *6*(4), 657–671.

Novy, J., & Colomb, C. (2021). Urban tourism as a source of contention and social mobilisations: A critical review. *Travel and Tourism in the Age of Overtourism*, 6–23.
Pastor, A., Ana, B., Casado, A., & Ivars, J. (2024). Smart tourism ecosystems and urban governance: A qualitative analysis of Barcelona. SMARTDEST Final Conference, Barcelona, 15–16 September.
Pechlaner, H., Beritelli, P., Pichler, S., Peters, M., & Scott, N. R. (2015). *Contemporary destination governance: A case study approach*. Emerald Group Publishing.
Pechlaner, H., Innerhofer, E., & Erschbamer, G. (Eds.). (2019). *Overtourism, tourism management and solutions*. Routledge.
Pechlaner, H., Kozak, M., & Volgger, M. (2014). Destination leadership: A new paradigm for tourist destinations? *Tourism Review*, *69*(1), 1–9.
Phi, G., & Dredge, D. (2019). Collaborative tourism-making: An interdisciplinary review of co-creation and a future research agenda. *Tourism Recreation Research*, *44*(3), 284–299.
Pizam, A. (1978). Tourism's impacts: The social costs to the destination community as perceived by its residents. *Journal of Travel Research*, *16*(4), 8–12.
Presenza, A., Del Chiappa, G., & Sheehan, L. (2013). Residents' engagement and local tourism governance in maturing beach destinations: Evidence from an Italian case study. *Journal of Destination Marketing & Management*, *2*(1), 22–30.
Romão, J., Domènech, A., & Nijkamp, P. (2021). Tourism in common: Policy flows and participatory management in the Tourism Council of Barcelona. *Urban Research & Practice*, *16*(2), 222–245. doi:10.1080/17535069.2021.2001039
Stompff, G., & Gerritsma, R. (in revision). Who is at Amsterdam's tourism policy making table: Co-designing tourism policies; who is (not) participating and how is this perceived among tourism stakeholders. *Journal of Design Studies*.
Thees, H., Pechlaner, H., Olbrich, N., & Schuhbert, A. (2020). The living lab as a tool to promote residents' participation in destination governance. *Sustainability*, *12*(3), 1120.
Valente, F., Dredge, D., & Lohmann, G. (2015). Leadership and governance in regional tourism. *Journal of Destination Marketing & Management*, *4*(2), 127–136.
Voorberg, W., Bekkers, V. J. J. M., & Tummers, L. (2014). *A systematic review of co-creation and co-production: Embarking on the social innovation journey*. Social Science Research Network. https://papers.ssrn.com/abstract=2444075
Wan, P. Y. K., & Pinheiro, F. V. (2014). Macau's tourism planning approach and its shortcomings: A case study. *International Journal of Hospitality & Tourism Administration*, *15*(1), 78–102.
Wonderful Copenhagen. (2017). *The end of tourism as we know it; towards a beginning of localhood strategy 2020*. Wonderful Copenhagen.
Wonderful Copenhagen. (2021). *Tourism for good*. Wonderful Copenhagen. www.e-unwto.org/doi/book/10.18111/9789284414024

17 Benefits of an Integrative Tourism Policy

On the Way to a New Tourism Culture Using the Example of South Tyrol

Mirjam Gruber, Elisa Innerhofer and Maximilian Walder

1 The Status Quo of (Over)Tourism in South Tyrol

Although South Tyrol's tourism industry is unquestionably a significant contributor to the local economy, the region's current tourism policies and its tourism situation are frequently criticised. The phenomenon of overtourism in South Tyrol has been discussed at an academic level (Carvalho et al., 2020; Erschbamer et al., 2018; Scuttari et al., 2019; Weiss, 2021) as well as in local and international media (Benedikter, 2018; Ebner, 2022; FF Media, 2021; Fischer, 2019; Mair, 2020; Rainews, 2022; Schwarz, 2023). Arrivals and overnight stays have risen steadily over the last 20 years (except for 2020 due to the COVID-19 pandemic) and the number of beds in the accommodation establishments has also increased. We have observed two main developments in this region: Firstly, the country is recording more and more arrivals and overnight stays, and secondly, the number of high-class hotels (four- and five-star establishments) has been increasing since 2010, with the majority of "new beds" falling into this category. Like many other regions, tourism in South Tyrol started with simple bed and breakfast accommodation (*Gasthäuser*) where the owners were the innkeepers (*Gastwirte*). In the meantime, one- and two-star establishments have become fewer (ASTAT, 2023; Windegger et al., 2022), although other types of smaller accommodation such as campsites, private accommodations and agritourism ventures are growing (Windegger et al., 2022). Currently, the dominant establishment type is the three-star hotel. This change in type of accommodation available has an impact on the local economy as smaller accommodations, which usually offer only bed or bed and breakfast, encourage guests to use the local infrastructure such as swimming pools or saunas for leisure activities, and to visit local restaurants, bars and cafes. Most hotels in the four- and five-star segment provide various services and leisure activities, following a concept that keeps guests in the hotel as much as possible. In fact, these hotels increasingly have large wellness facilities, swimming pools, restaurants, entertainment facilities, etc. In other words, the local economy benefits more from guests who do not limit their consumption to their accommodation, but from those who shop, eat out and take part in leisure activities locally.

DOI: 10.4324/9781003365815-24

The (over)tourism situation in South Tyrol exhibits significant variations across different valleys, areas and villages. For instance, certain villages are strongly focused on tourism, boasting numerous hotels, while others serve as popular day trip destinations, such as the Dolomites. Conversely, there are villages or zones with limited or no tourist accommodations. Consequently, the distribution of tourism must be approached with discernment. One notable aspect in South Tyrol is the considerable influx of traffic during holiday seasons, exerting strains on the local population and the environment. Indeed, one issue in South Tyrol regarding overtourism is the South Tyroleans' attitude towards tourism (*Tourismusgesinnung*). In a representative survey of households in South Tyrol conducted in 2020, 77 percent of all respondents stated that the advantages of tourism outweigh the disadvantages (de Rachewiltz et al., 2021). While in absolute terms this number is not low, a similar survey in 2018 showed the support at 95 percent. Although the questions about the South Tyroleans' attitude towards tourism in the two surveys are not identical, the number still indicates a downwards trend in the attitude towards tourism. Of note, younger people (between the ages of 14 and 18) and the elderly (over the age of 65) showed a less positive attitude than residents between the ages 25 and 65. Moreover, respondents with higher education and persons working in the food and accommodation sector valued tourism more than others. When asked about the future importance of tourism in the region, the respondents univocally stated that it will play a rather important or even very important role in South Tyrol's overall future development. Approximately 12 percent of the households polled expressed a desire for more tourism in the future, 23 percent preferred less tourism in the future and the remaining two-thirds stated that the current level of tourism should be maintained.

The need to steer touristic development is not new in South Tyrol. Indeed, the consideration of the level of tourism development as a basis for tourism regulation is reflected in the Decree of the Provincial Governor dated October 18, 2007, No. 55, on the "regulation on the expansion of hospitality businesses and the designation of zones for tourist facilities".[1] This divides municipalities into the categories of "heavily developed", "developed" and "structurally weak" based on the number of beds and the bed density. Municipalities in South Tyrol are at different stages of tourism development, and regulations certainly have to take this into account. In a more recent study on tourism development in the region – in which the authors of this contribution were engaged and which represents an important foundation for the present article – municipalities were divided into three categorizations based on their degree of development (Pechlaner et al., 2022). To determine the degree of touristic development, the authors of the study used the tourism exposure index, which is based on two variables of tourism intensity and the tourism density for each town. After calculating the tourism exposure value for each town, a ranking was created in which the bottom 25 percent of municipalities were classified as "touristic low developed", the top 25 percent as "touristic high developed" and the middle 50 percent as "touristic developed" towns. Accordingly, this classification could be used by politicians and administrators to establish clear guidelines and regulations for potential qualitative or quantitative expansions in the tourism sector for the different classifications.

In fact, in the media and in politics a complete bed freeze (*Bettenstopp*) was discussed, which could be introduced more restrictively or more loosely on the basis of this classification, depending on the tourism exposure of the municipalities (Ebner, 2022; Kofler, 2022; Südtiroler Landesverwaltung, 2022; SWZ, 2022).

However, we argue in this chapter that in South Tyrol the current concrete regulations are not enough and that a broadly diversified sustainable and well-thought-out tourism policy is also required, which should integrate various aspects of political, cultural and social life as well as different views from a variety of stakeholders. To this aim, in the next section we introduce tourism policy as a research field and concept. We explain the understanding of tourism policy and outline previous research on the topic in South Tyrol. As mentioned above, this contribution stems from an investigation on tourism in South Tyrol, carried out by the Eurac Research Center for Advanced Studies (Pechlaner et al., 2022). The authors are affiliated with the research team that actively participated in the study and the following reflects the outcomes derived from the aforementioned research endeavour. In particular, the third section where we propose a set of measures for action in South Tyrol, which could be part of a holistic tourism policy in the region, is based on that research.

2 A Framework of Tourism Policy

In this section we take a step back and explain the concept of tourism policy, why an integrated tourism policy is needed, and how it could look (see e.g. Haigh, 2020; Liasidou, 2019; Panasiuk & Wszendybył-Skulska, 2021). Tourism policy, like tourism science, is a cross-sectional issue that touches on a wide range of economic and social factors, as well as areas of public administration (transport, spatial planning, environmental protection and nature conservation, trade law, labour law, etc.) (MCI Tourismus, 2014).

Freyer (2006) highlighted the influence on tourism policy of various political departments (economic policy, financial policy, labour market and social policy, foreign policy, legal policy, regional planning and building policy, transport policy, technology and research policy, environmental policy, international policy), system divisions (social system, environmental system, economic system, legal system, social system, individual sphere, operating system) and sciences (geography, economics, sociology, political science, psychology, architecture, law). Moreover, Krippendorf (1976), defined indirect tourism policy as many measures that are not primarily motivated by tourism, but which have a significant impact on tourism. Indeed, tourism development is influenced by political actors even without the existence of an explicit tourism policy (Lun et al., 2014). This circumstance, combined with the fact that tourism has an impact on economic, social and environmental factors, necessitates the development of a tourism policy. Tourism and its design thus touch almost every aspect of life of the population of a tourism destination, making it all the more important to actively pursue and legitimise tourism policy.

But what exactly is tourism policy? In the literature we find numerous definitions. Tourism policy is defined by Berg (2021, p. 486) as "the development and alteration of framework conditions and instruments by state agencies to demand and control tourism at supranational, national, regional, and municipal levels".[2] Similarly, Freyer (2015, p. 449) described it as "the deliberate planning and shaping of tourism and its future by various organisations (state, private, higher-level)".[3] Kaspar (1991, p. 145) defined tourist policy as "intentional promotion and shaping of tourism by affecting the tourism-relevant reality of communities",[4] as opposed to Krippendorf's (1976), indirect tourism strategy. Even though political engagement in specific day-to-day occurrences is sometimes viewed negatively by tourism operators and entrepreneurs, tourism should not be viewed as a self-runner at the political level.

Lun et al. (2014) identified three distinct levels of general tourism policy. The definitions of framework conditions such as laws, constitutions and institutions are used as a starting point. The specific contents of tourism policy are found on the second level, where actions, goals, tasks and problem solutions are developed. The third level covers the processes of tourism policy design, which include actions that "lead to the implementation, abandonment, or compromise of substantive concepts"[5] (Böhret et al., 1988; Lun et al., 2014, p. 63).

According to Opaschowski (2002, p. 300), "tourism policy must create the social framework conditions for an optimal holiday experience, enable access to travel for as many sections of the population as possible, ensure an intact social and natural environment and help to improve the living conditions in the destinations".[6] Essentially, tourist policy serves an overall political, sociopolitical and environmental function. In addition, tourist policy has an instrumental character and can help to achieve regional policy objectives. As a result, the tourism industry is frequently regarded as a "start-up economy" for economically deprived regions (MCI Tourismus, 2014). Finally, tourism strategy is about producing and securing jobs, raising tax money and promoting an area, city or country as an appealing location for vacation or business travels (Berg, 2021, pp. 488–489).

Lun et al. (2014) have specifically dealt with the tourism policy of South Tyrol and underlined, among other things, the influence factor of governance (see also Bichler & Lösch, 2019). Three levels that are considered central to tourism governance in South Tyrol are the municipalities, the province and the state, with the latter receiving the least attention. At the local level, the municipalities are responsible for the following:

- Concrete planning and implementation of projects
- Coordination, financing and implementation of infrastructural construction activities
- Administration in tourism
- Guest registrations or the redistribution of financial and other resources
- Liaising between tourism enterprises and the regional administration
- Support of tourism activities, organisation of events and product development

The regional level, i.e. the Autonomous Province of Bolzano-Bozen, is responsible for the following aspects:

- Urbanism, spatial planning, hospitality regulations
- Laws and regulations that serve to guide tourism activities
- Tourism promotion criteria and destination marketing organisations
- Financial subsidies

The results of the study also indicate that the role of political actors in the tourism system may change in the future and should be adapted to current needs and situations. In addition to governance, Lun et al. (2014) have identified external and social factors as determining factors. The *Zeitgeist* (spirit of a time) and social structures of a region, which are in turn characterised by the expectations and attitudes of a population, play a role. Furthermore, the competencies of the political actors, legal (scope for action, regulatory system) and structural factors (tourism organisations, microstructures) have an influence on tourism policy in South Tyrol.

Based on the concept of tourism policy and the studies presented in this section, in the next section we introduce various measures for action which could help to establish a tourism policy in South Tyrol, explicitly also considering governance structures as well as the situation of overtourism.

3 Measures for Action in South Tyrol in the Context of an Integrated Tourism Policy

Due to the importance of tourism to South Tyrol, there is not only a separate administrative area for tourism at the level of the South Tyrolean provincial administration, but also a separate tourism department within the South Tyrolean provincial government. Nevertheless, the political challenge remains to illuminate the advantages and disadvantages of tourism development, because the cross-sectional function of tourism touches many areas of political responsibility.

At the political level it is important to be sensitive to the interrelationships of tourism development with other sectors and areas of public life to enable and promote sustainable tourism development. Therefore, a broader political understanding of the phenomenon of tourism is needed. In other words, politics and the public need to be sensitised to the necessity of transforming tourism into a sustainable development engine for South Tyrol (Pechlaner et al., 2022). We argue that this is not only a matter of identifying strategies and measures for spatially compatible tourism development, but also of thinking about how tourism policy can be designed (primarily at the provincial level). Authors' engagement in the work established by Pechlaner et al. (2022) serves as a basis for the subsequent paragraphs, where proposals for tourism policy interfaces and cooperation within the diverse realms of South Tyrolean political and administrative domains are presented: South Tyrolean small and medium-sized enterprises from the tourism sector should have access to financial support and advice in the development and implementation of innovations as well as in the application, filing and registration of intellectual property and trademarks. For this purpose, concrete criteria and conditions should be developed that regulate access to these subsidies and support measures. These support mechanisms should be communicated accordingly via the (tourism) associations.

- The possibilities for cooperation between tourism and agriculture should be expanded. To this end, regional partnerships should be increasingly developed beyond the individual farm level. For example, partnerships between various hotel cooperations such as Wanderhotels, Belvita Leading Wellness Hotels South Tyrol or Vitalpina Hotels South Tyrol with the South Tyrolean Farmers' Union or agricultural cooperatives can be promoted to further develop topics such as "regional delicacies" and create regions of enjoyment. This would also increasingly highlight local and regional products and integrate them into the hotel and gastronomy sector. In addition, cooperation potentials could also be realised between agricultural enterprises, ski resorts and alpine pasture gastronomy.
- The functional area of tourism should monitor the development of tourism in the region with the aim of achieving appropriate progress while complying with the defined capacity limits. Special attention should be paid to the protection and preservation of the natural environment, which is essential for the quality of life of the local population, but also for the quality of experience of the guests. In this context, environmentally friendly behaviour, e.g. environmentally friendly arrivals and departures, sustainable water use, eco-friendly activities (use of local public transport) etc., should be promoted among guests and locals as well as tourism entrepreneurs.
- It is of great importance to incorporate the benefits of strategic forestry management into tourism, with special consideration of the costs and revenues of forest management, investments in water protection, avalanche protection, etc., along with consideration of future developments around extreme events in the context of climate change (snowy winters or drought as an example, including an analysis of the effects). The background to such an initiative is the fact that tourism stakeholders are already sensitised in many ways to the interfaces with agriculture, but not to the same extent for forestry, which will undoubtedly have a massive impact on the Alpine landscape of the future in connection with climate change.
- In order to achieve the country's goal of becoming a "model region for sustainable Alpine mobility", measures should be expanded, and rail connections to Germany as well as to Italian cities and to major cities in the Netherlands, Belgium, Luxembourg, Austria and Switzerland should be further developed. Mobility plays a central and burdensome role in emissions. In the fight against climate change, the area of mobility thus has a corresponding role to play. The improvement of local public transport, the accessibility of valleys and rural areas as well as their connection to urban centres should be promoted – ideally with the help of alternative drives. Special efforts should be made for infrastructure projects that lead to traffic-calming or traffic-free centres. During the tourism-intensive months, public transport, i.e. bus and train connections, should be increased to guarantee a functioning and pleasant mobility for both

guests and locals. Innovative mobility concepts need to be developed in cooperation with research institutions and partners from industry.

- Health at the workplace should become more of a focus, especially for professions in tourism. To this end, concepts need to be developed around how employees and employers can consider, ensure and promote individual (and collective) health in various activities. The COVID-19 pandemic showed the close links between population protection, health and tourism. In this respect, it is necessary to continually review and act on the data to ensure the safety of guests and hosts through evidence-based decisions.
- Employment in tourism is regularly measured within the framework of monitoring and reviewed regarding the quality of jobs and the attractiveness of the industry. Issues such as seasonal work, part-time work, a five-day week, a shortage of skilled workers, and employment of locals and workers from abroad are addressed together with the hotel management schools, the associations (*Verbände*) and local tourism associations (*Tourismusvereine*). IDM, as the country's location agency, also plays an important role, especially in the area of communication. Concrete measures need to be developed together with the educational institutions of the province (e.g. regional hotel schools) and with the support of local associations. In this respect, it should be considered that – where possible – further education offers should be made available online so that they reach as many workers as possible and can be taken up flexibly.
- The aim is to raise awareness of the UNESCO World Natural Heritage Site "Dolomites" regarding the special protection of natural areas along with instruments for visitor guidance. The aim must be to raise awareness both in the local population as well as in the tourist markets. Visitor management can distribute visitor flows intelligently according to space and time, but it must also be able to point out weak and strong signals at capacity limits. UNESCO World Heritage should be the benchmark for this.
- Cooperation between tourism operators and nature and environmental protection associations along with cultural and art associations and actors should be promoted in a targeted manner. Initiated at the political level and driven by the associations (*Verbände*), these should be supported by digital and analogue platforms that serve to develop joint projects between tourism operators and nature conservation and environmental protection actors, and to address conflict issues.

In conclusion, it is imperative that the Department of Tourism assumes a leadership role in fostering collective responsibility among all areas involved in shaping the perception and acceptance of tourism in South Tyrol, utilizing their respective competences for its promotion. However, this undertaking is not without challenges, as tourism often garners high priority only when specific projects with significant implications for the responsible area are imminent. To address this, concrete measures can be implemented, such as organizing regular public events on crucial tourism policy decisions and industry developments, enabling the collection and documentation of public opinions and expectations. Additionally, monitoring instruments can be employed to gauge the satisfaction of the local population with tourism and its overall acceptance, with periodic evaluations conducted from the perspectives of various departments. To effectively implement tourism as a cross-sectional policy, it is essential to aspire towards establishing an "inter-departmental committee" that serves as a dedicated platform for ongoing exchanges on pertinent subjects.

Notes

1 "Dekret des Landeshauptmanns vom 18. Oktober 2007, Nr. 551. Verordnung über die Erweiterung gastgewerblicher Betriebe und die Ausweisung von Zonen für touristische Einrichtungen".
2 Own translation from German. Original quotation: "die Schaffung und Veränderung von Rahmenbedingungen und Instrumenten durch staatliche Stellen zur Forderung und Steuerung des Tourismus auf supranationaler, nationaler, regionaler und kommunaler Ebene".
3 Own translation from German. Original quotation: "die zielgerichtete Planung und Beeinflussung/ Gestaltung der touristischen Realität und Zukunft durch verschiedene Träger (staatliche, private, übergeordnete)".
4 Own translation from German. Original quotation: "bewusste Förderung und Gestaltung des Tourismus durch Einflussnahme auf die touristisch relevanten Gegebenheiten von Gemeinschaften".
5 Own translation from German. Original quotation: "zur Durchsetzung, Ablehnung oder zu Kompromissen von inhaltlichen Konzepten führen".
6 Own translation from German. Original quotation: "Tourismuspolitik die gesellschaftlichen Rahmenbedingungen für optimales Urlaubserleben schaffen, möglichst allen Bevölkerungsschichten den Zugang zum Reisen ermöglichen, für eine intakte soziale und natürliche Umwelt sorgen und die Lebensbedingungen in den Feriengebieten verbessern helfen".

References

ASTAT. (2023, February 3). *tourismus.qvw*. Beherbergungsbetriebe Und Betten Nach Kategorie. https://qlikview.services.siag.it/QvAJAXZfc/opendoc_notool.htm?document=tourismus.qvw&host=QVS%40titan-a&anonymous=truesmus.qvw

Benedikter, T. (2018, September 7). Touristische Überbelastung: was tun? [Tourism overload: what can be done?] *Salto.bz*. www.salto.bz/de/article/05092018/touristische-ueberbelastung-was-tun

Berg, W. (2021). "Tourismuspolitik" [Tourism policy]. In A. Schulz, B. Eisenstein, M. A. Gardini, T. H. Kirstges & W. Berg (Eds.), *Grundlagen des Tourismus*. De Gruyter Oldenbourg.

Bichler, B. F., Lösch, M. (2019). Collaborative governance in tourism: Empirical insights into a community-oriented destination. *Sustainability*, *11*(23): 6673.

Böhret, C., Jann, W., & Kronenwett, E. (1988). *Innenpolitik und politische Theorie: Ein Studienbuch* [Domestic politics and political theory: A study book]. Westdeutscher Verlag. www.amazon.de/Innenpolitik-politische-Theorie-Ein-Studienbuch/dp/3531114948

Carvalho, F. L., Guerreiro, M., & Matos, N. (2020). Overtourism. In M. Korstanje, C. Ribeiro de Almeida, A. Quintano, M. Simancas, R. Huete & Z. Breda (Eds.), *Handbook of research on the impacts, challenges, and policy responses to overtourism*. IGI Global.

de Rachewiltz, M., Dibiasi, A., Favilli, F., Ghirardello, L., Habicher, D., Laner, P., Omizzolo, A., Scuttari, A., Tomelleri, A., Trienbacher, T., Walder, M., Watschinger, S., & Windegger, F. (2021). *Die Beobachtungsstelle für nachhaltigen Tourismus in Südtirol (STOST). Jahreszwischenbericht – Ausgabe 2021*. Eurac Research. https://webassets.eurac.edu/31538/1643198519-de-insto-progress-report-2021.pdf

Ebner, S. (2022, July 11). Südtiroler ächzen unter den Touristen-Massen: Kommt der Bettendeckel? [South Tyroleans grunt under the tourist masses: will the bed stop come?] *Merkur. de*. www.merkur.de/welt/news-suedtirol-italien-tourismus-deutschland-betten-bozen-reisen-91657859.html

Erschbamer, G., Innerhofer, E., & Pechlaner, H. (2018). *Overtourism: How much tourism is too much?* Eurac Research. https://bia.unibz.it/esploro/outputs/report/Dossier-Overtourism-Wieviel-Tourismus-ist-zu/991006321295801241

FF Media. (2021, December 22). Südtirol will "begehrt und nachhaltig" sein – Panorama [South Tyrol wants to be "desired and sustainable"]. *ff – Das Südtiroler Wochenmagazin*. www.ff-bz.com/politik-wirtschaft/panorama/2021-51/suedtirol-will-begehrtnachhaltig-sein.html

Fischer, J. (2019). *Philosophische Anthropologie: Eine Denkrichtung des 20. Jahrhunderts* [Philosophic anthropology: A 20th century school of thought] (1st ed.). Alber Karl.

Freyer, W. (2006). *Tourismus: Einführung in die Fremdenverkehrsökonomie* (überarbeitete und aktualisierte Auflage). Oldenbourg Wissenschaftsverlag.

Freyer, W. (2015). *Tourismus: Einführung in die Fremdenverkehrsökonomie* [Tourism: Introduction to tourism economy] (11th ed.). De Gruyter Oldenbourg. www.degruyter.com/document/doi/10.1515/9783486857542/html?lang=de

Haigh, M. (2020). Cultural tourism policy in developing regions: The case of Sarawak, Malaysia. *Tourism Management*, *81*, 104166.

Kaspar, C. (1991). *Die Fremdenverkehrslehre im Grundriss. St. Galler Beiträge zum Fremdenverkehr und zur Verkehrswirtschaft* [A basic tourism outline. St Gallen contributions to tourism and transport] (Vol. 2). Institut für Fremdenverkehr und Verkehrswirtschaft Hochschule St. Gallen.

Kofler, M. (2022, July 24). Der Bettenstopp-Kompromiss – Die Neue Südtiroler Tageszeitung. *Die Neue Südtiroler Tageszeitung*. www.tageszeitung.it/2022/07/24/der-bettenstopp-kompromiss/

Krippendorf, J. (1976). Schweizerische Fremdenverkehrspolitik zwischen Pragmatismus und konzeptioneller Politik [Swiss tourism policy between pragmatism and conventional politics]. *Schweizerische Wirtschaftspolitik Zwischen Gestern Und Morgen. Festgabe Zum 65. Geburtstag von Hugo Sieber*, *25*, 444.

Liasidou, S. (2019). Understanding tourism policy development: A documentary analysis. *Journal of Policy Research in Tourism, Leisure and Events*, *11*(1), 70–93.

Lun, L-M., Pechlaner, H., & Pichler, S. (2014). Politik und Tourismus: die zukünftige Rolle von politischen Akteuren im Tourismus [Politics and tourism: the future role of political actors in tourism]. In R. Conrady & D. Ruetz (Eds.), *Tourismus und Politik: Schnittstellen und Synergiepotentiale*. Erich Schmidt.

Mair, G. (2020, December 2). "Neue Tourismuskultur" – Wirtschaft. *ff – Das Südtiroler Wochenmagazin*. www.ff-bz.com/politik-wirtschaft/wirtschaft/2020-49/neue-tourismuskultur.html

MCI Tourismus. (2014). *Analyse der Tourismuspolitik im Alpenraum 2013–2014*. https://docplayer.org/56734583-Analyse-der-tourismuspolitik-im-alpenraum.html

Opaschowski, H. W. (2002). *Tourismus. Eine systematische Einführung Analysen und Prognosen*. VS Verlag für Sozialwissenschaften. https://doi.org/10.1007/978-3-322-94948-6

Panasiuk, A., & Wszendybył-Skulska, E. (2021). Social aspects of tourism policy in the European Union: The example of Poland and Slovakia. *Economies*, *9*(1): 16.

Pechlaner, H., Innerhofer, E., Gruber, M., Scuttari, A., Walder, M., Habicher, D., Gigante, S., Volgger, M., Corradini, P., Laner, P., & von der Gracht, H. (2022). *Ambition Lebensraum Südtirol. Auf dem Weg zu einer neuen Tourismuskultur. Landestourismusentwicklungskonzept 2030+* [Ambition Living Space South Tyrol. Towards a new tourism culture]. Eurac Research. https://webassets.eurac.edu/31538/1651050857-ltek_de_final.pdf

Rainews. (2022, October 20). Overtourism: Wann ist es zuviel? [Overtourism: when is it too much?] *RaiNews*. www.rainews.it/tgr/tagesschau/articoli/2022/10/tag-Overtourism-Wann-ist-es-zuviel-192e91aa-36f4-47f4-b7f0-3ae1d071eb3a.html

Schwarz, H. (2023, February 16). Studie in Prags über das Leben mit Overtourism [Study in Prags about life with overtourism]. *Südtiroler Wirtschaftszeitung*. https://swz.it/studie-in-prags-ueber-das-leben-mit-overtourism/

Scuttari, A., Isetti, G., & Habicher, D. (2019). Visitor management in world heritage sites: Does overtourism-driven traffic management affect tourist targets, behaviour and satisfaction?: The case of the Dolomites UNESCO World Heritage Site, Italy. In H. Pechlaner, E. Innerhofer & G. Erschbamer (Eds.), *Overtourism: Tourism Management and Solutions*. Routledge.

Südtiroler Landesverwaltung. (2022, September 13). Landesregierung beschließt Regeln zur Bettenkontingentierung [Provincial government decides on bed allocation]. *News Presseamt*. https://news.provinz.bz.it/de/news/landesregierung-beschliesst-regeln-zur-bettenkontingentierung

SWZ. (2022, September 13). Die Bettenstopp-Regeln sind fixiert [The bed stop rules are done]. *Südtiroler Wirtschaftszeitung*. https://swz.it/die-bettenstopp-regeln-sind-fixiert/

Weiss, T. (2021). *Host community attitudes and overtourism: The case of the Puster Valley in South Tyrol, Italy*. http://urn.kb.se/resolve?urn=urn:nbn:se:uu:diva-455272

Windegger, F., Scuttari, A., Walder, M., Erschbamer, G., de Rachewiltz, M., Corradini, P., Weisel, Z. K., Habicher, D., Ghirardello, L., Wallnöfer, V., Garzon, G., & Moroder, P. (2022). *The sustainable tourism observatory of South Tyrol (STOST). Annual Progress Report – 2022 edition*. Eurac Research. https://webassets.eurac.edu/31538/1678895987-stost-2022_full-report.pdf

18 Resilience Agility in Tourism

A Strategic Approach to Mitigate Overtourism and Gain Sustainability

Anastasia Traskevich and Martin Fontanari

1 Introduction

COVID-19 emerged as a specific crisis challenge for tourism, since it developed not in the form of a scenario-based event with critical impacts but as an unknown and unpredictable process (Goessling et al., 2021; Hao et al., 2020; Lai & Wong, 2020; Li et al., 2021; Sigala, 2020). Therefore, the ongoing long-term study on resilience in tourism is currently focused on grounding new dimensions, implementation factors and development opportunities of resilience, also taking into account the tremendous insights which have been discovered and tangibly experienced by the tourism business during the pandemic crisis (Hall et al., 2022; Hoffmann et al., 2023; Islam et al., 2022). Furthermore, the COVID-19 crisis can be placed in a broader context with the intense ongoing discussion on overtourism and its consequences (Fontanari & Traskevich, 2021a; Fontanari & Traskevich, 2022; Fontanari et al., 2021b). Both issues concern a more sensitive and responsible use of local recreational resources; as well as involvement of residents and travellers in the tourism system. Furthermore, it entails a newly oriented cooperation of service providers in the sense of a resilient destination management (Hoffmann et al., 2023).

Integrative and synthesising approaches to conceptualising resilience in tourism are only emerging in recent years (Butler, 2017; Colmekcioglu et al., 2022; Gretzel & Scarpino-Johns, 2018; Innerhofer et al., 2018; Filimonau & DeCoteau, 2020; Traskevich & Fontanari, 2021b). These approaches are based on the converged inductive evolution of the core characteristics of ecological, corporate and technological resilience, which emphasises its significance for the community within destinations (Chen et al., 2020; Sheller, 2020). Thus, the current research challenge, which contains numerous gaps, is the holistic application of resilience knowledge to the multi-dimensional nature of tourism within the systemic context of an enterprise, a community and a destination as an open networked system (Cochrane, 2010; Hall et al., 2018). The present research gap concerns the interrelated dimensions of resilience against numerous kinds of vulnerability in the global tourism industry (Calgaro et al., 2014; Sharpley, 2012), in particular overtourism (Fontanari & Berger-Risthaus, 2020; Fontanari & Traskevich, 2021a; Fontanari & Traskevich, 2022; Fontanari et al., 2021b). This gap is addressed through the

DOI: 10.4324/9781003365815-25

lens of sustainability paradigm (Espiner et al., 2017; Font et al., 2021; Holladay, 2018; Kato, 2018; Nunkoo, 2017; Ruhanen, 2008; Sausmarez, 2007) to advocate for resilience as an advanced applied concept for sustainable tourism management (Traskevich & Fontanari, 2021b).

2 Resilience Agility as a Strategic Approach for Transformative Tourism Development

The postulates of crisis resilience that were previously studied (Cushnahan, 2004; Jiang et al., 2019; Ritchie, 2008, 2009) are extended in the present contribution to highlight that the theory of resilience could also have worked preventively for the COVID-19 crisis (Fontanari & Traskevich, 2022; Traskevich & Fontanari, 2021b) and the critical issues of overtourism (Butler & Dodds, 2022; Cheer et al., 2019; Gonzalez et al., 2018). The present strategic considerations provide insights into new components of resilience to be identified for future transformative tourism development in the context of the latest findings of Hall (2022) and Sharpley (2022). Furthermore, Hoffmann et al. (2023) introduce the inductive approach to resilience building which suggests iterative inductive accumulation of resilience knowledge considering all the levels of resilience: from corporate resilience towards community and network resilience and ultimately towards the systemic level of destination resilience.

From this perspective, the reactive business experiences achieved in tourism business in the course of critical conditions, like the COVID-19 crisis or overtourism, require further theoretical analysis and conceptualisation. This is focused on defining ways of applying resilience knowledge in the long term, not only in relation to crisis scenarios but also in relation to transformation for prosperity, sustainability and well-being. The present conceptual approach suggests that such positive perspectives for resilience building focus on cooperation and networking (Hoffmann et al., 2023), resilient leadership, resilient technological advancement (Bethune et al., 2022; Fontanari & Traskevich, 2022), and innovative value creation for self-sufficiency (Traskevich & Fontanari, 2018, 2021b), in particular through partnerships with local leisure sharing economies.

The strategic gaps in resilience building are identified in the domain of regional networking and cooperation (Hoffmann et al., 2023), particularly: confidence in network communication for the efficient data exchange and knowledge transfer; mutual help and support in the destination to attain flexibility in the implementation of revised plans in crisis; locally based value co-creation. Building on resilience of the human capital appears an advanced field of management for the tourism business. In the future, the imperatives of transformative tourism development (Ateljevic & Sheldon, 2022; Dredge, 2022) will require tourism management to address more carefully the issues of financial stability of the employees, as well as to contribute actively to the health, wellbeing and psychological resilience of the staff.

In applying resilience knowledge in tourism management, it appears relevant to introduce a participative approach to finding solutions within the framework of quality management. Knowledge management structures should be more actively addressed by tourism stakeholders for establishment and application of resilience knowledge. The present integrative approach discloses the lack of managerial resilience awareness in terms of stability of the resource procurement for business continuity (Hoffmann et al., 2023). Proactive approach to resilience building should further embrace considerations on the issue of self-sufficiency and alternative use of the physical infrastructure within a plan "B" for critical situations (Fontanari et al., 2021a). In this instance, future-oriented structural changes in management and related investment measures dedicated to resilience development are required on the corporate level, within the network and at the destination.

Hoffmann et al. provide a revised integrative definition on resilience (2023, p. 2) which postulates proactive and positive treatment of the challenge of crisis and vulnerability of tourism systems. In this case, tourism business units are seen not solely within the domain of business continuity in crisis. The positive definition of resilience also implies expanding business opportunities for further development into multidimensional directions: business, products, infrastructure, knowledge-management, organisational structures and leadership (Jiang & Wen, 2020; Li et al., 2021; Prayag, 2023; Ritchie & Jiang, 2019; Traskevich & Fontanari, 2021b). From this perspective, the multidimensional understanding of **re**silience **kno**wledge (ReKo) was elaborated (Hoffmann et al., 2023, p. 2; Traskevich & Fontanari, 2023).

These multidimensional directions bring the notion of entrepreneurial agility which is required for integrative resilience building. First definitions of agility applied to tourism concern supply chains (Ku, 2022; Mandal & Saravanan, 2019) and niche product development (Mengoni et al., 2009). However, the fundamental and holistic understanding of entrepreneurial agility is elaborated within the IT industry. Therefore, the present research addresses the complex agility components from applied cybernetics with the aim of outlining a definition of resilience agility in tourism. For the purpose of proactive resilience development, the following innovative components of agility should be addressed: lean portfolio management; agile organisational design; enablement teams; technology agility; agile business framework; strategic business planning; agility metrics for added value, quality and flows; agile leadership and corporate culture; agile talent management; agility for discoveries and validation (Lyytinen & Rose, 2006; Wang et al., 2003).

These conceptual transdisciplinary considerations allow for the following definition of resilience agility for tourism business units:

> Resilience agility is understood as a proactive and initiating change management that relies on a broad and deep resilience know-how of residents and tourism stakeholders and thus builds on a complementary interaction of all units participating in tourism in order to increase the sustainable quality and adaptability of tourism in established regional networks.

3 The Implementation of Resilience Agility Strategies

This chapter presents the elaborated conceptual approach towards the application of resilience knowledge in tourism and presents the existing theoretical basis for the further development of the relevant strategies on resilience agility. Resilience agility as a strategic approach is presented in Figure 18.1. Based on the present theoretic approach, it is composed of complementary and interdependent models of resilience thinking and the model application in the design of systems, organizations and products in tourism development (Fontanari & Traskevich, 2022; Hoffmann et al., 2023; Traskevich & Fontanari, 2018, 2021b, 2023). What all the models have in common is that the underlying understanding of resilience is clearly differentiated and synthesized from big ideas elaborated in the fields of crisis management (Cushnahan, 2004; Faulkner, 2001; Ritchie, 2008, 2009; Tajeddini et al., 2023; Zhai & Shi, 2022); business continuity management (Buzzao & Rizzi, 2023; Herbane, 2010; Namdar et al., 2021; Nguyen et al., 2022; Saad & Elshaer, 2020; Thees et al., 2022); corporate social responsibility (Battisti et al., 2022; Font & Lynes, 2018; Pereira & Anjos, 2021; Wong et al., 2021); sustainability management (Espiner et al., 2017; Font et al., 2021; Holladay, 2018; Kato, 2018; Nunkoo, 2017; Ruhanen, 2008; Sausmarez, 2007); and environmental social governance (Fafaliou et al., 2022; Hassan & Meyer, 2022; Nunkoo, 2017). Furthermore, the resilience agility as a strategic approach implies elaboration on a strategic mindset. This mindset entails aligning and designing resilient structures in tourism management, and – at the same time – creating new, innovative, marketable and competition superior approaches to network-based business development.

The idea of applying transdisciplinary resilience knowledge to tourism in the direction of long-term integrative resilience and sustainability dates back to the first sufficient empirical results on the ongoing resilience research (started in 2017) (Fontanari & Kredinger, 2017; Innerhofer et al., 2018). These findings were correlated with the first wave of COVID-19-related resilience awareness in tourism business achieved in 2021 (Traskevich & Fontanari, 2021b). The first conceptual considerations to apply resilience knowledge to integrative advanced management of resilience for tourism businesses were grounded (Traskevich & Fontanari, 2021b). Since then, the ongoing research on tourism resilience has been dedicated to creating integrative theoretical models which are to be applied to diverse business aspects of tourism development to show the most efficient, competitive and responsible performance of tourism business units in all case scenarios of the development of their micro- and macro-environment, as well as local and global socioecological and geopolitical situations (Traskevich & Fontanari, 2021a, 2021b; Fontanari & Traskevich, 2022; Hoffmann et al., 2023).

The fundamental Conceptual Integrative Model of Resilience in Tourism was elaborated in 2019 (Traskevich & Fontanari, 2021b). It was designed with regard to the new framework conditions and imperatives created by COVID-19 for the tourism sector. It was done with the aim to prioritise the concept of resilience in tourism. The model is based on the fields that affect strategic positioning and product policy of tourism companies and destinations based on local resources

Figure 18.1 Strategic approach to resilience agility

Source: The authors

and competencies. The model also incorporates the components of personal resilience, mental wellbeing and spirituality for cooperative expansion of the tourism value chain. The research shows that addressing the model of resilience ensures an increase of tourist attractiveness and competitiveness of the product portfolio, as well as innovative business models for tourism enterprises. Further steps of research have also shown the effectiveness of the concept of resilience to overcome the negative impacts of mass tourism, like overtourism (Fontanari & Traskevich, 2021a), and for local communities and tourists to achieve a meaningful life. Finally, the application of resilience knowledge in tourism management is directed to mitigate overtourism with the help of innovative technologies (Fontanari & Traskevich, 2022). The conceptual model suggests integrative application of intelligent solutions within tourism cooperative networks and destination governance.

The ReKo-Model (Traskevich & Fontanari, 2023) is an integrative theoretical model of resilience knowledge application for business units in tourism. The model describes dimensions of resilience knowledge application. It encompasses an implementation framework which indicates factors of resilience knowledge application and offers a range of business implementation tools, instruments and methods. The ReKo-Model offers a holistic integrative approach to implement the concept of resilience in tourism business practice. The empirical application of the ReKo-Model (Fontanari & Traskevich, 2021b; Fontanari & Traskevich, 2023b; Traskevich & Fontanari, 2021a) demonstrates the applied perspective on building resilience awareness, resilience performance and resilient cooperative commitment in tourism for its transformation and flourishing.

Besides the aforementioned integrative models of resilience in tourism, we have elaborated complementary models that are focused on the most crucial and unexplored strategic dimensions of resilience building. Resilience development in the context of business model innovation (Fontanari et al., 2021a) for tourism companies is theoretically grounded. Furthermore, the research considers a conceptual approach which would direct cooperative initiatives in tourism and hospitality (Bhat & Milne, 2008; Cai, 2002; Chang et al., 2019; Guo et al., 2013; Nguyen et al., 2022; Zhang et al., 2022) to proactive resilience building. The ongoing research on resilience agility suggests addressing internal factors of tourism service production as objects of resilient cooperation. This strategic approach incorporates the design framework for tourism cooperative networks for resilient value co-creation of tourism enterprises based on the synthesis of corporate, networked and destination resilience (Hoffmann et al., 2023). In this case the strategic approach in resilience building reiterates the embeddedness of tourism stakeholders in destinations through the resilient business model with further transition towards destination resilience. The resilience agility is realised through inductive components of self-sufficiency, resilient product development, alternative accommodation and catering services, leading to new experiential knowledge and design of meaningful livelihoods (Traskevich & Fontanari, 2021b). In this direction, the ongoing theoretical elaboration of resilience in tourism also integrates the concept of the leisure sharing economy for its application to tourism acceptance by local communities and strategic priorities for gaining stronger resilience and sustainability. In this context, resilience agility also describes the elements and mechanisms for the positive impact of the leisure sharing economy on community wellbeing, fair inclusion of local residents, innovation, authenticity and the touristic dissemination of local knowledge, values, traditions and skills.

In the context of the present discussion on overtourism, the presented strategic approach to resilience agility incorporates the indicative model for statistical complementary elements of socio-cultural tourism satellite accounts (Fontanari et al., 2021b) which is introduced to advocate resilience as an advanced managerial concept against the crucial vulnerabilities of the global tourism industry. In this case, resilience agility is called for in order to balance mass tourism development for responsible and sustainable business orientation of tourism enterprises in compliance with claims for tourism acceptance and community benefits. This way, the challenges and impacts of overtourism can be mitigated in the long term for transformative tourism development in the post-COVID-19 era.

At the systemic level of destination resilience, the present strategic approach to resilience agility in tourism considers the development of criteria for different forms and characteristics of resilient destinations with regard to niche tourism themes – on the basis of which these destinations can be assigned to different targets or clusters. The different resilience clusters imply various development strategies of the destination in order to strive for a differentiated, holistic and thus competitive development and to design profile-forming tourism offers for the native tourism companies. In this field, the conceptual and assessment model is elaborated (Traskevich & Fontanari, 2019) implementing the statistical tool of fuzzy-analysis to correlate the unique prerequisites of tourism regions with the core requirements for resilient tourism product development.

Last but not least, the present strategic approach to resilience agility can be implemented also towards the applied field of tourism product development. The Convergent Development Model of Mental Wellness Within Resilient Destinations, Code-Red Model (Traskevich & Fontanari, 2018), is elaborated for this instance and applied to the design of new mental wellness products, that contribute to the destination's resilience. Directly derivable products can be implemented in rural areas, thus leading rural stakeholders to a marketable supply quality and allowing for sufficient distribution of critical tourism masses away from the present hotspots. This is how overtourism can also be mitigated by means of resilient product development. The product approach comprises self-sufficiency measures and social interactions within a fulfilling Mental Wellness paradigm. The integrative framework for multi-level Mental Wellness products is elaborated to sustainably incorporate the natural habitat and indigenous knowledge of the destination, in order to interpret recreational resources authentically, to ensure an environmentally oriented identity, and to initiate co-operative approaches between stakeholders. Besides, we elaborated a design framework for resilient product development (Fontanari & Traskevich, 2023a) on the basis of the Code-Red Model.

4 Discussion: The Benefits of Resilience Agility

It can be concluded that the already elaborated theoretical models cover inductively all the key fields of resilience agility in tourism. The integrative models create the relevant basis of resilience knowledge and managerial expertise for building on resilience agility for tourism and hospitality companies, suppliers, mobility providers and all the other tourism-relevant stakeholders. At the same time, supplementary focused modules provide the range of contributions, as follows:

- Advance resilience orientation in the fields of personal resilience of the employees and tourism communities
- Integrate private providers for networked resilience building
- Disclose the essence of cooperative approaches to long-term resilience building
- Create methodologies for resilient product development

Thus, the higher common goals for regional sustainable and resilience-oriented development can be achieved.

As there is no empirical study that examines resilience agility in thinking and acting (Prayag, 2023), the contributions of such a strategic approach to corporate success and mitigation of global vulnerabilities of tourism, like overtourism, still present a research gap for future investigation. However, the presented approaches of individual models (Figure 18.1) allow for drawing a theoretical consideration on contributions of resilience agility as a strategic paradigm (Figure 18.2).

5 Conclusion

The presented conceptual study on resilience agility in tourism brings up a discursive formation on the issue of clearly dispersed and antagonistic professional and academic attitudes towards the phenomenon of resilience. This comes in line with other recent conceptual studies on resilience application in tourism (Prayag, 2023). The present study postulates two levels at which the concept of resilience can be applied in tourism research: reactive resilience awareness and proactive resilience agility. These research subjects can be synthesised to construct a paradigm that develops from reactive crisis management to a proactive long-term approach for resilience knowledge application, complete with the imperatives of innovativeness, meaningful and responsible development and sustainability within the new framework of the "disruptive twenties". This further emphasises the need to study resilience as a broad interdisciplinary concept by defining the new integrative dimensions of resilience knowledge in tourism with the key components of supply chains, business networks, technology, community, personality and ecosystems as a research focus. These complex theoretical dimensions require both qualitative and quantitative empirical investigation. The quantitative approach could further evaluate the range and depth of resilience awareness and resilience agility in tourism management through the system of benchmarks that characterise innovation, cooperation, networking, leadership competence, biodiversity and the resilience of ecosystems, and engineering resilience in the context of destinations.

Contributions by applying to resilience agility

Advantages of "Autarky"-oriented approach to destination resilience	
• *local entrepreneurial activity in the field of tourism*	stakeholders know the specifics of local recreational resources and hospitality competence; they are bearers of uniqueness and authenticity of the destination
• *efficiency of the regional and local supply system*	stability, high quality and low cost of products and services
• *strengthening and vertical extension of the value-chains*	USP tourism products are competitive in the global tourism market
• *growing attractiveness of the product portfolio*	no significant growth of the costs thanks to the synergies created by partnerships in the destination

Implementation advantages	
• *growth of investment in-flows and their long-term orientation*	infrastructure, training of stakeholders & local community; recreational resources
• *design of new tourism products*	diversification of the product portfolio; USP-products
• *improve-ment of the regional destination management*	

Benefits for a destination against existing vulnerabilities and overtourism	
• *diversification of entrepreneurial risks*	extension of the destination portfolio with experimental resilient regions
• *minimization of the tour operator's insurance risks*	safety of tourists in emergency and crisis situations are insured
• *loyalty of the local stakeholders and the regional tourism administration*	investment of the tour operator into the projects in the destination
• *deeper focus on modern trends in tourist demand*	priority of safety, uniqueness and authenticity, individual approach, niche tourism themes, trust and loyalty to the brand, sensitivity to the factor of sustainability
• *qualitative expansion of the client base and branding activities*	a tour operator can generate a significant segment of "resilience-sensitive consumers"; the portfolio of resilient destinations as a powerful branding factor

Figure 18.2 Contributions by applying to resilience agility strategies

Source: The authors

Thus, the new dimensions of resilience in tourism should be further conceptualised in order to create a new integrative framework of resilience agility for tourism management. This resilience agility, which is based on innovative, value oriented and sustainable approaches, will be iteratively communicated to real business and tourism markets with their specific requirements. The transfer of integrative resilience knowledge will be achieved through evolving stakeholder understanding aimed at long-term proactive resilience agility in tourism management. Furthermore, institutional consolidation and coordination of research activities in the field of resilience in tourism will be beneficial in achieving this long-term goal.

References

Ateljevic, I., & Sheldon, P. J. (2022). Guest editorial: Transformation and the regenerative future of tourism. *Journal of Tourism Futures*, *8*(3), 266–268. https://doi.org/10.1108/JTF-09-2022-284

Battisti, E., Nirino, N., Leonidou, E., & Thrassou, A. (2022). Corporate venture capital and CSR performance: An extended resource-based view's perspective. *Journal of Business Research*, *139*, 1058–1066. https://doi.org/10.1016/j.jbusres.2021.10.054

Bethune, E., Buhalis, D., & Miles, L. (2022). Real time response (RTR): Conceptualizing a smart systems approach to destination resilience. *Journal of Destination Marketing & Management*, *23*, 100687. https://doi.org/10.1016/j.jdmm.2021.100687

Bhat, S. S., & Milne, S. (2008). Network effects on cooperation in destination website development. *Tourism Management*, *29*(6), 1131–1140. https://doi.org/10.1016/j.tourman.2008.02.010

Butler, R. W. (Ed.). (2017). *Tourism and Resilience*. CABI.

Butler, R. W., & Dodds, R. (2022). Overcoming overtourism: A review of failure. *Tourism Review*, *77*(1), 35–53.

Buzzao, G., & Rizzi, F. (2023). The role of dynamic capabilities for resilience in pursuing business continuity: An empirical study. *Total Quality Management & Business Excellence*. doi:10.1080/14783363.2023.2174427

Cai, L. A. (2002). Cooperative branding for rural destinations. *Annals of Tourism Research*, *29*(3), 720–742. https://doi.org/10.1016/S0160-7383(01)00080-9

Calgaro, E., Lloyd, K., & Dominey-Howes, D. (2014). From vulnerability to transformation: A framework for assessing the vulnerability and resilience of tourism destinations. *Journal of Sustainable Tourism*, *22*(3), 341–360.

Chang, Y-W., Hsu, P-W., & Lan, Y-C. (2019). Cooperation and competition between online travel agencies and hotels. *Tourism Management*, *71*, 187–196. https://doi.org/10.1016/j.tourman.2018.08.026

Cheer, J. M., Milano, C., & Novelli, M. (2019). Tourism and community resilience in the Anthropocene: Accentuating temporal overtourism. *Journal of Sustainable Tourism*, *27*(4), 554–572.

Chen, F., Xu, H., & Lew, A. A. (2020). Livelihood resilience in tourism communities: The role of human agency. *Journal of Sustainable Tourism*, *28*(4), 606–624.

Cochrane, J. (2010). The sphere of tourism resilience. *Tourism Recreation Research*, *35*(2), 173–185.

Colmekcioglu, N., Dineva, D., & Lu, X. (2022). "Building back better": The impact of the COVID-19 pandemic on the resilience of the hospitality and tourism industries. *International Journal of Contemporary Hospitality Management*, *34*(11), 4103–4122. https://doi.org/10.1108/IJCHM-12-2021-1509

Cushnahan, G. (2004). Crisis management in small-scale tourism. *Journal of Travel & Tourism Marketing, 15*(4), 323–338.

Dredge, D. (2022). Regenerative tourism: Transforming mindsets, systems and practices. *Journal of Tourism Futures, 8*(3), 269–281. https://doi.org/10.1108/JTF-01-2022-0015

Espiner, S., Orchiston, C., & Higham, J. (2017). Resilience and sustainability: A complementary relationship? Towards a practical conceptual model for the sustainability–resilience nexus in tourism. *Journal of Sustainable Tourism, 25*(10), 1385–1400.

Fafaliou, I., Giaka, M., Konstantios, D., & Polemis, M. (2022). Firms' ESG reputational risk and market longevity: A firm-level analysis for the United States. *Journal of Business Research, 149*, 161–177. https://doi.org/10.1016/j.jbusres.2022.05.010

Faulkner, B. (2001). Towards a framework for tourism disaster-management. *Tourism Management, 22*(2), 135–147.

Filimonau, V., & DeCoteau, D. (2020). Tourism resilience in the context of integrated destination and disaster management. *International Journal of Tourism Research, 22*, 202–222.

Font, X., & Lynes, J. (2018). Corporate social responsibility in tourism and hospitality. *Journal of Sustainable Tourism, 26*(7), 1027–1042. doi:10.1080/09669582.2018.1488856

Font, X., English, R., Gkritzali, A., & Tian, W. S. (2021). Value co-creation in sustainable tourism: A service-dominant logic approach. *Tourism Management, 82*, 104200. https://doi.org/10.1016/j.tourman.2020.104200

Fontanari, M., & Berger-Risthaus, B. (2020). Problem and solution awareness in overtourism: A Delphi study. In H. Pechlaner, E. Innerhofer & G. Erschbamer (Eds.), *Overtourism: Tourism management and solutions*. Routledge.

Fontanari, M., & Kredinger, D. (2017). Risiko- und Resilienzbewusstsein. Empirische Analysen und erste konzeptionelle Ansätze zur Steigerung der Resilienzfähigkeit von Regionen [Risk and resilience awareness. Empirical analyses and first conceptual approaches to increase the resilience of regions]. ISM Working Paper, 9.

Fontanari, M., & Traskevich, A. (2021a). Consensus and diversity regarding overtourism: The Delphi-study and derived assumptions for the post-COVID-19 time. *International Journal of Tourism Policy, 11*(2), 161–187. doi:10.1504/IJTP.2021.10039110

Fontanari, M., & Traskevich, A. (2021b). The concept of resilience for organisational units in tourism: Application for advanced sustainable management. Paper presented at the Clusters and Competitiveness Conference, Session "Resilience and Change Management for Cities, Regions and Firms", Ruhr Universität Bochum, Geographical Institute, Bochum, Germany, 23–24 September.

Fontanari, M., & Traskevich, A. (2022). Smart-solutions for handling overtourism and developing destination resilience for the post-Covid-19 era. *Tourism Planning & Development*. doi:10.1080/21568316.2022.2056234

Fontanari, M., & Traskevich, A. (2023a). Nature-based and wellness tourism in resilient destinations: Methodological approaches and pilot applications for product development in experimental destinations. In F. Niccolini, I. Azara, E. Michopoulou, J. R. Barborak & A. Cavicchi (Eds.), *Nature-based tourism and wellbeing: Impacts and future outlook*. CAB International.

Fontanari, M., & Traskevich, A. (2023b). The integrative concept of crisis resilience in tourism and hospitality: First empirical insights from the hotels in Germany and Austria. In K. Andriotis (Ed.), *Innovative sustainable practices in travel and tourism. Recovery and resilience: Book of abstracts of the International Conference on Tourism (ICOT2023), Nicosia, Cyprus*. International Association for Tourism Policy; European University Cyprus.

Fontanari, M., Traskevich, A., & Kutsch, H. (2021a). Corporate resilience within tourism enterprises. In M. Valeri, A. Scuttari, H. Pechlaner (Eds.), *Resilienza e sostenibilitá:Dinamiche globali e risposte locali* [Resilience and sustainability: Global dynamics and local responses]. Giappichelli.

Fontanari, M., Traskevich, A., & Seraphin, H. (2021b). (De)growth imperative: The importance of destination resilience in the context of overtourism. In K. Andriotis (Ed.), *Issues and cases of degrowth in tourism*. CAB International.

Goessling, S., Scott, D., & Hall, C. M. (2021). Pandemic, tourism and global change: A rapid assessment of COVID-19. *Journal of Sustainable Tourism*, *29*(1), 1–20.

Gonzalez, V. M., Coromina, L., & Gali, N. (2018). Overtourism: Residents' perceptions of tourism impact as an indicator of resident social carrying capacity – case study of a Spanish heritage town. *Tourism Review*, *73*(3), 227–296. https://doi.org/10.1108/TR-08-2017-0138

Gretzel, U., & Scarpino-Johns, M. (2018). Destination resilience and smart tourism destinations. *Tourism Review International*, *22*(14), 263–276.

Guo, X., Ling, L., Dong, Y., & Liang, L. (2013). Cooperation contract in tourism supply chains: The optimal pricing strategy of hotels for cooperative third party strategic websites. *Annals of Tourism Research*, *41*, 20–41. https://doi.org/10.1016/j.annals.2012.11.009

Hall, C. M. (2012). Island, islandness, vulnerability and resilience. *Tourism Recreation Research*, 37(2), 177–181.

Hall, C. M. (2022). Tourism and the Capitalocene: From green growth to ecocide. *Tourism Planning & Development*, *19*(1), 61–74. doi:10.1080/21568316.2021.2021474

Hall, C. M., Prayag, G., & Amore, A. (Eds.). (2018). *Tourism and resilience: Individual, organisation and destination perspectives*. Channel View Publications.

Hall, C. M., Safonov, A., & Naderi Koupaei, S. (2022). Resilience in hospitality and tourism: Issues, synthesis and agenda. *International Journal of Contemporary Hospitality Management*. https://doi.org/10.1108/IJCHM-11-2021-1428

Hao, F., Xiao, Q., & Chon, K. (2020). COVID-19 and China's hotel industry: Impacts, a disaster management framework, and post-pandemic agenda. *International Journal of Hospitality Management*, *90*, 102636.

Hassan, A. S., & Meyer, D. F. (2022). Does countries' environmental, social and governance (ESG) risk rating influence international tourism demand? A case of the Visegrád Four. *Journal of Tourism Futures*. https://doi.org/10.1108/JTF-05-2021-0127

Herbane, B. (2010). The evolution of business continuity management: A historical review of practices and drivers. *Business History*, *52*(6), 978–1002. doi:10.1080/00076791.2010.511185

Hoffmann, S., Deppisch, T., Fontanari, M., & Traskevich, A. (2023). Creating cooperative value for destination resilience. *Tourism Management Perspectives*, *48*, 101160. https://doi.org/10.1016/j.tmp. 2023.101160

Holladay, P. J. (2018). Destination resilience and sustainable tourism development. *Tourism Review International*, *22*(11), 251–261.

Innerhofer, E., Fontanari, M., & Pechlaner, H. (Eds.). (2018). *Destination resilience: Challenges and opportunities for destination management and governance*. Routledge.

Islam, M. S., Kabir, M., & Hassan, K. (2022). Resilience strategies of tour operators during the uncertainty of COVID-19: Evidence from Bangladesh. *Tourism Planning & Development*. doi:10.1080/21568316.2022.2144429

Jiang, Y., & Wen, J. (2020). Effects of COVID-19 on hotel marketing and management: A perspective article. *International Journal of Contemporary Hospitality Management, 32*(8), 2563–2573.

Jiang, Y., Ritchie, B. W., & Verreynne, M. L. (2019). Building tourism organizational resilience to crises and disasters: A dynamic capabilities view. *International Journal of Tourism Research, 21*, 882–900.

Kato, K. (2018). Debating sustainability in tourism development: Resilience, traditional knowledge and community: A post-disaster perspective. *Tourism Planning & Development, 15*(1), 55–67.

Ku, E. C. S. (2022). Technological capabilities that enhance tourism supply chain agility: Role of E-marketplace systems. *Asia Pacific Journal of Tourism Research, 27*(1), 86–102. doi:10.1080/10941665.2021.1998162

Lai, I. K. W., & Wong, J. W. C. (2020). Comparing crisis management practices in the hotel industry between initial and pandemic stages of COVID-19. *International Journal of Contemporary Hospitality Management, 32*(10), 3135–3156.

Li, B., Zhong, Y., Zhang, T., & Hua, N. (2021). Transcending the COVID-19 crisis: Business resilience and innovation of the restaurant industry in China. *Journal of Hospitality and Tourism Management, 49*, 44–53.

Lyytinen, K., & Rose, G. M. (2006). Information system development agility as organizational learning. *European Journal of Information Systems, 15*(2), 183–199. doi:10.1057/palgrave.ejis.3000604

Mandal, S., & Saravanan, D. (2019). Exploring the influence of strategic orientations on tourism supply chain agility and resilience: An empirical investigation. *Tourism Planning & Development, 16*(6), 612–636. doi:10.1080/21568316.2018.1561506

Mengoni, M., Germani, M., & Mandorli, F. (2009). A structured agile design approach to support customisation in wellness product development. *International Journal of Computer Integrated Manufacturing, 22*(1), 42–54. doi:10.1080/09511920802326233

Namdar, J., Torabi, S. A., Sahebjamnia, N., & Pradhan, N. N. (2021). Business continuity-inspired resilient supply chain network design. *International Journal of Production Research, 59*(5), 1331–1367. doi:10.1080/00207543.2020.1798033

Nguyen, V. K., Pyke, J., Gamage, A., Lacy, T., & Lindsay-Smith, G. (2022). Factors influencing business recovery from compound disasters: Evidence from Australian micro and small tourism businesses. *Journal of Hospitality and Tourism Management, 53*, 1–9, https://doi.org/10.1016/j.jhtm.2022.08.006

Nunkoo, R. (2017). Governance and sustainable tourism: What is the role of trust, power and social capital? *Journal of Destination Marketing & Management, 6*(4), 277–285. https://doi.org/10.1016/j.jdmm.2017.10.003

Pereira, T., & Anjos, S. J. G. (2021). Corporate social responsibility as resource for tourism development support. *Tourism Planning & Development*. doi:10.1080/21568316.2021.1873834

Prayag, G. (2023). Tourism resilience in the "new normal": Beyond jingle and jangle fallacies? *Journal of Hospitality and Tourism Management, 54*, 513–520. https://doi.org/10.1016/j.jhtm.2023.02.006

Ritchie, B. W. (2008). Tourism disaster planning and management: From response and recovery to reduction and readiness. *Current Issues in Tourism, 11*(4), 315–348.

Ritchie, B. W. (2009). Crisis and disaster management for tourism. Channel View Publications.

Ritchie, B. W., & Jiang, Y. (2019). A review of research on tourism risk, crisis and disaster management: Launching the annals of tourism research curated collection on tourism risk, crisis and disaster management. *Annals of Tourism Research*, *79*, 102812.

Ruhanen, L. (2008). Progressing the sustainability debate: A knowledge management approach to sustainable tourism planning. *Current Issues in Tourism*, *11*(5), 429–455. doi:10.1080/13683500802316030

Saad, S. K., & Elshaer, I. A. (2020). Justice and trust's role in employees' resilience and business' continuity: Evidence from Egypt. *Tourism Management Perspectives*, *35*, 100712. https://doi.org/10.1016/j.tmp. 2020.100712

Sausmarez, N. (2007). Crisis management for the tourism sector: Preliminary considerations in policy development. *Tourism and Hospitality Planning & Development*, *1*(2), 157–172.

Sharpley, R. (2012). Tourism and vulnerability: A case of pessimism? *Tourism Recreation Research*, *37*(3), 257–260.

Sharpley, R. (2022). Tourism and development theory: Which way now? *Tourism Planning & Development*, *19*(1), 1–12. doi:10.1080/21568316.2021.2021475

Sheller, M. (2020). Reconstructing tourism in the Caribbean: Connecting pandemic recovery, climate resilience and sustainable tourism through mobility justice. *Journal of Sustainable Tourism*, *29*(9), 1436–1449.

Sigala, M. (2020). Tourism and COVID-19: Impacts and implications for advancing and resetting industry and research. *Journal of Business Research*, *117*, 312–321.

Tajeddini, K., Gamage, T. C., Tajeddini, O., & Kallmuenzer, A. (2023). How entrepreneurial bricolage drives sustained competitive advantage of tourism and hospitality SMEs: The mediating role of differentiation and risk management. *International Journal of Hospitality Management*, *111*, 103480. https://doi.org/10.1016/j.ijhm.2023.103480

Thees, H., Störmann, E., & Pechlaner, H. (2022). Business modeling for resilient destination development: A multi-method approach for the case of destination Franconia, Germany. *Tourism Planning & Development*. https://doi.org/10.1080/21568316.2022.2121313

Traskevich, A., & Fontanari, M. (2018). Mental wellness in resilient destinations. *International Journal of Spa and Wellness Industry*, *1*(3), 193–217. doi:10.1080/24721735.2019.1596656

Traskevich, A., & Fontanari, M. (2019). Fuzzy-clustering approach for strategic development of resilient destinations for mental wellness. In K. Andriotis (Ed.), *Tourism into the new decade: Challenges and prospects*: *Book of abstracts of the International Conference on Tourism (ICOT2019), Braga, Porto, Portugal*. International Association for Tourism Policy.

Traskevich, A., & Fontanari, M. (2021a). *Thinking about regional and corporate resilience in the tourism industry: Models, empirical research and first insights by the German hotel industry*. Paper presented at the Tourism Naturally Conference, Deggendorf Institute of Technology, Germany, 14 October.

Traskevich, A., & Fontanari, M. (2021b). Tourism potentials in post-COVID-19: The concept of destination resilience for advanced sustainable management in tourism. *Tourism Planning & Development*. doi:10.1080/21568316.2021.1894599

Traskevich, A., & Fontanari, M. (2023). Resilience knowledge application in tourism and hospitality: A theoretical ReKo-Model. In K. Andriotis (Ed.), *Innovative sustainable practices in travel and tourism. Recovery and resilience*: *Book of abstracts of the International Conference on Tourism (ICOT2023), Nicosia, Cyprus*. International Association for Tourism Policy; European University Cyprus.

Wang, Y. H., Yin, C. W., & Zhang, Y. (2003). A multi-agent and distributed ruler based approach to production scheduling of agile manufacturing systems. *International Journal of Computer Integrated Manufacturing*, *16*(2), 81–92. doi:10.1080/713804987

Wong, A. K. F., Köseoglu, M. A., & Kim, S. S. (2021). The intellectual structure of corporate social responsibility research in tourism and hospitality: A citation/co-citation analysis. *Journal of Hospitality and Tourism Management*, *49*, 270–284. https://doi.org/10.1016/j.jhtm.2021.09.015

Zhai, Y., & Shi, P. (2022). The evolutionary characteristics, driving mechanism, and optimization path of China's tourism support policies under COVID-19: A quantitative analysis based on policy texts. *Current Issues in Tourism*, *25*(7), 1169–1184. doi:10.1080/13683500.2021.1972942

Zhang, T., Chen, Y., Wei, M., & Dai, M. (2022). How to promote residents' collaboration in destination governance: A framework of destination internal marketing. *Journal of Destination Marketing & Management*, *24*, 100710. https://doi.org/10.1016/j.jdmm.2022.100710

Excursus

The Concept of Destination Resilience to Prevent Overtourism

Anastasia Traskevich and Martin Fontanari

The growing problem of overtourism (Dodds & Butler, 2019) often relates to the soft aspect of a low level of tourism awareness and tourism acceptance at destinations (Fontanari & Traskevich, 2021). Thus, the challenges of reorientation of tourism development and mitigating overtourism raise the question of the basic understanding and guiding philosophy of tourism. To overcome subjective critical perceptions on the side of local communities, it is essential to draw the attention and sensitivity of residents to the positive tourism effects which touch and enrich everyone. This requires creating a stronger integration and interaction between tourists and residents. This way, social-cultural barriers and critical stresses can be reduced.

At the same time, these soft components create the fundamentals of resilient community structures (Bec et al., 2015). They are constituted by a strong orientation towards resilience within the authentic local lifestyle at the destination (Traskevich & Fontanari, 2018). Furthermore, community and destination resilience is achieved by the sustainable co-inhabitation of residents and guests (Chen et al., 2020; Kato, 2018; Traskevich & Fontanari, 2021). This is how the concept of resilience in tourism is introduced as a bridging solution to combat the negative impacts of overtourism (Fontanari & Traskevich, 2021, 2022).

From this perspective, the consideration of tourism as a cross-sectoral industry across many other economic sectors and markets is to be emphasised and amplified (Stylidis & Dominguez-Quintero, 2022). Tourism can be seen in its cross-sectoral function as an instrument of regional development, especially in relation to the design of infrastructure, projects, themes and products that help authentic tourism potentials to become attractive to tourists and to be thematised internally and externally in marketing. Infrastructure measures and leisure services provided by tourism development also increase the locational quality of a municipality. This positively impacts the attractiveness of the native place for a large number of non-tourism enterprises, which benefit directly from measures of tourism development and the associated advancement of the life quality in the region. This is also true for the local population, whose leisure time, livelihood and employment benefit directly and indirectly from tourism structures.

Conversely, overtourism has perceived negative impacts on quality of life (Lai et al., 2020). To address these, the dimensions of socio-cultural and

DOI: 10.4324/9781003365815-26

socio-psychological resilience of both locals and guests have to be strengthened (Hassan et al., 2017). This approach would allow for dealing with existing impairments in such a way that powerful changes in lifestyles and personal attitudes can be introduced (Traskevich & Fontanari, 2018). From this perspective, the instrumental function of tourism design for regional development becomes broader and deeper. It raises the question of further orientation and positioning of destinations, also in the context of overtourism.

A growth-oriented yet more sustainable tourism that absorbs the negative impacts of overtourism must, therefore, attain a new quality (Hall, 2022). This implies tourism that has a strong impact on destinations by increasing local social prosperity and cultural sustainability. Furthermore, this means tourism that safeguards the living conditions of residents and guests. All these overarching objectives are reflected in the development approach based on the multidimensional concept of resilience in tourism (Butler, 2017; Fontanari & Traskevich, 2021; Shin et al., 2022; Traskevich & Fontanari, 2021; Zacher et al., 2022). Thus, a "development bridge" from overtourism to resilience-oriented tourism development (Fontanari & Traskevich, 2022) can be designed as shown in Figure Ex5.1.

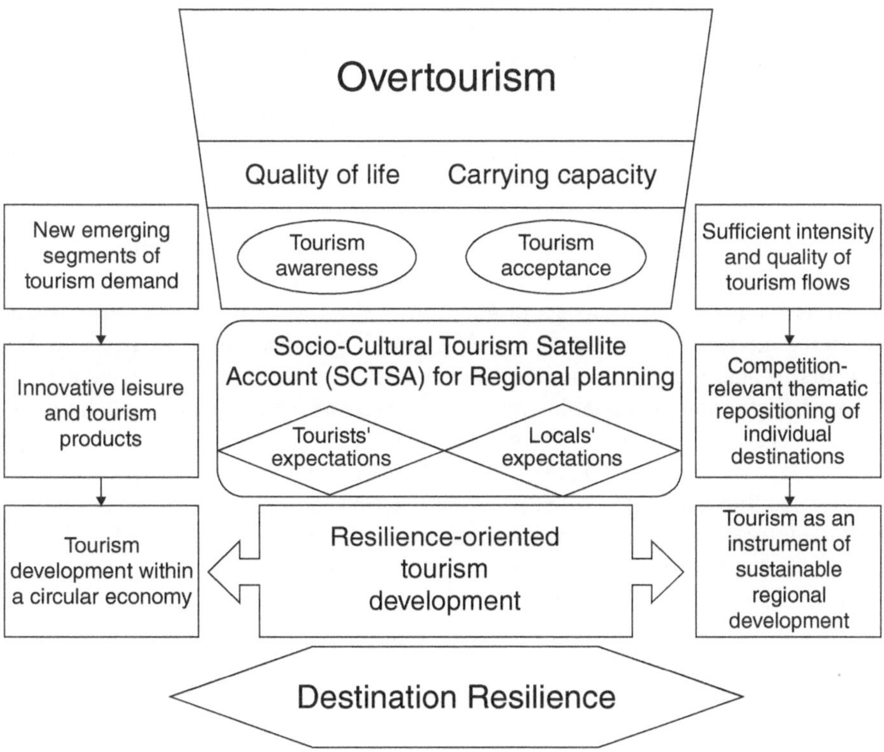

Figure Ex.1 Resilience-oriented framework to handle overtourism: The approach to tourism awareness and quality of life

Source: The authors

The framework presented postulates the following guiding considerations:

- Destinations enjoy loyal and new emerging segments of tourism demand providing sufficient intensity and quality of tourism flows
- Tourism offerings range between old "carefree" and current "new-normal" safety-oriented leisure and tourism products
- Tourism development focuses on the competition-relevant thematic repositioning of individual destinations in the context of sustainability and differentiation

Strategic resilience orientation would then be the overarching goal and consistent development paradigm towards sustainability (Hall et al., 2018; Pechlaner et al., 2022). In this context, tourism awareness and tourism acceptance play the central foundation for strong communities, agile networks and cooperative structures (Thees et al., 2022). Intensive interaction and communication with guests can be designed as an awareness-raising process that addresses the potential opportunities and enrichments of communicative exchange. Tourism activities should be channelled more strongly in the direction in which the locals can add value contributing to both monetary and to social-cultural revenues of tourism. In tourism planning, this means that locals are more involved in different supply structures of the destination and enrich local value creation. Thus, guests enter into more intensive economic and socio-cultural contact with locals. This should have immediate positive effects on the economic and socio-cultural interactions of tourists and residents, which will contribute both to mitigating overtourism and to implementing destination resilience (especially, the self-sufficiency aspects of regional supply in the context of a circular economy).

This involves development and implementation of the measurement instruments of a survey and a system of key performance indicators that can be continued as long-term monitoring. In addition to the economic added value, such long-term monitoring allows the socio-cultural added value of tourism to be measured. Thus, tourism awareness and tourism acceptance can be evaluated in a more holistic and sustainable way. The framework of application of the concept of destination resilience is to be applied within strategic planning (Fontanari et al., 2021a; Innerhofer et al., 2018) through the instruments of the socio-cultural tourism satellite account (SCTSA). SCTSA forms the basis for the quantitative and qualitative evaluation of the prerequisites and the perspectives for building a resilient destination, with a focus on the psychological resilience of residents and guests (Fontanari et al., 2021b).

In summary, the core idea for overcoming the soft, perception and relationship-based aspects of overtourism lie in making tourism development strongly oriented towards the integration of value-creating contributions of the locals for authentic tourism products. This also implies the design of meaningful leisure-related socio-cultural contributions which are executed as diverse elements of basic supply targeted at both locals and guests. For this purpose, resilient supply structures must be implemented holistically in tourism development planning. At the same time, activities on awareness building and monitoring measures which are focused on the positive socio-cultural contributions of tourism are required. Applied resilience knowledge serves as a development bridge to mitigate overtourism and gain destination resilience.

References

Bec, A., McLennan, C., & Moyle, B. D. (2015). Community resilience to long-term tourism decline and rejuvenation: A literature review and conceptual model. *Current Issues in Tourism, 19*(5), 431–457.

Butler, R. W. (Ed.). (2017). *Tourism and resilience*. CABI.

Chen, F., Xu, H., & Lew, A. A. (2020). Livelihood resilience in tourism communities: The role of human agency. *Journal of Sustainable Tourism, 28*(4), 606–624.

Dodds, R., & Butler, R. (Eds.). (2019). *Overtourism: Issues, realities and solutions*. De Gruyter.

Fontanari, M., & Traskevich, A. (2021). Consensus and diversity regarding overtourism: The Delphi-study and derived assumptions for the post-COVID-19 time. *International Journal of Tourism Policy, 11*(2), 161–187.

Fontanari, M., & Traskevich, A. (2022). Smart-solutions for handling overtourism and developing destination resilience for the post-Covid-19 era. *Tourism Planning & Development, 20*(1), 86–107. doi:10.1080/21568316.2022.2056234.

Fontanari, M., Traskevich, A., & Kutsch, H. (2021a). Corporate resilience within tourism enterprises. In M. Valeri, A. Scuttari & H. Pechlaner (Eds.), *Resilienza e sostenibilitá:Dinamiche globali e risposte locali* [Resilience and sustainability: Global dynamics and local responses]. Giappichelli Editore.

Fontanari, M., Traskevich, A., & Seraphin, H. (2021b). (De)growth imperative: The importance of destination resilience in the context of overtourism. In K. Andriotis (Ed.), *Issues and Cases of Degrowth in Tourism*. CAB International.

Hall, C. M. (2022). Tourism and the Capitalocene: From green growth to ecocide. *Tourism Planning & Development, 19*(1), 61–74.

Hall, C. M., Prayag, G., & Amore, A. (Eds.). (2018). *Tourism and resilience: Individual, organisational and destination perspectives*. Channel View Publications.

Hassan, A., Ramkissoon, H., & Shabnam, S. (2017). Community resilience of the Sundarbans: Restoring tourism after oil spillage. *Journal of Hospitality and Tourism, 15*(1), 1–18.

Innerhofer, E., Fontanari, M., & Pechlaner, H. (Eds.). (2018). *Destination resilience: Challenges and opportunities for destination management and governance*. Routledge.

Kato, K. (2018). Debating sustainability in tourism development: Resilience, traditional knowledge and community: A post-disaster perspective. *Tourism Planning & Development, 15*(1), 55–67.

Lai, H. K., Pinto, P., & Pintassilgo, P. (2020). Overtourism: Impacts on residents' quality of life – evidence from Macau. In C. R. de Almeida, A. Quintano, M. Simancas, R. Huete, & Z. Breda (Eds.), *Handbook of research on the impacts, challenges, and policy responses to overtourism*. IGI Global.

Pechlaner, H., Zacher, D., & Störmann, E. (Eds.). (2022). *Resilienz als Strategie in Region, Destination und Unternehmen: Eine raumbezogene Perspektive*. [Resilience as a strategy in region, destination and company: A spatial perspective]. Springer Gabler. https://doi.org/10.1007/978-3-658-37296-5

Shin, H. W., Yoon, S., Jung, S., & Fan, A. (2022). Risk or benefit? Economic and sociocultural impact of P2P accommodation on community resilience, consumer perception and behavioral intention. *International Journal of Contemporary Hospitality Management*. https://doi.org/10.1108/IJCHM-12-2021-1561.

Stylidis, D., & Dominguez-Quintero, A. M. (2022). Understanding the effect of place image and knowledge of tourism on residents' attitudes towards tourism and their word-of-mouth intentions: Evidence from Seville, Spain. *Tourism Planning & Development*, *19*(5), 433–450. doi:10.1080/21568316.2022.2049859.

Thees, H., Störmann, E., & Pechlaner, H. (2022). Business modeling for resilient destination development: A multi-method approach for the case of destination Franconia, Germany. *Tourism Planning & Development*, *20*(2), 212–235. doi:1080/21568316.2022.2121313.

Traskevich, A., & Fontanari, M. (2018). Mental wellness in resilient destinations. *International Journal of Spa and Wellness*, *1*(3), 193–217. doi:10.1080/24721735.2019.1596656.

Traskevich, A., & Fontanari, M. (2021). Tourism potentials in post-COVID-19: The concept of destination resilience for advanced sustainable management in tourism. *Tourism Planning & Development*, *20*(1), 12–36. doi:10.1080/21568316.2021.1894599.

Zacher, D., Philipp, J., & Pechlaner, H. (2022). Destination resilience. In D. Buhalis (Ed.), *Encyclopedia of tourism management and marketing*. Edward Elgar Publishing.

Index

Note: **Bold** page numbers refer to tables; *italic* page numbers refer to figures; ***bold italic*** page numbers refer to boxes and page numbers followed by "n" denote endnotes.